THE FEEDBACK LOOP
OF THE MYSTIC

REVELATIONS ON THE SCIENCE OF THE SELF & ITS PROTECTIVE AND TRANSFORMATIVE POWERS

John Brighton with Christine Brighton

Order this book online at www.trafford.com
or email orders@trafford.com

Most Trafford titles are also available at major online book retailers.

Print information available on the last page.

ISBN: 978-1-4907-9357-3 (sc)

Library of Congress Control Number: 2019932774

Trafford rev. 02/28/2019

 www.trafford.com
North America & international
toll-free: 1 888 232 4444 (USA & Canada)
fax: 812 355 4082

DEDICATION

Towards the establishment of a love-and-understanding, rather than a fear-and-defensive, based world. May the logos be with us in body, mind and spirit to guide the way.

ACKNOWLEDGMENTS

☆ To my wife Christine, for her love, research, and intellectual contributions, which made this book possible.

☆ To my daughter Joelle Morales, who offered valuable suggestions supported by her background in mathematics and biology.

☆ To Karen Nalbone, for her many years of friendship and dedication to the Fourth Way and the compassionate counseling of others, and for commenting on the manuscript, particularly in matters of physics and mathematics.

☆ To our psychologist friend Jim Strohl, for the many years of friendship and collaboration in matters of spiritual psychology.

☆ To our friend Laurel Leland, a psychological counselor, for her emotional support, and for providing relevant research and links to others whose work have affinities to psiology.

☆ To Teri and Larry White, for their friendship, encouragement, and generous financial and loving support.

☆ To numerous other contributors, too many to mention, who directly or indirectly continue to have a positive influence on our intellectual and spiritual development.

TABLE OF CONTENTS

INTRODUCTION

We are given a precious jewel called the mind. Our personal freedom depends on learning how it can become an inner-directed servant rather than an other-directed master. This requires learning how it functions, especially how it modulates itself in a self-referential manner. Not learning about its nature and how to control it for personal and social benefit makes it quite likely that someone else, or some self-serving group, will do so. Given the technical nature of our time, coupled with the powerful force of "group think" and the hidden powers that control society, the wherewithal for this to take place on a very large scale is almost a foregone conclusion. However, it's never too late to begin. Why not start now?

We go through life not realizing that our consciousness is teased by a constant barrage of stimuli emanating from the inner world of memories, physiological processes, and the outer world of constantly fluctuating events. Now and then some inner process "chooses" from this plethora of possibilities, and brings it into a state of focus, where everything else is suppressed. It could be about physical or psychological suffering, a car accident, or finding a one hundred dollar bill on the sidewalk. It doesn't matter whether the mind fools us into becoming aroused by something that appears as something else, but these moments are "real" to the observer, and our bodies are at the mercy of the mind's interpretations.

Then there are what I call "signature moments" that live in our memory like beads strung on a necklace of time; these moments are not ordinary ones. They stand out because of their personal uniqueness, for we intuitively know that despite their common elements, they could not have happened to anyone else in the same way. Usually it takes a dramatic event around which the beads gather in thematic relationships.

For me these relationships are rooted in numerous psychic experiences that began with an out-of-body event (OBE) when I was just three years of age. These are the events that led me into the counterintuitive journey I describe in this book.

To fully express the roots of this journey I felt that it was important that I try not to devalue them; while they are special I did not want to think of them as something superior that would set me apart from other people, for I intuitively knew we all share such potentials. While not fully knowing how they came into being I kept notes, thinking that someday I would be able to explain or enrich them from a scientific perspective.

Personal history and modern science do not easily blend, even though much of the insights that come to the creative mind in any field emerge from the depths of the personal spirito-psychophysiological history of individuals in a social context. It is always a holistic process that is not isolated to the individual but includes the interactive time and place where the insights occur. I refer to them as moments of "relative unity" because each one was riveting in its own way, uniting me into a wholistic system of mind, body, and spirit.

Often I would re-read my notes only to find that they were actually forming a long statement about the universe being expressed through me. Each incident was like a sentence of a paragraph stretched out through many years. It was as if each event was changing my mind and life, one gestational step at a time.

After many decades of research and working on myself to attain spiritual benefits, I experienced the emergence of what I thought to be a Jungian "Symbol of unity" that not only altered my level of being and consciousness but also began "organizing" my notes in counterintuitive ways.

At that moment I felt that "secrets" about our nature and the universe were being revealed to me in symbolic visions, and finally via the Symbol, by some Cosmic Intelligence, which I eventually realized was mediated by my Self in a spiritual mode of being. Exhilarated by it all, I took its suggestions and began to reorganize my numerous notes in order to write this book. What follows is what came from that intuitive relationship mediated by the Symbol.

GLIMPSES OF HIDDEN REALITIES

> "Instead of searching for the proof of truth, which you do not know, go through the proofs you have of what you believe to know. You will find you know nothing for sure – you trust on hearsay. To know truth, you must pass through your own experience." - Nisargadatta Maharaj

The quest began with, of all things, the need to write a short story for a course in creative writing I was taking during my sophomore year in college. I had never written a short story before, and my language skills were rather wanting. Doubting that I would be able to complete the assignment, I began to worry.

ZEUS DID IT

In the midst of my growing anxiety I decided one day to lie in bed to relax and shift my mind from these negative expectations as the deadline neared. As I began to drift off I experienced a sudden rush of energy throughout my body that culminated with an electrical discharge in my head. In that instant the first paragraph of a story entitled *Zeus Did It* popped into my awareness, accompanied by a sustained and all-enveloping feeling of joy.

> To fall in love with a statue is a blissful thing. Statues of women are beautiful, especially when you take into consideration the dexterity that went into making beauty on a large scale. I wonder who posed for the artist. It must have been a tart. It must have taken a special chisel to carve those delicate nates and Venus's breasts, and the smoothness of the marble is another thing. The round hard curves that lured men and beguiled them. Fools.

I immediately got up and began to type it, and the rest of the story, just as surreptitiously, flowed out as I typed.

What emerged was a first-person tale of a violent psychopath who falls in love with a large marble statue of Venus (the Roman version of the Greek goddess of love) housed under a domed ceiling in a New York City museum. Throughout the story he is constantly committing murder of those who represent the social figures and voices that repress the instinctive motivations that propel him to find love in the statue of Venus. Yet he is constantly blaming Zeus for his atrocities, as signified in the title.

Periodically he visits the museum to fulfill his fantasies. His obsession with the statue compels him to find ways to remain in the museum after it closes, allowing him to emerge from hiding and approach it with amorous intent. However, in the balcony overlooking the sun-or-moonlit Venus stands a "threatening" statue of Zeus, who views with disdain the amorous advances of this mortal creature. In the end he encounters the wrath of the god, who destroys him with a bolt of lightning, or so it seems. Reality and illusion intermingle in this brief first-person plot so that one is indistinguishable from the other.

While there's an obvious antagonism between the protagonist and Zeus, in some instances he identifies with the god when he reaches into his mysterious "bag" to draw a presumptive "gun" to kill his victims with thunderbolts. Obviously he wants his power. Yet in his desperation to find love he taunts Zeus to kill him. In the final scene the protagonist, quite suicidally, returns to the museum to take his chances, right below Zeus's gaze, to climb onto the huge and cumbersome thighs of Venus with erotic intent, only to taunt Zeus to use his lethal thunderbolt.

All the statues seemed to have their attention on Venus, and Zeus looked down from the balcony with red veins in his bulging eyeballs. Philandering hypocrite. As I slowly walked my heels clicked echoes. Adagio wound itself up in the rhythm of the click. Daaa da da click daa dada click dee dum dee dum click dee dum dum de de de click. There she was, as close as breath. I reached out and touched her big toe. I slipped my hand up her leg. It was a cold marble leg, long and slender but substantial. Her legs gracefully flowed from her wide hips. I moved like molasses with my two hands up around her hips and on to her waist.

It is going to happen soon. They will come alive and speak to me. I was dead sure: the white figures were turning their heads lifelike. Venus, having a taste of my affection, was beginning to soften. She was so big. The stone was pure flesh now. I stretched my arms as far as they could go, trying to reach her shoulders, but all was in vain. She was beyond my reach. I treated myself to the idea of a crimson kiss. My shoes slipped off her kneecap and I went stumbling to the ground. Raising myself and forming two fists, I took a deep breath and yelled, "You damn lousy tart!" I reached down for my bag. A voice froze me as I stooped over. Zeus. "Hey, you crazy man, what..." I pulled it out of the bag and pointed straight at Zeus. A bolt of light sounded. I felt a stabbing acute pain in my chest as the ground pushed the bag up to my stomach. I could see a dim glow in the museum through the slits of my eyes. The museum must have been dawning with light. I could hear faintly the shuffling of feet all round me, on all sides: the lesser gods. I ought to tell them.

It felt like years had passed before I finally built up enough energy. I raised one arm painfully and pointed toward the statue up in the balcony. As I heaved what seemed to be my last breath, the statue smiled and became blurry. Before my eyes began to roll back, I whimpered "Zeus di . . . di . . ."

When I handed it in I had no idea of its quality or whether it would pass muster, but, to my surprise the professor loved it, read it to the class, and decided to publish it in the university's literary journal *Between Worlds* (Neimann 1960).

As a token prize he gave me two books that were to have a strong influence on how I began to view the mind: Paramahansa Yogananda's *Autobiography of a Yogi* (1946), and the *Larousse Encyclopedia of Mythology* (1960).

Not everyone, however, was elated by the story. Another professor of English thought my sanity was questionable, and I felt somewhat embarrassed and a bit disappointed that it had such a public airing. As a matter of fact, I too was quite shocked that I harbored such a horrendous tale within me.

Realizing that I had not consciously created it, I assumed that I had drawn it from somewhere. From where? I was clueless. As I saw it at the time, the mythological theme probably came from a confluence of readings of Roman mythology and Edith Hamilton's *The Greek Way*

3

(1930), and the insane character was quite likely a figure of my pent up anger, a kind of symbolic rage of the verbal and physical violence that frequently took place between my parents when I was a child.

There's a tendency to attribute such occurrences to the subconscious (a.k.a., unconscious) mind, but even though they are the result of unknown processes, this often turns out to be the resolution of a mystery with another one. However, at the time I couldn't come up with a different hypothesis. Because of the involuntary nature of the act I thought of it as occurring in a trance state of automatic writing (although on a typewriter).

Fascinated by the mystery of how it came about and the emotional affect it had on me, at the end of the semester I switched my major from economics to psychology. I resorted to extracurricular readings in depth psychology, particularly the works of Sigmund Freud and Carl Jung, to help me understand the psychological meaning of the story.

What was of interest to me was how these theories explore the relationship of the conscious and non-conscious levels of the mind through the interpretation of symbols, manifesting in dreams, art, myths, and other representative areas of life. The clinical aim is, as Freud averred, "To make the unconscious conscious" as the means by which therapy can foster a more harmonious psychosocial state of being.

Generally speaking, it puts forth the idea that within each person there is a constant psychodrama mainly taking place in a non-conscious domain (i.e., preconscious, subconscious, unconscious, personal unconscious, collective unconscious), involving a struggle between:

1. A pluripotent energy that manifests as innate instinctive impulses (i.e., food, safety, sex, etc.) seeking unbridled satisfaction (Freud's Id), or as universal archetypes (i.e., Jung's cosmic complements to gender: the animus for the male, and the anima for the female) seeking symbolic union.
2. Constraints posed by social rules and conformity (i.e., Freud's "superego").

As a result of this conflict there are unresolved, anxiety producing, memory constructs (MCs) in the non-conscious known as complexes.

It is the role of a reconciling principle, referred to as the ego, to ameliorate the anxiety through defense mechanisms like repression, denial, rationalization, and sublimation (the socially-friendly expression

4

of "taboos"—Freud), or through transformation of opposing archetypical forces into symbolic states of unity (Jung).

There was no doubt in my mind that the story was describing a powerful complex that affected every aspect of my life.

It is interesting to see how the "mind" uses art to allow one to peer into taboo relationships of one's psychic life in the virtual arena of apparent inconsequence.

Every once in a while I reread it, and would discover new meanings condensed in the symbolism that would gradually reveal other still-unfolding dimensions of an emerging personal history.

From my indeed crude self-analysis I realized that I never identified with my father, who abandoned the family during my early childhood. I rarely saw love or tenderness between my parents; if there was some of that it was eclipsed by their frequent verbal and physical violence towards each other. Most of my adult life, I struggled with feelings of anger towards him, eventually leading me to change my last name. This explains the protagonist's antagonistic attitude towards Zeus, and, more pointedly, the blaming of him for his violence (i.e., the defense mechanism of projection).

THE ZEUS COMPLEX

Even though I became quite conscious of the "Z Complex," as the story eventually was dubbed, I was not able to fully self-resolve it after several years of psychotherapy simply by being cognizant of its existence and reasoning about it. I painfully became more conscious of how it still lurked within me, or the traces it left in my persona.

It was so evident in almost every nuance of my behavior, particularly my approach to other people, always tinged by fear of not being wanted. To compensate I resorted to being witty and humorous in my conversations. It brought forth many relationship problems with women and authority figures. I could see it in my posture, and even in the tone of my voice—how it became subdued and at times faltered. There appeared a fruitless search for a father figure by how I sought the friendships of older, highly educated men with whom I enjoyed intellectual conversations and being admired by them.

While the Freudian interpretation to the story seemed quite literal and the Jungian one introduced mystical elements of integral causation, there was yet another aspect of the event that was equally, if not more, intellectually interesting: the overwhelming *energetic effect* the story had on my entire being when it emerged.

Transcending the Z Complex with the Z Complex

As I eventually discovered, the energetic effect would also come forth whenever I read the story aloud to myself. Strangely enough, it was only in that state that I felt freed from the Z complex. It was as if I had removed a major barrier to my natural way of being by inoculating myself with the nature of the problem. In many instances the transcendental effect would last for hours, at times for days on end.

After having periodically performed this ameliorative, indeed mysterious exercise for many years, the Z complex eventually lost its all-powerful grip on me, as I saw it at the time. However, I could later still identify vestiges left over from its days of power.

Why, I wondered, can a psychopathic tale, full of blood, murder, and disaster, yield such a wholesome, indeed spiritual effect? Was this a princess-kissing-the-frog syndrome, or William Blake's *The Marriage of Heaven and Hell*?

PSYCHIC HOMEOPATHY?

Years later I saw analogies of this repetitive process to a healing modality known as *homeopathy*, simply because it uses the cause of the problem to dispel it. In other words, it repeats the condition to itself in a self-referential manner.

Homeopathy is the treatment of disease by repetitive administration of highly diluted doses of natural substances that in a healthy person would produce symptoms of a particular disease.

For example, exposure to poison ivy and poison oak often results in the skin breaking out with red, swollen, intensely itching blisters, sometimes followed by oozing or crusting. By using a remedy with substances that produce these symptoms, like *Rhus Toxicodendron, Croton Tiglium,* or *Xerophyllum,* the condition can be remedied. The more one dilutes the formula (i.e., the subtler it becomes and cannot be detected by scientific instruments) the greater is the potency.

An important characteristic of using homeopathic remedies is the nature of the healing process. At some point the organism will undergo a "healing crisis," which occurs when the disease shifts from the inner organs towards the periphery, which in psychological terms would mean its emergence into consciousness.

During this period the symptoms become more aggravated, reaching a maximum before the condition is resolved. In mysticism this would be akin to the "dark night of the soul," which I interpret to be the emergence of one's contradictions into consciousness.

While the relationship to homeopathy was a compelling analogy, it did not explain why or how this weakens the complex by expelling it to the surface (i.e., the analogy of the conscious mind to the skin).

Moreover, no one, as far as I knew, has ever convincingly explained how homeopathy works, except perhaps that it causes the metabolic rate of the immune system to become more sensitive to the disorder and accelerate its rejective activity.

Why, I wondered, does iterating the cause of a problem evoke the organism to be more cognizant of the problem and accelerate its healing function?

The issue of repeating the dose (i.e., the rereading/re-thinking of the story) began to take on more importance, for it suggested that it had some amplifying effect, that is, bringing it from a subtle (i.e., subconscious) level of causation into a visible manifestation, called the "healing crisis" (i.e., consciousness of the complex). Maybe it had something to do with changing the energy level or frequency in which the disease/complex exists, and that might have something to do with how it manifests into awareness. This possibility brought to mind a statement Yogananda made in his book about the rate at which the mind operates.

THE RATE OF MENTAL OPERATION

There was so much in that book that I couldn't understand. It all seemed beyond the grasp of a mind immersed in conventional notions of reality. Almost every chapter told of people who transcended the "known" laws of our world. They all seemed to be intentionally altering reality in mind-boggling ways. There were, for example, yogis who never slept, rarely ate, levitated, exuded the sweet scent of various flowers, and could disappear and reappear at will (all caught on camera). Due to their extraordinary ability to concentrate their minds, indicating the degree of attentional focus, they were able to perform these transformative acts. I had never read anything like that before. It all seemed miraculous to me.

Fascinated by it all, I wrote to the Self Realization Fellowship to inquire about their home study course, which I eventually began, but never got into with the fervor needed to succeed. After awhile I abandoned the whole thing.

Yet, there was a passage in the book that impressed me by its scientific implications. It made me think of the mind as a vibrational system operating at a certain rate, which could be modified by its degree of attention. This higher rate of operation could be instrumental in acquiring higher states of consciousness involved in creative and healing activities of the mind. Yogananda wrote:

> "The infinite potencies of sound derive from the Creative Word, AUM, the cosmic vibratory power behind all atomic energies. Any word spoken with clear realization and *deep concentration* has a materializing value. Loud or silent repetition of inspiring words has been found effective in Coueism and similar systems of psychotherapy; *the secret lies in the stepping-up of the mind's vibratory rate*" (1946, Footnote 1-12. Italics added).

The term Coueism refers to the French psychologist Émile Coué who introduced a mantra-like method of self-improvement based on optimistic autosuggestion, like: "Every day, in every way, I'm getting better and better" (see Coué 1922).

In this case the repetition involves a message that is suggested to the "subconscious" (he referred to it as the "unconscious"), which he equated with "imagination." The idea of repeating an objective to oneself brought to mind the formation of memory and habits, in this case the habit of being healthy.

While Coué mainly recommended his self-help therapy for alleviating medical problems by making positive autosuggestions, Yogananda's method is about the transformation of "being" and "consciousness," apparently a much broader agenda.

While both methods differ in scope they overlap in the use of focused repetition in order to modify, in Coué's approach, the imagination, and in Yogananda's as the means for evoking a superconscious state. One would therefore assume that the subconscious and conscious "minds" differ in intended context and vibratory rate.

According to this line of reasoning, these levels of mind would be determined by the contextual intent and the rate at which they vibrate, which also implies the amount of time they take to process energy and information.

It's important to note that Yogananda is not, in the common parlance of yoga and other systems of spiritual development, alluding to "being out of time", but to accelerating a temporal aspect of the mind, namely the amount of time it takes for a vibration to complete a cycle (i.e., a return to a point of origin).

What's so interesting about his assertion is its basis on time as a mental-and-conscious modifier and, of equal importance, our ability to alter it. A hypothesis began to develop.

> The information process would be one and the same at all levels of operation, but what is subject to change is the particular context and the energy-fueled rate and scope of operation, determined by the degree to which the mind is concentrated; as the rate increases, the psychological system becomes more powerful, intelligent, and transformative.

The act of repeating a context to oneself beyond the limit of habituation (i.e., the diminishing of a physiological or emotional response to a frequently repeated stimulus) requires a certain voluntary dedication to the task, hence, a degree of will and concentration. But, how I wondered, does concentration make the mind go faster?

Attentional focus and iterative mental activity seemed to be the key factors in answering this question, but I made notes of the issue and set it aside for the future to answer. There was so much on my plate at the time with academics and my commitments to playing basketball. As I pursued my commitments, the Zeus story and this question faded into the background.

QUICKENING

Part of accepting the sports scholarship was also the agreement to take required courses in religion, which used the King James Holy Bible as the text. As I read the Bible the notion of speeding up the mind began to take on a scientific as well as a spiritual dimension. Speed is a scientific concept when thought of in terms of mathematical equations. However, the two came together when I encountered the word "quickening" in the Bible that appeared in several places. For example in Psalms:

> "[Thou,] which hast shewed me great and sore troubles, shalt *quicken* me again, and shalt bring me up again from the depths of the earth" (71:20. Italics added).

Later in St. Paul the term appears again, but there it is given a more evolutionary purpose in regards to the resurrection of the spirit from the confines of the body.

> "But some [man] will say, How are the dead raised up? and with what body do they come? [Thou] fool, that which thou sowest is not quickened, except it die: And that which thou sowest, thou sowest not that body that shall be, but bare grain, it may chance of wheat, or of some other [grain:] But God giveth it a body as it hath pleased him, and to every seed his own body. All flesh [is] not the same flesh: but [there is] one [kind of] flesh of men, another flesh of beasts, another of fishes, [and] another of birds. [There are] also celestial bodies, and bodies terrestrial: but the glory of the celestial [is] one, and the [glory] of the terrestrial [is] another. [There is] one glory of the sun, and another glory of the moon, and another glory of the stars: for [one] star differeth from [another] star in glory. So also [is] the resurrection of the dead. It is sown in corruption; it is raised in incorruption: It is sown in dishonour; it is raised in glory: it is sown in weakness; it is raised in power: It is sown a natural body; it is raised a spiritual body. There is a natural body, and there is a spiritual body. And so it is written, The first man Adam was made a living soul; the last Adam [was made] a quickening spirit. Howbeit that [was] not first which is spiritual, but that which is natural; and afterward that which is spiritual. The first man [is] of the earth, earthy: the second man [is] the Lord from heaven. As [is] the earthy, such [are] they also that are earthy: and as [is] the heavenly, such [are] they also that are heavenly. And as we have borne the image of the earthy, we shall also bear the image of the heavenly" (1 Corinthians, 15:35-49).

Not having the wherewithal to fully process what this more specifically meant at the time, I left it to mercy of the future. I found a job, got married, and one day while I was driving home from a small mountain town in the wee hours of the morning, I encountered the second incident that brought my mind into a state of extreme focus, further challenging my agnostic leanings, and bolstering the notion of quickening in regards to that of *psychological space-time*.

VW VERSUS MACK TRUCK

It was dark and misty, and as I was taking a curve on a narrow serpentine mountain road with my Volkswagen (VW) I saw a huge truck coming towards me with very little room to spare. I braced for impact. Seconds before hitting its large chrome bumper I felt a surge of energy rush through my body, and I went into a hyper-accelerated state of mind where everything I witnessed was occurring in slow motion. Except for the VW, myself, and the truck, everything else was totally excluded. It seemed like it took twenty minutes for the collision to occur, giving me enough time to make a life-saving maneuver that avoided a dead-center impact.

The VW hit the edge of the bumper, glanced off to the other side of the road, the wheels still spinning, and returned back towards the side of the rig and collided with a refrigerator unit attached to its underside. Then it glanced off again and landed behind the rig with its front end considerably crumpled. When the car came to a halt I proceeded to step out through the open driver's side window, a feat I later realized was an impossible escape for a six-foot-five-inch two-hundred-pound body.

I ended up on a slope overlooking the accident at the other side of the road. People started to come down from their homes on the mountainside to check out the accident. I thought they would see me and ask questions. Surprisingly, as they passed by they were not aware that I was there. I looked at the VW and I could see my body sitting slightly slumped in the driver's seat. The onlookers thought I was critically injured or dead, and proceeded to call for help.

Strangely enough, I had no sense of urgency, fear, or emotional attachment while out of the body.

As I looked down on this scene a modicum of incredulity entered my mind, and I started to squeeze my arms to see if I had substance. My arms felt solid! In the blink of an eye, I returned to my physical body and I examined myself for injury and discovered that I had only a scratch on the bridge of my nose. Seeing me move, some onlookers came over to see if I was all right, and as I got out of the car fully intact and speaking their worst expectations were dispelled.

Later on it became clear that I was saved by the hyper-focused and accelerated state of mind. As time of the reality around me "slowed down" I was able to make crucial decisions on the spur of the moment. It seemed like the so-called passage of time was modified by the *frequency of mental function*: the rate of vibrations.

It was obvious that attention has degrees of focus and rates of operation that could be triggered intentionally through repetition, as Yogananda averred, and by moments of anxiety and great danger.

There was something similar to what occurred with the Zeus story, but in this case what stands out is not the synthesis of a psychodrama but the synthesis of another level of being, as if the information of the body (the physical aspect of the Self) contained in the nucleic memory (DNA) of trillions of cells was energetically transduced into a "spiritual" body. The scope of the incident seems to have far exceeded that of the energetic conditions that produced the Zeus story. However, both affected my entire being.

The accident ignited flashbacks, and reassessment, of my first OBE that took place one Christmas night when I was three years old while living with my parents in the lower apartment of an old two-story frame house in the Bronx. Powerful and unforgettable as it was, for years I thought of it as a dream, and it therefore had little significance for me. However, this time there was a witness who many years later helped to dispel that notion.

THE RED ROCKING HORSE

On that day our family and friends were having a festive time after a sumptuous dinner. Some time after midnight, after the dancing, laughter, and lively conversation subsided, I could tell the group was waning and getting ready to call it a night.

By that time, my sister Angela (a few years my senior) and I were exhausted from the festivities. Our leaden eyelids gave us away as the guests departed, and we were scooted off to two twin beds situated between the living room and the dining room. My parents put out the lights and went to sleep. Yet I could not fall asleep since my heart and vigilance were glued to the red rocking horse gift I had received that morning. Angie, as she was called, was also waiting for me to keel over with sleep so she could rock on it.

From my bed I could see the fireplace in the living room. Embers let out a subtle scented glow throughout the rooms. Of special interest to me was the appearance of a bright Christian cross hovering over, yet untouched by the fire. I had no idea then what it was but later it held a special significance for me.

Despite my exhaustion, I struggled to keep my eyes open as I watched the horse at the side of the room. I knew Angie was also watching. It was becoming increasingly impossible to keep my body from falling asleep. As I struggled to stay awake I felt myself slip out of my body.

Once out I ran over and mounted the wooden horse. It did not seem odd to see my inert body on the bed as I gently rocked. This is how a child unencumbered by social conditioning would see it. Then I saw my sister also leave her body and go towards the basement door where she grasped the left hand of an unknown female-looking person. They went down the basement steps. Out of curiosity I dismounted the horse and ran over to the open door, only to see her scoot past me and back into bed and enter her physical body before I returned to mine. She didn't seem interested in the horse.

Over forty years had passed with no mention of the incident between us. However, one day during a Thanksgiving dinner at her house, she brought up the subject, and without any prompting from me, she described the entire scene just as I had remembered it. By the way, she said the person who held her hand was an angel; this too stood in my memory with special significance.

The fact that two people shared an OBE moment was for me a breakthrough, for it indicated that it was not just a figment of my imagination or a dream. In this event our separate identities remained intact as we pursued our little agendas in the invisible space that surrounded us.

While OBEs appear to be a projection from the physical body, I discovered that a similar state can occur in what appeared to be in the opposite direction.

MICROCOSMIC INTROJECTION

A major reason for going to college in Puerto Rico was based on receiving a full athletic scholarship to play basketball. In addition to playing basketball for the Inter American University, I also played for the town of San Germán, where the university was situated.

One hot summer evening the San Germán team played the team from Arecibo, and we won a very close game. I recall playing a highly focused game. After the game I was utterly exhausted, but I could not get my mind to stop thinking about the game during the two-hour trip back home. It kept replaying the entire game over and over.

We arrived in San Germán around two in the morning, and instead of going another half-hour up to my mountain home in Maricao I decided to stay in a local hotel. My body was exhausted, but my mind had remained in a highly focused state, and time slowed down to where the outer world seemed to freeze.

This time I experienced my consciousness coming to a point in my forehead and dissociate from the body as it fell into deep sleep. However, instead of projecting it went "inward." One might say that it introjected (as an inverted vector, not in the psychoanalytic notion of the unconscious adopting the ideas or attitudes of others). Perhaps the mind can "expand" in scope via the positive powers of the number scale (X^{+n}) or shrink via the negative ones (X^{-n}) into hyperspace.

I entered a vast and endless darkness, and I could see a profusion of very tiny specks of brilliant light, like a night sky. Once again, I entered into that now familiar state of absolute *stillness*, giving me the impression that I was an immovable presence, watching motion. It was probably a steady state in which energy inflow and outflow were balanced.

Each time I focused on a point of light it would blossom out into a lucid scene. Each one was a totally different world, so to speak. All the while I was fully aware of myself, my sleeping body, and the room around me, bathed in a translucent black luster. It felt that I was observing from a 3D sphere.

The starry scene was convincingly similar, with some exceptions, to what Jung wrote about to describe the archetypical nature of the chaotic unconscious as consisting of a "multitude of little luminosities."

> "The hypothesis of multiple luminosities rests partly, as we have seen, on the quasi-conscious state of unconscious contents,

and partly on the incidence of certain images which must be regarded as symbolical. These are to be found in the dreams and visual fantasies of modern individuals, and can also be traced in historical records. As the reader may be aware, one of the most important sources for symbolical ideas in the past is alchemy. From this I take, first and foremost, the idea on the scintillae—sparks—which appear as visual illusions in the 'arcane substance.' Thus the *Aurora consurgens*, Part II says: '*Scito quod terra foetida cito recipit scintillulas albas*' (Know that the foul earth quickly receives white sparks). These sparks Khunrath explains as '*radii atque scintillae*' of the '*anima catholica*,' the world soul, which is identical with the spirit of God. From this interpretation it is clear that certain of the alchemists had already divined the psychic nature of these luminosities. They were seeds of light broadcast in the <u>chaos</u>, which Khunrath calls '*mundi futuri seminarium*' (the seed plot of a world to come)" (De Laszlo 1959, 60).

Years later, this brought to mind a similar experience of the ambient space reported by the neurologist James Austin in his book *Zen and the Brain: Towards an Understanding of Meditation and Consciousness* (2000).

"Suppose you enter a light tight closet and close your eyes. In the dark, you'll still "see" some grayness. Grayness represents the background noise in the visual system. One might think of it as comparable to the faint hum in the background of your hearing. Internalized absorption penetrates beyond this, reaching *absolute blackness*. Absolute blackness is eye-catching, and it commands attention in itself. The Apollo astronauts, in their quest for the moon, looked out and were awestruck by the "blacker-than-black" intensity of outer space. That which permeates the depths of *inner* space casts a spell no less enchanting. In the mystical literature of the West, another term for the black void is "the Divine Darkness." We must be careful about the word "darkness." It is one more metaphor (like "light") which mystics sometimes stretch to include other states" (2000, 481).

As Jung suggested, I believe I had come in contact with dormant memory constructs embedded in the timelessness of an "unconscious" field: perhaps related to Jung's "collective unconscious." Strangely enough, I could sense that I was an integral part of the field, for I would lose my sense of definition as if I had become spatial, while attention unfurled the different worlds locked in apparent frames of timelessness.

When I thought about the experience of time slowing it brought to mind a psychological interpretation of Einstein's theory of relativity. He pointed out, in terms that anyone can understand, how his theory would explain why in positive moments time "subjectively" slows down and while in negative states it speeds up. He said,

> "Put your hand on a hot stove for a minute, and it seems like an hour. Sit with a pretty girl for an hour, and it seems like a minute. That's relativity."

It seems that what the attentional process is dwelling on affects the metabolic rate at which the mind, that is the organism as a whole, processes information and how it affects our experience of time.

From a psychological perspective this would have something to do with resonant and dissonant states of mind. Experiencing a resonant moment would mean that expected and perceived outcomes are flowing in sync, and are in phase. Synchrony would have a pleasant effect on our sense of time. Dissonant moments would be those that are asynchronous and out of phase, breaking the smooth flow of time, causing it to drag.

The concept of concentrated attention started to take on a broader psychological meaning. It was not just selective awareness. It also had other transformative potentials that seemed to connect with an increase in metabolic rate (time), with particular memory constructs (context) embedded in a cosmic field and, at the same time, is able to amplify them into consciousness.

A similar metabolic effect seemed to be at work when playing basketball.

THE ZONE

Playing basketball provided other experiences I thought had relevance to some of these states. On many occasions, while playing highly focused and intense games, I would enter into what today is called the "zone," or "flow" (see Csikszentmihalyim 1991, for an interesting psychological analysis).

During these episodes I could sense my mind-body system merge into a kind of oneness in which, just like in the VW-versus-Mack-Truck incident, it would speed up and everything around me occurred in slow motion. It's important to note that there was no separate sense of thought followed by a muscular response. Thought and response registered to be one and the same.

Even though I was in a competition I felt linked not only to all the players but also to the audience and the physical environment as well. It wasn't so much that I lost my sense of Self; the scope of my identity appeared to expand and become more inclusive. There was no sense of "body" versus "spirit." Just me.

Strangely enough, I was competing with no sense of urgency or emotional attachment while merged with the creative intelligence that guided, or actually made, my every move. My best games occurred while in this state.

Certain conditions seem to lead into the being-in-the-zone effect. The first began when I was reaching the edge of an energy threshold. At that point I would experience my body kind of quitting on me. If I ignored it and persisted I would acquire a "second wind," and while this began to occur I would take deep breaths to slow my breathing down while maintaining the same physical activity. As the breathing and heart rate appeared to synchronize the zone effect would kick in, and I could play for hours within this expanded state of consciousness.

Obviously, there is an energy component that, I presumed, was sufficient to "fuel" a wider connection to all of the potentials (cellular, biochemical, electromagnetic) within the Self. There was no doubt that more of me was involved.

As these potentials manifested I experienced a synchrony with the activity at hand; I would feel being electrical, and I assumed it was this state that obviously connected me to the wider environment.

A BOY'S QUEST FOR WHOLENESS

After graduating I began to pursue a livelihood in the "real" world, and I took on a part-time instructorship in psychology at the university, which became a four-year interlude without any major "paranormal" events to speak of.

To keep my mind from wandering into my area of interest I simply tracked my lectures to the textbook. About a year later I was offered the position of Director of Admissions at the San Juan campus of the university.

This brought a longed for change of pace now that I was ready for a new environment, even though that type of work was not my first choice.

While there I took a course in educational anthropology from Claude Crawford, a visiting professor in the graduate program, and it was his sparkling spirit and love of his subject that rekindled my interest in epistemology and the subjective aspects of the mind.

As the course was winding down, he invited me to join the teaching staff in an experimental education program at an elementary school, based on the "open-classroom" concept he was heading in Saugatuck, Michigan, which I excitedly accepted.

This would bring me back to the mainland where I would be able to find out more about the burgeoning so-called "New Age" movement that I had been reading about, humanistic psychology, and new experiments that were going on with biofeedback related to mental focus.

The open classroom experiment lasted one year. The community either was not fully prepared to accept such a conceptual shift or they were overwhelmed by the lack of rigid classroom order to which they were accustomed. My disappointment had to be tempered by the fact that the citizens of this small town on the edge of Lake Michigan had at least taken a courageous experimental leap. I later learned that they went back to the "closed classroom" concept and seemed to be pleased with the results. The experimental school was a source of many insights about how children learn, which is always linked to their attentional abilities. I then set my sights on graduate school, but I was broke.

Sweetening the down moments in my life were a series of fortunate serendipitous events that Jung would have described as "synchronicity." My sister Angie would often say that I had a guardian angel looking over me, and I started to believe her. This thought was prompted by the events that unfolded from the collapse of the open-classroom experiment. A number of people involved with its formulation were looking for other involvements once it ended.

One such person was Byron Antcliff, the superintendent of the Saugatuck school system. He went on to another district and was also involved with the University of Michigan's Graduate School of Education. He had connections with people doing research in progressive forms of education, and they were looking for research fellows to participate in some of their projects. He highly recommended me as a candidate, and soon after I was invited into The Child Development Consultant Program.

In the summer of 1968 I was admitted to the program, which began with an in-residence counseling role at Fresh Air Camp for emotionally disturbed boys, mainly from the inner cities. The camp was situated in a beautiful rural Michigan setting with ample acreage beside Patterson Lake, near a small town with the ironic name of "Hell."

Because of the emotional instability of the children it was necessary to have a counselor with them at all times. We planned outings, games, canoeing and row boating, had three meals daily with them at the main lodge (a stately log building with a huge stone fireplace), and slept in their cabins at night.

While this was a special treat for these kids, who might have never traveled beyond the city streets, their emotional states often seemed to eclipse the healing potential of the natural setting, and it seemed that every minute was a challenge to curtail the verbal and physical violence, a syndrome one of the professors, David Wineman, addressed in his book, coauthored with Fritz Redl, *Children Who Hate*.

Each group of six boys lived in a cabin supervised by two counselors on an alternating schedule. The counselors arrived at the camp earlier and were trained by the staff to work with the children using various psychological intervention methods. Also, courses were provided periodically during our time away from the boys. We remained there for two intensive summer sessions with a different set of children each time. At the end of each period we were required to write a case history for each boy. This is where I learned more about the emotional dimensions of attention.

One boy, whom I shall call David to protect his privacy, was of special interest to me because he was diagnosed with severe attention deficit disorder (ADD), the inability to sustain attention required in learning new tasks; however, I noticed behavior which took place at night when the other children were asleep that showed that he could be extremely focused.

David was nine years old. It was bedtime, and, as usual, I read a story, which he listened to periodically as he would fidget, before I turned off the lights. I got into my bed and waited till everyone was quietly asleep before I got ready to sleep. A shade of moonlight came through the window. As I began to doze off I noticed from the corner of an eye that he raised his head slightly to look around. Curious to what was going on, I feigned being asleep, but kept my eyes slightly ajar. I could sense his energy sweep across the cabin space. He slowly uncovered himself and stealthily moved out of his bed, and like a panther in the night, began to move towards another cabin mate's foot locker, gently opened it and took something out. Then just as stealthily he returned to his bed and fell asleep.

Out of curiosity, I allowed this behavior to continue for several nights before I dealt with the issue. The dissonance between David's diagnosis of ADD and the very focused act of stealing of which he was capable raised academic questions.

What I subsequently learned was that he was not interested in the things he stole. It was more of an emotional charge that I believed came from the precariousness of the moment: the dark room, the possibility of being caught, the up-beating heart and the upsurge of adrenalin and neurotransmitters. There was little doubt that David sought these out-of-pattern acts because they brought an elated feeling of wholeness, a feeling that was often drowned out by his condition and the rigid routines of life.

The entertainment industry is the primary provider for the general population of such virtual moments of emotional intensity, and the much-desired holistic response of being one with the moment. However, these periodic dramas are not enough to sustain an enduring sense of heightened presence for which, I felt, everyone deep within their psyche longs.

When this sense of wholeness is not brought on by the world around one, or from fortuitous events, there is the possibility that moments of great danger, which depend on the subjective belief structure of each person, will be sought.

These involve the thrill seekers, which of course has many variations. While the physiological response to *out-of-pattern events* (OOPEs) may be standard, in all cases the contextual aspect imposed by one's belief system and personality plays a key role in how it will manifest.

SELF-HEALING & THE INNER VOICE

Once Fresh Air Camp closed for the season I moved to Ann Arbor into a campus apartment, and plunged into graduate studies and practicum work as a consultant at various elementary schools. While living there, I went through a divorce that brought powerful emotional energies to surface. Because of the stress that I was harboring, I became ill one day with what I felt was a ruptured appendix, a life-threatening condition. Something told me to not go to the university hospital. Why I trusted this "voice" I cannot say.

As the pain increased I could feel that the source of the voice was guiding me, and I went to bed without any fear of the possible consequences. I simply followed its intent. At one point the room became ice cold, and I wound a wool blanket around myself and fell into deep sleep.

The next day I awoke, free of pain, feeling completely healed! To make sure, I pressed the left side of my abdomen and quickly let go (a diagnostic procedure I learned from a medical book); there was no pain. Again, the illness brought me into a highly focused state in which all I could think about was the pain; when I decided to trust the inner voice all my anxiety faded, and the subconscious mind, I presume, took over the healing process. It was that "letting go" factor again.

A pattern was starting to form. What seems to have occurred was similar to the surrender that preceded the Zeus story: I let the voice take over. The subconscious mind, when given charge of the situation is willing and able to help, regardless of the situation. It seems to have the ability to create an integral response with specificity to what is of deep concern. When needed, it can produce a short story, it can alter space-time, integrate the Self and project it into the ambient space, and it can heal the body. It wasn't simply a "servo mechanism" as has often been suggested in some of the literature.

Even though I felt weak and famished from it all, I could feel a state of elation rippling throughout my body while, in the background, I could experience that state of *stillness* watching. Before I could get some food I started to hear voices.

SPOOKY ACTION AT A DISTANCE

I could hear my mother's voice in my head calling me and asking if I was all right. She lived in New York City. Seconds later the phone rang and it was her asking if I was OK. Chills rippled throughout my body.

Later on I could hear the voice of an old college friend that I had not seen or heard from for several years, and her phone call inquiring about my health came soon after.

I didn't let them know about hearing their voices before the calls.

What's so interesting about these episodes was the extreme focus and relaxation that lingered in the aftermath of the illness. It seems that may have been instrumental in connecting with a non-local energy field through which such communication could occur.

Only the people that had some deep positive emotional relationship with me were the ones that experienced the connection and called. The principle of *resonance* and *quantum superconductivity* seemed to be involved. Skeptical of the findings of quantum physics, Einstein referred to this notion as "spooky action at a distance."

THE ALERT HYPO-METABOLIC STATE

During the 1960s Eastern meditative practices were becoming popular amongst the generation of college students that formed the so-called "New Age" movement; while I was chronologically and culturally out of sync with that ethos, in 1970 I was swept up by the trend. Along with a number of friends and acquaintances I began to meditate, using the techniques of Transcendental Meditation (TM) created by the Maharishi Mahesh Yogi, the grey-bearded guru brought to fame by the Beatles.

Even though I was interested in the expansion of consciousness and the health benefits of meditating espoused by the Maharishi, supported by studies at the University of Chicago and Harvard, I also felt this would be an opportunity to see if this process could help to understand, and perhaps intentionally produce, the magical unbidden states that somehow came to me.

The technique was quite simple. I went through a brief initiation and was given a personal mantra to silently repeat to myself as I comfortably sat closed-eyed for at least twenty minutes twice daily.

Once a week I would go to the TM center to talk with an advisor about any problems I might be having. Almost invariably I was told to not let the context of what was arising in my mind take me away from the mantra ("letting go"), which was a sound, as far as I could tell, without any known contextual meaning.

After several weeks of meditation it was obvious that the notion of *feedback* was essential to the process of focusing my mind. Until then I hadn't realized how chaotic my mind could be in the absence of some pending urgency, and how difficult it was to control it to suit my purpose.

This was an interesting personal discovery because it revealed my attentional style: why I needed a sense of urgency to focus my mind on academic tasks, and why personal interest in something generated greater states of enduring attention. As David's case revealed, I too showed reliance on a strong emotion to accelerate my metabolic rate in order to generate mental focus.

Feedback is a cyclic information process that oscillates between an intended outcome and the behavior to achieve it. It involves steering the mind towards a certain goal, which in this case was the mantra.

It was also clear that silently repeating the mantra to myself produced the paradoxical result that simultaneously drew my body towards a sleep-like restfulness while keeping me awake. There is a midpoint between these two extremes in which I had to learn to dwell.

Often times I would succumb to the urge to sleep and lose consciousness. Occasionally, I was able to experience the paradoxical presence of being awake while asleep, as had occurred in a number of "paranormal" events I had experienced. It was also quite like being in the zone. The Maharishi referred to it as an "alert hypo-metabolic state" (a.k.a. "alert-relaxed" state). Somehow I knew that all my unusual experiences were rooted in this paradoxical state.

BIOFEEDBACK

I became very excited when I learned about the biofeedback breakthroughs that were occurring at the University of Chicago with the psychologist Joe Kamiya in 1958. Kamiya began experiments on brainwave frequencies by training subjects to respond to feedback signals from an electroencephalograph (EEG). It opened a new way to view the transformative powers of attention.

The interesting part of these experiments was the amplifying effect of what one aimed to modify in order to complete the feedback loop with the observer. Kamiya was essentially amplifying into awareness electromagnetic pulsations emitted from neurons with the EEG, otherwise involuntary and non-conscious processes, and learning to alter them in certain ways. Using this technology it was possible to calm the mind-body system down into a relaxed alpha state or ramp it up into the high beta-plus ranges where greater specified focus reigns. This was done by interacting with a signal (e.g., a sound, a light) to indicate the desired result.

It was thus established that people could control brainwaves, which had been thought to be involuntary states. This was the beginning of the brain wave biofeedback revolution. The *Psychology Today* magazine did an article on Kamiya in 1968 and the concept became a household term. Kamiya opened a door to a new way of understanding the mind-body interface.

Tapping this breakthrough, the biofeedback industry started to produce instruments for use by the general public. Believing that this technology could provide answers to many of my questions, I purchased a galvanic-skin-response (GSR) meter to measure electrical dermal activity (EDA) due to emotional tone, and an *Autogen* EEG machine, about the size of a shoebox, to perform almost daily experiments on mental focus in the basement of my home, which I set up as a study and personal laboratory.

Over time, I learned that meditating and practicing with the biofeedback instruments produced the "alert-hypo-metabolic" effects about which the Maharishi had spoken. The secret to avoiding negative forms of stress was to imbue the memory of relaxing with a positive or neutral context. The hypo-relaxed state seemed very suggestible while the context of the alert state could modify it. Again, all this relied on the ability to focus the mind intentionally for a certain period of time.

Using feedback from the GSR to learn to relax allowed me to enter deep states of relaxation. In going deeper, memories of the "past" would crop up, and I simply needed to "let them go" if I wanted to advance the process. The conscious-relaxed state correlated with the alpha frequency detected and amplified by the Autogen.

Eventually, I could use either device to produce the same state. At some point, just the thought of doing so evoked this paradoxical response. This was the most interesting breakthrough, for it implicated Pavlovian conditioning to some neutral stimulus.

LUCID DREAMING

Influenced by Jung's writings on dreams, I decided to keep a dream journal. I used autosuggestion just before falling asleep, indicating that I would be aware of my dreams.

In keeping with my hypothesis, the autosuggestion is an iterative function that imposes my goal (will) on the subconscious (encoded in the alpha-theta-delta frequency domain) through quantitative iteration. The key was the highly motivated act of focusing, the proposed psycho-accelerative effects of attention, which at this point had become a habit. The focusing continued as the body began to doze off.

There were moments when I could view my brilliant dreams like being in a movie, which were so delightful and mysterious that it was difficult to part from them in the morning. Yet I could continue where they left off the following night; I simply made a mental note that this is what would happen. This was the first time that I noticed I could intentionally affect my dreams.

Here, as you by now may have surmised, the gap (i.e., reaction time) between attention and the psycho-cybernetic wizard (the non-conscious) had considerably shrunk. However, while there was a modicum of control of the dream scenarios, I was only aware of the results, never the cognitive processes that produced them. What I learned from these experiences is that it is possible to be intentional with dreams when sustained attention is enfolded into the alpha-theta band of frequencies. From my lucid dreams I discovered seven factors:

1. There were two modes of consciousness that worked in an either-or fashion: the inner or the outer. When the outer one was active the inner one became inactive, and *vice versa*. Therefore, each would become the "non-conscious" of the other. During the waking hours one is not conscious of the "inner world," and vice versa while sleeping.
2. Being aware of the subconscious after falling asleep required some frontal activity, the specifics of which I was not able to identify.
3. I was aware of the dream environment.
4. I sustained a sense of identity within that context, which I equated, as the Maharishi averred, to being alert in a hypo-metabolic state.
5. I was able to critically think about what I was observing.
6. It was possible to have some control over what I dreamed.
7. Dreams have greater degrees of freedom when compared to the physical world. In this regard, dreaming and imagination seem closely related, and our creativity is intimately linked to this potential.

25

In some instances, especially at the start, I would emotionally assert to myself, "This is a dream!" I could pick a dream sequence, but I could not specifically produce the kind of dream I wished to have. I thought the reason for this was that the dream world is always a psychosynthesis of all that is impinging on the wave structure of the mind. This would include physiological, sensory, and memory inputs that have salience in terms of personal and hardwired instinctive pathways. As I thought about the "dream wave," as I called it, I could see that the cognitive processes that were generating it were one and the same, minus the sensory inputs that eclipse them during the waking hours. In a sense we are constantly dreaming!

My take on it all was that there is no set non-conscious mind per se, but only processes that we are not aware of, and this is constantly shifting. What is subliminal at one moment could become conscious in another one. I would not say that it is like a stream of impressions but more like a dynamic collage of information from which a symbolic-like definition is synergistically extracted. There was something apperceptive about it all. The mind is always synthesizing a story, the story of our lives. The writer James Joyce and the surrealist artistic movement captured its essence quite well. The Impressionists captured its pre-focal qualities. The Zeus Did It story was a strange compression that I still see unfolding to this day.

All this, I felt, had a relationship to brainwaves as carrier waves. This is how I conceived it. In deep delta-dreamless sleep (.5-4 Hz) all information is distributed in a field (perhaps the quantum field or the aether) in a power spectrum that we define as 1/f noise (pink noise). I suspect that this mysterious noise is not the ordinary notion of chaos, but a strange order of all the information of the universe from which we extract our existence.

In quiet introspective moments we can hear it in the head as a soft hissing or hushing sound (not tinnitus in the ears). As the organism shifts towards the faster rhythms detected by the EEG it "selects" from this repertoire in specific ways. It seemed that attention begins to become more specifically resolute in the beta (14-30 Hz) range, due perhaps to the modulation of these slower waves.

HOLOGRAPHIC VIEW OF THE BRAIN

One night after focusing on an imagined dark spot just a few inches outside the area between the eyes, my body went into a "rigid-like" state, and I was surprised to be able to see my entire brain like a translucent holographic image in full color floating before my forehead. I had not planned to do this. Perhaps, I thought, I was projecting this part of my body by simply amplifying it with attention since I was dwelling on the sensation of "fullness" in my head before the image appeared.

MOLECULAR BODY

One night after generating the dark spot a few inches from my forehead I could feel a subtle vibration throughout my body, and an energy bubble developed at the forehead, which I intuited was the cosmic egg. It took awhile for me to begin to doze off.

A bit later I could sense waves rippling throughout my body and surging towards my forehead, and in a matter of seconds I projected out of the body in what I later thought was a molecular body because it ("I") "floated" up toward the ceiling and got stuck in the upper corner of the room, from where I could see my physical body on the bed. It seemed to be trapped like a perfume scent in a bottle.

THE CONVERGENCE OF SCIENCE & SPIRITUALITY

During the 1970's, Eastern forms of mysticism were being exported to the United States in great number, forming part of the *New Age* movement. The term supplanted the one used by the hippies in the 1960's called the *Age of Aquarius*. This was the ambiance in which I joined the TM organization and learned to meditate.

However, there was so much that attracted me to this broad eclectic movement, encompassing religion, philosophy, mysticism, health, holism, ecology, science, especially the quantum-relativistic paradigm in physics, and the so-called "occult" (see Taylor 1999; and Chandler 1993, for how it manifested in the USA). While there were numerous "New Age" movements throughout the world's history, the one manifesting in the U.S. took on a more scientific ethos.

In the summer of 1968, Sheila Ostrander and Lynn Schroeder visited the Soviet Union, Bulgaria, and Czechoslovakia, to explore the groundbreaking research the Soviets were performing on psychic phenomena. In their book *Psychic Discoveries Behind the Iron Curtain* (1970) they revealed to the West for the first time the astonishing scientific breakthroughs and the key personalities who were performing and investigating these unknown human potentials.

Absolutely awed by what I was reading, I felt that it supported the intriguing events that had already transpired in my life. For example, the Soviets mastered "artificial reincarnation," trained animals, and induced hypnosis by distant telepathy, and filmed numerous feats of telekinesis. Even though the "cold war" was in full bloom I could not desist from respecting the courageous work the Soviets were doing while most North American scientists were in denial of the psychic realm.

Nevertheless, as the movement progressed in the USA the rigid barriers separating the spiritual and scientific communities began to soften. The tolerance reached an interesting confluence, especially in cities where large research universities are situated. Groups of students, including many professors, were already primed by their resistance to the Vietnam War to deviate from the fear-based group-think that led to that conflict. Not only was social disobedience in the air but also a willingness to allow old ways of thinking to merge with new visions in science.

Two iconic figures of the time were the Harvard researchers Richard Alpert (now known as Ram Dass) and the physician and psychoanalyst Timothy Leary, who were performing controversial studies on LSD (lysergic acid diethylamide), the potent hallucinogenic drug. Both were fired from Harvard after they espoused the spiritual and therapeutic uses of the drug.

After traveling to India to practice yoga under the mentorship of the guru Neem Karoli Baba (a.k.a. Maharaj-ji), Alpert was transformed into a new being and given the stature of sainthood and the name of Ram Dass by his guru, after sharing an OBE with him. In 1971 he shifted away from the use of drugs as the means to spirituality in his famous book *Be Here Now.*

Leary continued, to the chagrin of the powers-that-be, with the chemical/entheogenic approach, and in 1977 he developed a therapeutic and spiritual system he called *Exo-Psychology,* an eightfold path of development that was based on his interactions with other explorers of consciousness, such as the Bolivian philosopher Oscar Ichazo, whose work had much affinity with the system of human development developed by Gurdjieff.

Leary's prolific research led to explorations with a sensory-deprivation flotation tank, which he describes in his book *The Center of the Cyclone: An Autobiography of Inner Space* (1972). Reading it led me to consider the possibility that psychotropic chemicals, critical *out-of-pattern events* (OOPEs), and various forms of meditation can produce psychophysiological effects that "separate" the electromagnetic dimension of the Self from its sensory systems.

This occurs naturally when we transition towards sleep, but the purpose of meditation is to link, as the Maharishi averred, the wakeful state of mind with the more relaxed one. Floating the body in a tank filled with salt water allows extreme sensory isolation to take place, and by adding LSD to the equation, a chemically supported inner alertness produces the paradoxical state.

It was known by then that "uniform stimuli," any safe recurring event, causes the system to habituate, leading to boredom and sleep. Sensory deprivation, in its various forms, is the method used for centuries by transformative spiritual teachings to exclude temporal clues from the "outer world," allowing the mind to go inward to explore the inner nature of the Self (to "know thyself").

Psychotropic chemicals appear to also have this effect via their relationship to chemical structures like the neurotransmitter serotonin, involved in the synthesis of melatonin, the sleep molecule metabolized in the pineal gland. When the physiology transitions towards the recuperative benefits of deep sleep, a dissociation takes place from the outer-directed sensory system.

There was so much to choose from in this holistic, yet chaotic, cornucopia of information. No one, even to this day, has been able to define the New Age; it was a totally open manifestation without a central figure, except that one could glean a general purpose towards which it seemed to gravitate.

While many of its critics decried it as cultish, a satanic conspiracy devised by the anti-Christ, and even drawing parallels to Nazism, proponents saw it as a "revolution in consciousness" (see Guiley 1991, 406). It contributed immensely to the "new thought" movement, namely that one can intentionally alter the nature of one's personal reality.

This challenged the notion of determinism in orthodox religion and deterministic science.

Intention formed a conceptual nucleus around which gyrated a sense of being responsible for not only oneself but also for the social as well as the physical environments. The notions of caring, cooperation, and love were its emotional mandates. Within these rather broad ideals science was held to its humanistic roots in serving humanity and its habitats.

The fact that so much was simultaneously occurring drove many into decision-making conflicts: what to practice, who or what to follow? The New Age was rich in information and poor in how to harness this cornucopia towards a finite purpose. I too was caught up in the kid-in-the-candy-store syndrome. My inclination leaned towards its scientific offerings.

The merger between spirituality and science was echoed in the hippie term "vibes" to describe how a person was energetically coming across to another. Rooted in the spiritual notions of Dass and the radical Leary, the ethos of the most recent "New Age" was the raising of energy mediated by psychotropic biochemicals and meditative techniques.

While I did not indulge in psychotropic drugs, I deduced from all the unusual events I experienced, and my readings in neuropsychology, that they altered not only the rate at which the mind processes energy (biochemical, electromagnetic) and information but also how the energy was configured, by reducing its resistance to editing structures situated in the center of the brain, the alleged brain structures of the "ego."

In 1975 the physicist Fritjof Capra published *The Tao of Physics*, a best seller that correlated Eastern spiritual teachings with quantum physics. Soon to follow on the best seller listings was Gary Zukav's *The Dancing Wu Li Masters: An Overview of the New Physics* (1979). These exciting reads added a more technical dimension to this historical convergence.

Not so iconic, though not of less importance, was British psychologist Maxwell Cade (1918-1985), who performed research on biofeedback and its correlates to higher states of awareness (1979). What added an extra element of significance to his work was his background; he was not only a psychologist, but also a Zazen meditator and teacher of Raja Yoga, as well as a research scientist and distinguished physicist. He pioneered a new way of simultaneously measuring the EEG power spectrum from both hemispheres of the brain by devising an instrument he called the "Mind Mirror®."

With this instrument he discovered that the mental state of individuals is determined by hemispheric ratios of the EEG spectrum. In probing the states of highly developed gurus and mystics, he found that their brainwave patterns were consistently more balanced in both hemispheres than those of people not practicing a transformative teaching. His work added a holistic perspective in viewing the mind in a more practical and scientific way.

However, other than Kamiya's studies, there were no further insights into the scientific nature of attention, especially its psycho-transformative powers, the central theme of my quest. It mostly revealed what attention could "do" but little on what it is and how it works.

The only link that I could discern that brought science and the transformative tradition closer was the notion of vibrations. After all, vibrations are so powerfully experienced by every individual, which we tend to express in somewhat prosaic language. Science broke from this limitation by how it rendered vibratory phenomena into finely measured wave oscillations: the speed and frequency of sound, the frequencies of the electromagnetic spectrum, the rhythmic biological parameters like the heartbeat, and so on.

In keeping with the New Age ethos, in 1974 I began reading works by the Russian theoretical mathematician and mystic Peter I. Ouspensky. A number of these books were based on the system of spiritual development he learned from his mentor, the Greek-Armenian mystic George I. Gurdjieff (1866?-1949). Ouspensky dubbed the system the *Fourth Way*, and held that it was a resurrected form of a "forgotten science."

Because of its aim to foster transmutations of the physical body into higher embodiments (a.k.a. astral, mental, divine bodies, etc.) of being and consciousness, I thought of it as *transformative psiology*, and the systems that used it as the *transformative tradition*.

During that period I joined a Fourth Way school (which I will subsequently refer to as the "School") and after a year of dedicated practice to the teachings I experienced two OBEs, which I determined were clearly electromagnetic in nature. I will briefly describe in the following section some key tenets of the system, and what these OBEs were like, however I will postpone the more technical aspects of the Fourth Way and incorporate it in the chapter *Principles & Practice of Transformative Psiology*.

ELECTROMAGNETIC PROJECTIONS OF THE SELF

I joined The School at the Cleveland, Ohio center, and travelled from Ann Arbor on weekends to attend meetings and workshops. Then I transferred to a more convenient, newly-opened center in a suburb of Detroit, housed in a modest mansion of many rooms.

After spending one year in Detroit, I moved to San Francisco to be closer to the main center and the Teacher, as he preferred to be called. When I first arrived I temporarily stayed in a teaching house where I was invited to use a basement room until I found a place to stay.

Before I got there I had secured a job with a clinical laboratory selling lab tests and servicing medical clients. Throughout the workday I made efforts to *remember my Self*, a method central to the Fourth Way system.

It was important for me to remember that this teaching suggests that memory is "real" and is transformationally consequential. For example, as I saw it, the physical body is a bio-construct held together by the memory of the Self in the molecules of DNA (deoxyribonucleic acid), and the mind emanates electromagnetically (i.e., detected as brainwaves) from these molecules, while the soul or spirit emanates from the mind as, quite likely, a plasma of sorts. These separations are viewed as phase transitions of a universal "substance," not any different in principle to how ice turns to water, water to vapor, and back again. This was in keeping with the various natures of my OBEs.

SELF REMEMBERING

Self remembering has two branches: a contextual one in which one seeks to remember all that one has learned about the teaching, and a behavioral one, of which one puts into practice what one has learned. Of this behavioral aspect, the non-expression of negative emotions is key, as well as a technique called *divided attention*.

MY FIRST MASTER MEMORY CONSTRUCT

Ouspensky defined Self remembering as "many lines of work converging on one point." While at the time I didn't quite understand what he meant by "one point," it had the ethos of the notion of "single pointedness" in Zen. However, it was obvious that the contextual aspect (i.e., thoughts, concepts, etc.) of Self remembering involves remembering a "new way of thinking" as the means by which the ego can be "reprogrammed" to affect one in a transformative way.

I had to keep in mind that "thinking" is a state of being mediating between the physical and the spiritual modes of the Self. Therefore, in *Proverbs of the King James Bible* we find the phrase

"As a man thinketh in his heart, so is he" (23:7).

According to this ancient wisdom, the heart is the vibratory interface between the mind and the body. Nowadays this interface can be interpreted with the parameters of the EEG, particularly involving the slow frequencies of deep sleep (i.e., epsilon/delta <.5-.5-4 Hz) with those of being awake in the alpha-to-gamma+ range, forming the basis of a "science" that is poorly understood. This book helps to unravel the mystery.

The old way of thinking is referred to as "false personality" (FP: with affinity to the "ego") while the new way entails the formation of a "true personality" (TP). To maintain some contextual coherence with psychology I thought of this new personality as a *master memory construct* (MMC) that could be used to supplant or modify the MMC of the ruling network of FP with a more complementary evolutionary format.

What distinguishes the MMC of TP from FP is coherence, the degree to which the thoughts are connected in a meaningful way with one's Essence (a.k.a., soul, inner child, etc.). FP is based on a rather loose, often contradictory and disempowering, context that briefly coheres based on the evoking stimuli of the moment and inhibiting everything else. Hence, it is constantly changing, but rarely with the power to enact a phase transition of the Self beyond that of the ordinary use of the mind.

As I had experienced, occasionally critical out-of-pattern events (OOPEs) shock the entire system into a temporary state of wholeness in which false personality is bypassed. Otherwise, the divisive effect of FP would prevent this "holy" state from taking place. Though these were enlightening yet fortuitous events, transformative spiritual teachings seek to intentionally evoke them through certain attentional practices and remember their effects as part of the MMC.

However, to avoid false personality from usurping the developing MMC of TP, the system uses esoteric symbols to keep the two initially "separate," that is, until the MMC acquires some "gravitas" and is able to overpower the memory network of false personality and entrain it into its novel context. This, I deduced, was the purpose of the symbol of the enneagram in particular, and of esoteric symbols in general, like: the Latin cross (✝), yin-yang (☯), astrology (♒), and so on.

In this regard it was interesting to me how Gurdjieff organized his entire transformative system on the symbol of the enneagram. It represented a succinct model of the universe, existing in vibratory frameworks (planes, levels, worlds) indicated by the notes of the solfeggio musical scale (measured as an octave based on frequency doublings). According to Gurdjieff:

> "The enneagram is a universal symbol. All knowledge can be included in the enneagram and with the help of the enneagram it can be interpreted. And in this connection only what a man is able to put into the enneagram does he actually know, that is, understand. What he cannot put into the enneagram he does not understand. For the man who is able to make use of it, the enneagram makes books and libraries entirely unnecessary. Everything can be included and read in the enneagram. A man may be quite alone in the desert and he can trace the enneagram in the sand and in it read the eternal laws of the universe. And every time he can learn something new, something he did not know before" (Ouspensky 1949, 294).

Such symbols now can be used as a hub to which the many lines of the teaching can be connected. While this connectivity made much sense in terms of weighting the MMC with associations involving all of the psychic centers, there was no scientific rationale at the time to explain why it was such a powerful tool.

LINKING EVERYTHING TO THE ENNEAGRAM

Upon seeing it for the first time the enneagram held a deep fascination for me. Perhaps this was influenced by my interest in Jung's view of symbols as archetypes of the mind. Using esoteric symbols as a modality for thinking is not the same as thinking in literal representative terms.

While representational in function, it was obvious that these types of symbols transcend ordinary signs in scope and meaning by how they incorporate multiple levels of information into a single frame of reference. Because of this, I assumed, that when it was thought of or viewed it would act as a multi-dimensional stimulus in the form of a mnemonic that can evoke a broader and more integral *psychophysiological* response, including the "conjunction of opposites" (the "marriage") that can unify the Self into a higher being body. In this regard it could be thought of as a master memory construct (MMC) that could be the force for a new level of being and consciousness. However, I was not sure how all this would take place in the neurological system.

Taking Gurdjieff at his word, I decided to express everything I learned about the system, and anything else (regardless of how subjective it may have been), with the vibratory dynamics of the enneagram.

To facilitate doing so I ordered a rubber stamp of the enneagram so that I could easily reproduce it on a page and add my comments to it on note cards.

The use of the enneagram in this way proved to be a very efficient *integrating mnemonic*, allowing me to remember, translate, and connect large amounts of information. In this regard, the MMC is constantly evolving. It helped me to think in a synthetic manner able to conjoin oppositional ideas. Eventually I would include various aspects of physics, neurophysiology, psychology, and other scientific theories.

All in all, I felt this integrative practice was instrumental in clearing the way for the OBEs that took place while I was involved in the School. Yet there were questions about the nature of the integration, which of course involved a "quantitative" factor I had not put my finger on, but which subsequently came to light when I learned more about neural synchrony and the nature of attention.

Eventually, I came to realize that it is not only the effect of what one believes but also the coherent nature spread throughout one's being of that particular context that provides the wherewithal of the transformative powers of the Self.

THE DARK NIGHT OF THE SELF

While this is happening the Self undergoes a powerful contradiction between the "true" and "false" personalities (referred to in some teachings as the "dark night of the soul"), which creates "the warring factions within." If one is poised to endure it, it will eventually resolve, when the MMC based on TP acquires "psychic mass." It will then be able to incorporate (i.e., entrain) the contents of the ego and assimilate it into its new frame of reference.

While more details on this aspect of transformative methodology will be discussed in the chapter *Principles & Practice of Transformative Psiology* in the section *Esoteric Symbols as Psycho-Transformative Tools* here I will briefly touch on the method, for it forms the roots of my discovery, and is key to understanding the theme of this book.

NON-EXPRESSION OF NEGATIVE EMOTIONS

It must be kept in mind that energy and context are closely related. For example, the mind configures energy into contexts, which are conserved as encodings in the non-conscious field. A pattern was starting to form that indicated that attention (fortuitously or intentionally evoked) draws information from the non-conscious and amplifies it into the vibratory threshold of awareness by how it alters its frequency.

DIVIDING ATTENTION

According to this hypothesis, attention is a selective and amplifying agent that allows for broad or specific windows of awareness and behavior. Therefore, information is *energy configured* in a certain way, which means that the non-conscious field is brimming with enormous amounts of potential energy-and-information (i.e., encoded memory constructs) of which, during ordinary states of consciousness, we expend in a non-transformative manner (e.g., by excessive talking, aimless movements, and especially in negative ways).

This is why in transformative teachings expressing it in unnecessary ways is discouraged, and why inner and outer "silence" (Peace) is encouraged. Hence, one needs to know what's going on inside in order to be aware of the motivations that precede the behavioral expressions.

The purpose of not expressing negative emotions is to conserve energy (in the School it was expressed as different gradients of energy, Gurdjieff referred to as "hydrogens": ranging from H-3 to H-96…) for the evolutionary process. However, dividing attention requires turning attention inward to observe what is going on inside, while at the same time to also observe what is going on in the external environment.

Both goals challenged the habitual way I would be aware of myself and of the world. Ordinarily, I would be responding to it in a mechanical, one-directional, stimulus-response fashion. While Self remembering, I had to be aware of the old habits and respond intentionally to circumstances as they came up.

Usually, I would find it taxing to do both simultaneously. The literature in psychology claimed that it was not possible to divide attention in this way. However, with a bit of practice it seemed that I could pull it off.

An interesting thing about dividing attention is that one can choose the *scope* of what is being attended to. That day I had decided to experiment by extending the scope of what I was experiencing in the external world. For example, I made an effort when interacting with a client to also see that person in the context of the room we were in. With the next client I expanded to what I recalled of the building, including the parking lot. Initially, I had doubts that I could think and speak clearly while performing these mental gymnastics. Strangely enough, my mind became clear and I could express myself quite well.

STARRY BODY

The sun had set when I headed home after a grueling schedule that left me exhausted. Yet, I could feel a pleasant humming throughout my body as I drove along the highway.

When I got to my basement room I turned off the light and crashed on the bed, still in my work clothes. Since the room had no windows, it was pitch black, adding a degree of visual deprivation.

As I started to fall asleep my attention went to the point on my forehead, a practice that had become habitual since my days with TM, and I became very still. Shortly after, I could feel an electrical humming throughout my body before I popped out in a starry body that made a sizzling sound like a Fourth-of-July sparkler, and the room took on a subtle luster of light. I hovered over my tired body for what seemed to be just a few minutes before entering it and going to sleep. When I woke up I realized that this OBE was more electrical and fiery than the others I had experienced.

With this OBE I deduced that all subtle manifestations of the Self had a consciousness pattern: the higher can see the lower, but not the inverse.

This incident brought to mind one that Ostrander and Schroeder had described in their book *Psychic Discoveries Behind the Iron Curtain*. It involved a Kirlian picture, shown by the Kirlians (Semyon and Valentina, the inventors of the technique), to the Leningrad surgeon Dr. Mikhail Kuzmich Gaikin, of a man's body that looked like a carnival of light. When a Kirlian photograph was taken of Gaikin:

> "[H]e saw an unbelievable display of "fireworks" in his own hands—great channels of violet fiery flashes blazed turbulently. There were silent yellowish-red and blue lights like dwarf stars" (1970, 2110).

Gaikin wondered if these flares were being produced by electricity from the nerve endings in the skin. However, he was informed that these same flares could also be seen pouring out of plants, which have no nerves.

Curious of what the energy was, he remembered an incident in which he witnessed an acupuncturist in China heal a man of what was deemed an "incurable" disease. When he inquired of his Chinese hosts about the nature of the energy, they informed him that it consists of:

> "An energy we call life force or Vital Energy [that] circulates through the body on specific pathways" (Ibid, 211).

They described two energy systems in the organism: a bioelectric one and one of vital energy, which the Russian scientists spoke of as "bio-plasma."

Ostrander and Schroeder referred to it as the "X force" and correlated it to the various terms used since antiquity: Prana, Mana, Munis, Magnale Magnum, Animal Magnetism, Odic force, Motor force, N-rays, Etheric force, Bioplasmic Energy, Psychotronic Energy, and so on.

The Soviet research asserts that this "force" appears to be the doppelgänger, or etheric double, of the physical body (in spiritist circles it is referred to as the *perispirit*: see Kardec 1857). And, as the Chinese meridian theory proposes, it not only provides the "vital" energy source but also is the main sensorium of all the unedited energy and information entering the organism, in addition to its ability to connect, communicate, and create with the universe at large.

REMOTE VIEWING

Months later, I found a room to rent in another teaching house. Before going to sleep I felt my entire body humming with energy again. I simply let myself passively witness what was going on, and I gradually faded into a deep sleep.

Just before waking in the early morning the humming had accelerated to a higher pitch, and I projected in what could have been another electronic body, because I went right through the walls, like a radio wave (although I did not experience the penetration of the house). I hovered over the Golden Gate Bridge with an incredible clear view of the bay, the cars crossing the bridge, the morning sun scintillating off the rippling water. Because of its ability to tunnel through the walls I thought of it as consisting of quantum scalar waves (more on this later).

THE RAINBOW

One day, after working many continuous hours in a very focused manner on finishing a building project in the basement, I went to bed exhausted with a sharp headache, when suddenly I heard a loud sound of rushing water and became merged, aware in an absorbed manner, with a beautiful rainbow accentuated against an emerald green sky.

The sound may have been the amplified hum (deep pink noise) I could hear in my head during peaceful meditation. It seemed that in this instance sound was transduced into light. Spaced equidistantly apart, I could see the hooded heads of monks embedded along the curve of the rainbow. It seemed to be "telling me" that the spiritual mind is able to control the electromagnetic spectrum.

I use the term "aware" as if I were observing it separately, but I must emphasize that it was one of those integral states where subject and object do not manifest. I was enveloped in the field, yet paradoxically, I had a distinct sense of awareness, like the oneness of being in "the zone."

Rainbows have a paradoxical symbolic significance given that they are conjunctions of sunlight and rain, fire and water.

SYMBOLIC MIND

On many occasions while in an integral state I would experience my entire system become unusually energetic, and when I focused on the energetic sensations symbolic material would appear as visions or lucid dreams. The high-energy states suggested that the cosmic egg was working with large amounts of information, which I assumed the conditioned consciousness of the ego could handle only in compressed symbolic formats.

Some of these symbols were quite interesting to interpret from scientific and psychological points of view. Yet I felt that they were "messages" of some aspect of my Self (probably the higher mental center connected with a "Higher Mind") trying to tell me something about the universe and my inherent nature.

THE MULTI-COLORED BIRD

Occasionally I would have a symbolic dream with a personal message. One took place in a brightly colored scene in which I was a naked boy in a desert walking towards a tall stone circular wall. I decided to climb it to see what was inside. When I got to the top I noticed that the wall encircled a large brilliant multi-colored bird that I felt a strong connection to. It seemed that the bird had no idea that it could fly out of this predicament.

There was no doubt that the dream or vision was about my trapped soul (Self). The wall was my ignorance, but the solution seemed obvious: don't believe in the wall.

The wall could also be interpreted to mean horizontal resistance, and the opening as a pathway towards vertical freedom, transcendence.

IMMERSED IN THE GOLDEN LIGHT

One night, after falling into a deep yet vigilant sleep, I projected into a luminous scene of "golden light." There was nothing else but this enormous enveloping glow. It emanated a warm and loving feeling such as nothing I've ever experienced before that I wanted to stay there and never leave its midst.

A LADDER TO THE SUN

While living in a small apartment in Brooklyn Heights, NY, I began to awaken one morning and suddenly I entered a lucid vision in which I begin to climb up a ladder to reach the sun. As I climb, the noise from the city gets progressively quieter. On reaching the sun I experience an incredible silence and a loving golden warmth.

THE COSMIC EGG

One night just before falling asleep I projected what I construed to be the electromagnetic structure of the pineal gland. Focusing as I had always done, I saw it before my forehead, appearing as a three-dimensional oval, like an egg, animated by a soft pattern of different brilliant liquid colors *weaving in-and-out* of diamond shaped windows.

It did not appear like a material object, but more along the lines of being comprised of light waves, like a hologram.

At the time I did not know that the pineal gland was an organ involved with light and the hormone melatonin, used to regulate the sleep-wake cycle and the sexual system. In attempts to understand this particular experience, I often wondered if this gland was a loom for weaving (transducing, binding) threads of light into imagery in order to connect with physiological systems, and/or perhaps a resonant system for getting in touch with other worlds organized into a spectral arrangement, which I thought were vibratory milieus. Was this how the mind binds disparate phenomena into a seamless state of consciousness?

If we were indeed made from the light (photons) of the world, as mystics as well as physicists proclaim, this gland would play a central role in the nature of human life. The lay scientist in me wondered if this gland was the interface for certain forms of organic life with the electromagnetic spectrum.

By dint of its dynamic self-referential structure I thought of it as a *cosmic egg*, the binding mechanism of the mind. The fact that I was able to perceive it suggests that its role is to synthesize memory constructs in the non-conscious field, or the field itself, into objects of perception. But how, I wondered, was it aware of itself?

THE MARRIAGE AND THE VORTEX

Months later while living in Brooklyn Heights, NY, I went through a rough emotional period and, as I was about to get out of bed one morning, I experienced a vision of exquisite techno-symbolic content involving a vortex, which emanated a strange relationship with the cosmic egg, which at that time I couldn't figure out. For now, let's focus on the marriage and the vortex.

It was about a marriage. The groom was dressed in a black tuxedo and his bride was enveloped in glowing white veils. They were at the base of a spiral stairway enclosed in a vortical wrought-iron cage, as they were about to ascend towards the distant apex, where they could see a point of brilliant light.

Here I abstract the essential features of the vision: the couple, the vortex, the direction they must travel, and the point of light. Let's address each one separately as to what it symbolically conveys.

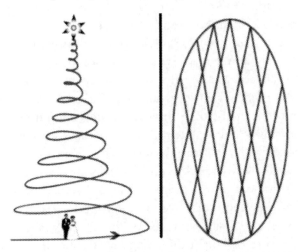

Figure 1: Marriage and the Vortex (left), and Cosmic Egg (right)

The Couple

As I saw it, the couple represented the complementary yet dual nature of the Self. In its physical mode of being it suggests the active and more passive nature of the autonomic nervous system: the sympathetic is symbolized by the male while the parasympathetic is symbolized by the female. The Marriage indicates equilibrium.

The Vortex

Though not of diminished importance is the conical wrought-iron cage, winding in a continuous and gradually tightening curve encasing the spiral of steps.

The entire structure directly suggests in spatial terms degrees of the attentional process. Each step upwards indicates a higher degree of focus and non-identification with other attractors.

After all, attention is essentially a "dissociative" process which by some means "separates" the relevant "signal" from the plethora of "noise" entering the system at all times. The vortical dynamic seemed to model that quite well.

The Direction of Travel

Starting at the base of the vortex indicates ascent in a counterclockwise direction, towards greater degrees of focus towards a smaller vortical configuration. Awareness at the base would be more general, like that of a forest, while each step upward leads towards greater specificity, like that of a tree, then a branch, and perhaps a leaf. Or: maybe everything just shrinks in scale and merges into a synthesis.

Because there's a defining movement involved, the spiraling direction suggests *time*. As such, it provides a temporal dimension (i.e., fourth dimension) that is expressed as frequency, the speed at which the ascent or descent takes place (i.e., the rate of vortical gyration).

Musing on this a bit further, we can envision the arrow going in the opposite direction such that it leads towards the base and a "slower" experience of space-time. Hence, movement towards the apex leads to a "faster" experience of the same.

Because these are either-or alternatives, we can assume that this is the basis of the dualistic state of mind. One direction leads to a more specified alertness while in the other to a more relaxed state of awareness.

In trying moments nature speeds up the process while in peaceful moments it slows it down. In the former it expends more energy while in the latter it conserves it.

If our psychology follows the law of *energy conservation*, a key principle in physics, it would of course be frugal with its use of energy and define our comfort zone as one in which there is minimal expenditure of energy. This would be defined by a threshold based on interpretations of survival. However, there's a catch. What would happen if the psycho-organism were prompted to co-exist in both states at the same time?

This is precisely the oppositional state of mind the Maharishi's spiritual system prompts his followers to practice: the "alert relaxed or hypo-metabolic" state of being and consciousness. This, of necessity, would entail parallel processing (as has been put forth by modern neuropsychology) in order to encompass upward and downward spirals to simultaneously co-exist in the mind. In this scenario the frequency equation would symbolically read: ><.

Jungian psychology would quite likely see this as the conjunction of opposites (*conjunctio oppositorum*) that forms the precursor to his "individuation" process. According to wave interference in physics the >< effect would theoretically produce a cancelling effect in the form of a standing wave. This leads to the "marriage" consummated as a star.

The Marriage

The notion of a marriage is common in spiritual themes (e.g., Shiva/ Shakti, male/female, Yang/Yin, etc.). As mentioned, it also plays a key role in Jungs's psychology, especially that aspect derived from his study of alchemy. According to the symbolist Cirlot:

> "In alchemy [marriage is] a symbol of 'conjunction', represented symbolically also by the union of sulphur and mercury—of the King and the Queen. Jung has shown that there is a parallel between this alchemic significance and the intimate union or inner conciliation—within the process of *individuation*—of the unconscious, feminine side of man with his spirit" (1983, 204. Italics added).

From a neurological perspective we might quite easily translate the marriage as coming from a conjunction of *parasympathetic*—the passive relaxing neural branch of the autonomic nervous system—with *sympathetic*—the active neural branch of the autonomic nervous system— giving birth to the symbol in the neocortex. The analogy suggests a triune relationship.

The Star

All I could deduce technically about the star was that it was a maximum compression of information streaming up on the legs of the vortex and being "bound" into a moment of perception.

This strongly suggests a "mechanism," if you will, by which information is drawn from the non-conscious field (i.e., any encoded memory in the psycho-organism that is not in awareness at any one time) into a phenomenal state of consciousness.

Once it reaches the peak there would be an implosive (i.e., fusion) generation of "light" (i.e., the electromagnetic field), where energy converges to produce a *wave interference pattern* we perceive as an event, and then, upon reaching a maximum of compression, reverts to progressively sink back into the long-waves of the non-conscious domain of signature-encoded information (i.e., memory). The up-down movement between the conscious frequency range describes a dynamic oscillation.

THE MYSTERY OF ATTENTION REMAINS

While the exercise of dividing attention proved once again that the hyper-focused attentional process seems to be a "causative" factor in accelerating my "mental" metabolism, indicated by the subtle electrical vibrations, and generating a duplicate of my Self into another phase of being and consciousness, I still had many questions about the nature of attention. How does it trigger these accelerative processes, and why is my identity and awareness not lost in these phase transitions?

Even though the Fourth Way system had compelling metaphors to implicate the vibratory nature of the Self to transform its modes of being and consciousness, it was still explained in somewhat prosaic terms. Nevertheless, by the 1980's neuroscientists took up the issue of attention, and as I began to read the literature I could sense they were nearing a great breakthrough. However, it would take some years before I came to realize what it actually was.

In the meantime the transformative processes that I had exposed my Self to all these years were still germinating inside me.

THE GROWING PRESENCE OF SELF

For many years hence after I left the School I continued to meditate, following the twice-daily TM protocol, and to also remember my Self as much as I could throughout the day. Both modalities seemed quite compatible in bringing the alert-hypo-metabolic equation into a state of augmented Presence.

Recall that an important aspect of meditation and Self remembering involves using attention in a paradoxical manner. Both employ alertness (one to the mantra while the other to the relaxed body and/or the outer world) and deep relaxation.

I deduced that this simultaneous-dual activity evokes a balanced neuroendocrine response in which intent, mediated by the central nervous system, evokes a conjunction of opposites in which the sympathetic (alert) and parasympathetic (restful) branches of the autonomic (non-conscious) nervous system are equally active or equilibrated.

Ordinarily, both branches are constantly shifting in ratios to one another in response to sensations. In evoking both simultaneously, and sustaining the ratio (which I assumed could be the golden mean ratio: 1.618...) for a period of time, a certain oppositional tension is experienced as a strong sense of solidity and inner stillness, an indication that the polarities of the Self are merging and transmuting into a more steady state. At some point I could feel the equilibration as a subtle tension producing a rather distinct joyful Presence inside of me.

GELLING A PERMANENT 'I'

Gurdjieff referred to this growing feeling of Self as preceding the formation of a more enduring sense of Self or "permanent 'I'." There were times when the Presence was quite deep; on other occasions it hovered around a rather peaceful yet common state of awareness. What determined how long it would endure was based on how long I could sustain (i.e., will) the simultaneous attention span (i.e., consciousness of) to what was going on inside of me and the outer world, especially when I was moving about while working on chores, an indication that the movement was necessary to increase my metabolic rate.

THE "ZONE"

This wasn't new to me, since I experienced what occurs when the physiological thresholds are intentionally exceeded while playing basketball, or running (see Rohé's *The Zen of Running* 1975). If one ignores the urge to stop, takes a deep breath to calm the system, and continues with the activity, an energy reserve is tapped. This is known in sports as "second wind," and one feels invincible. It is the precursor of going into the "zone." Gurdjieff spoke of the reserve as involving two accumulators, which are tapped when the organism exhausts its normal supply. This dynamic is also evident in other systems of spiritual development.

In yoga, for example, the energy reserve is called kundalini, situated in the root chakra (the *muladhara*), which is tapped when the polarity of the system is brought into balance (i.e., when *ida*, parasympathetic, equals *pingala*, sympathetic).

Any continuous exercise that puts a metabolic demand on the physiology would evoke the same process. This would include the use of difficult postures (such as yoga asanas), like bending, kneeling, working on demanding tasks, or taking long walks (as the Australian Aboriginal people do). Obviously, shamanic dancing would come under this category. The possibilities are endless.

SHAMANIC BIKING

About six years ago I performed a shamanic-like experiment, using a recumbent bike, to see if I could extend the levels of presence by accelerating my metabolism in a very relaxed manner in order to increase the levels of sympathetic-parasympathetic conjunction.

I hypothesized that if I continued pedaling beyond signs of "exhaustion" and maintained an equal level of poise, the level of consciousness would expand exponentially.

When I pedaled the bike to reach a point of exhaustion I would then continue for a few minutes or so and would stop to totally relax and just feel the energy surging through me. In those instances the energy flow would shift from the musculature inwardly and linger quite intensely.

In that state, everything around me appeared "static." Yet, I could feel the walls of the room and the sounds that normally came from the outer world suddenly freeze. For example, when I looked at the second hand of my watch, it literally "stopped." However, I knew that it actually didn't cease to function, but appeared so to my accelerated mind.

This correlated quite nicely with the VW-vs-Mack-Truck event, and the Microcosmic-Introjection one. Altering the metabolic rate appears to affect our experience of time, which also alters how we experience ourselves, and the world around us.

Theoretically, we can commune with different dimensions of reality by changing our vibratory rate.

This suggests that attention accelerates, and thereby amplifies a certain context into a state of perception. In the VW-vs-Mack-Truck case it amplified and integrated my perception of the oncoming truck, the road, and my Self in the VW. The Microcosmic-Introjection incident showed me attention could do the same, altering the scope and depth of different, fortuitously or intentionally, "selected" contexts. From the parsing of these experiences I had no doubt that attention is an *amplifying-selective-psycho-accelerant*. But, what was it amplifying, and how does it do this?

FIXING A RELATIVELY STABLE SENSE OF "I"

The bike exercise involved remembering my Self, with particular frontal-lobe focus, as I pedaled and listened to relaxing New Age music, for at least thirty minutes or more a day.

The aim was to see if I could extend and fix the memory of the state to every cell and atom in my being by increasing my metabolic rate through the continuous movement. The distributing (i.e., transmitting) mechanism would, of course, be the heartbeat with its broad electromagnetic field.

Not only did I want to spread the information more globally but also to connect it to the observing point at the center of my brow as the triggering center. Focusing on the frontal lobes would then bring the memory of the entire process to the executive function, which would allow me to evoke the holistic effect by intentionally recalling it.

A MORE PERMANENT 'I', THE SELF

After a month of this "shamanic biking" I could feel the Presence settling in my being with a more intense characteristic vibration that persisted whenever I would relax.

By the third month it took hold; only its intensity varied, which I associated with the duration of the biking. The longer I was able to bike the greater its intensity and robustness.

At one point it persisted in the background even as I watched a video or shopped for groceries. There was no doubt that I had to some considerable degree congealed the "permanent 'I'," which Gurdjieff had set as one of the goals of the Fourth Way (using the term "crystallized").

It was the Self reborn in a different, more inner, vibratory format (i.e., phase), one in which my meager consciousness had made contact with its non-conscious frequency. Whenever I wanted to amplify it I would simply divide attention, meditated from the center of the brow for a brief second or so, and it could reach a peak where I would be overwhelmed with orgasmic-like feelings. Gradually, my sensory systems became more sensitive, and there were moments when stimuli seemed unbearable.

DORMANT DARK NIGHTS OF THE SELF

Yet, there were costs. Accessing my entire system in this manner evoked dormant memories that I thought I had resolved. The release process caused me to dramatically face, especially in dreams, many issues connected with fears related to the dangerous environments of my childhood and remnants of the Z-Complex. I knew this was the "purification" process to which many of the teachings in the transformative tradition refer. While some were difficult to endure I felt assured that if I continued to accept and "let go", whatever came up I would be on the right path.

REQUESTS OF THE SELF

At night, just before going to bed, I took to "connecting" with this more permanent aspect of my Self and asking in an auto-suggestive way to give me "greater consciousness, health, compassion, and wisdom."

Many health issues were resolved from these requests. For example, I was able to lose about fifty pounds, gained after I ceased basketball activities, to help reduce the effects it had on my type II diabetes. The diabetic indicator A1C dropped from 11 to a "normal" for my age. Strangely enough, my sweat lost its pungent odor, a sign I took to mean that I was metabolizing food and repairing tissue more efficiently.

CONVERSATIONS WITH THE PRESENCE

Soon after, I explored its conceptual powers by asking it questions just before going to sleep. To my delight, answers in the form of visual and/ or lexical formats would be forthcoming in the early morning as I woke up, and I would quite excitedly get up and add them to the extensive notes I kept on 4x6 cards.

It was as if I had discovered the genie inside me who was behind all of my paranormal events. What is interesting about the process is how my entire body-mind-soul is participating, not just my brain. Every part of me is conscious, thinking, feeling, sensing, and contributing to the "response."

For the most part the questions focused on my ongoing search for scientific and metaphysical insights to the unusual events I had experienced. Thus far, it had integrated what they shared in common, suggesting a self-referential, feedforward-feedback (F-F) process, involving the following three transformative stages:

1. A triggering event, challenging my habituated psychophysiological thresholds.
2. The release of energy (molecular, electromagnetic) affecting the entire mind-body complex.
3. The rise in energy "fuels" the acceleration of psychophysiological processes, producing a unified sense of being, accompanied by an expansion of awareness, a rise in ability to remember, intelligence, creativity, and conflict resolution.

As had become routine, answers led to other questions. I was curious to know more about how this self-referential process functioned, since it seemed too mechanical to generate such rich creative effects. There was something missing.

While I was expecting a more straightforward response, that morning I was reminded of the cosmic egg and the marriage-spiral vision in a rather implicit manner. It came as a thought followed by an image of the infinity symbol (∞), suggesting that they all had a relation.

EMERGENCE OF
THE SYMBOL

"Science cannot solve the ultimate mystery of nature. And that is because, in the last analysis, we ourselves are a part of the mystery that we are trying to solve." - Max Planck

What did the infinity symbol have to do with the marriage-spiral and the cosmic egg? I tried to logically figure it out, but this was not successful and generated a dissonant feeling. Obstinate in my inquiry, I asked the same question again, expecting a lexical response, but this time I was given the same imagery. Realizing that I couldn't force it to fulfill my expectations, I gave up.

Surrendering was the key. The next morning I came out of a deep sleep with no dreams or visions to speak of, and to my surprise I felt that familiar electrical sensation building inside, and in a matter of seconds the following inscribed Symbol flashed into clear view.

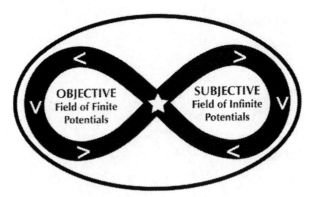

Figure 2: The Symbol

Just as occurred with the *Zeus Did It* story, I got up, bristling with mystery and excitement, and recorded what came to mind.

A SYMBOL OF UNITY

I had no doubt that it was, as Jung had averred, a "symbol of unity," confirming my efforts to come to grips with myself, particularly in regards to the Z-Complex, the powerful memory construct dictating in the darkness of my being, together with other divisive contexts.

Symbols of unity are the result of the process signaling "individuation," the culmination of "wholeness" in the individual. Jung states:

> "The phase of the process is marked by the production of the symbols of unity, the so-called mandalas, which occur either in dreams or in the form of concrete visual impressions, often as the most obvious compensation for the contradictions and conflicts of the conscious situation" (in DeLaszlo 1959, 458).

For the first time I know what Jung implied by his notion of the "conjunction of opposites" (*conjunctio oppositorum*), and that

> "There is no consciousness without discrimination of opposites." (Ibid, 347).

The integral nature of which was evident in the Symbol's equal emphasis on the objective and subjective potentials.

At the same time the Symbol emerges as a personalized expression of my Self coming to terms with its contradictions (involving psycho-fragmentation, neurosis, and other pathological conditions).

A BASIC UNIT OF CREATIVE INTELLIGENCE

While "symbols of unity" are idiosyncratic expressions unique to each individual, they share universal archetypal truths. Not only am I embracing the state of being it represents but also note that it gives back an integrative response whenever I ponder it for awhile. They are interactive, and in this sense I received a strong affirmative response when I thought it represented the dynamics of a unit of basic intelligence.

I immediately knew that all my queries would be answered by meditating on it. I also knew that I was no longer going to only be given direct "answers", but an integrating symbolic means by which they would be forthcoming.

After a few rounds of contemplating the image it settled in my forehead. For weeks I could not get it out of my mind, regardless of what I was doing. Then one day it suddenly vanished, leaving behind the Presence.

LIKE A WAVERING CANDLE FLAME

I could now sense-feel it as a subtle ever-present undulation (like a wavering candle flame) at the center of the brain, extending as a subtle joyous vibration throughout the cortex and the body, that emanated more intensely from a point that could be felt either at the center of my forehead or at the top of my head.

At times the two points would merge at the center of my brain, producing intense ecstasy. Initially I would feel a bit dizzy, but I realized that it indicated a new kind of psychophysiological functioning I had to get used to.

A DYNAMIC INTERACTIVE STILL POINT

I could immediately feel that its dynamic nature was interacting with every nuance of thought and behavior. Above all, it was arresting my attention to that still point, from which the emanations came forth. It was as if I were constantly in the zone. I was privy to the multiple streams of influence in the forming of every thought, feeling, and sensation; yet, at the time, they had no affect on me.

For example, I reach for a cup of coffee and I immediately know that the cup and the liquid in it are objectified potentials of infinite subjective possibilities. A strange cascade of what is "truly" involved with the simple object of a cup of coffee floods my awareness, and for the first time I know I have experienced a multi-dimensional form of consciousness.

> Clay-wet-earth-fire-ovens-pottery-ancient-peoples, defining moments of creativity-history-wars-water-agriculture-coffee-beans-domestication-of-animals, ice-cream—caffeine-molecules-atoms-particles, mental driving, wave patterns, "empty" space, all dancing before a sentient silence…

OBJECT & SUBJECT ARE THE DYNAMICS OF ONE

Questions I had regarding the nature and relationship of the objective and subjective fields are "intuitively" answered with the associative-connectivity of that seemingly irrelevant incident, and I'm led to know that I'm becoming privy to the dynamic interconnectedness of all things and non-things. Everything is important!

Each emerging point of interest is a doorway towards a whole new perspective. I deeply feel compassionately related to every thing and everyone and, strangely enough, to their infinite subjective potentials. In that perspective, the symbol is the 'name of the transformative game', where anything is possible.

CONVERGENCE AT THE STAR

The flow comes from every direction and culminates at the point, which on the Symbol is the conjunction, the *Star* window through which the LOGOS/GNOSIS is being revealed to me as an all-enveloping feeling of being in LOVE: the MARRIAGE.

THE PLURIPOTENT UNIFIED FIELD: PUF

However, I learn that the Symbol is not the ultimate reality; it instantiates an auto-kinetic (AK) aspect of something even more mysterious, from which it emerges as the means by which the inexhaustible creativity of subjective and objective potentials are enacted as "standing waves." I label this rather unnamable notion as the *Pluripotent Unified Field* (PUF).

The PUF is a wave-producing medium bristling with "pure potential." In its resting state the PUF's potentiality cannot be deduced or described. When it becomes modified by some means, it develops there dynamic wherewithal for phenomena and functionality to exist. Therefore, the quantum field, the zero-point energy field, the Higgs field, and so on, are secondary effects of the PUF. They are at least one step removed from being Primal and Omnipotent.

Theories that come closest to the PUF are those put forth by the astrophysicist Milo Wolff and his "wave-structure of matter theory" (2008) and by the Turkish engineer Hakan Egne and his theory of matter based on the notion of a primal "superfluid" (2012a, b). There are others, but these have been given special emphasis by the Presence via the Symbol.

WAVES & VORTICES CREATE PHENOMENA

The waves and vortices created by the PUF are both cognitive and creative. They form the basis of all the variety of form and function that we see all around us. They are responsible for the creation of the phenomenal cosmos, the galaxies, the suns, the planets, and all the subsystems that pertains to life on them. They form the very basis of systemic behavior, involving inputs, throughputs, and outputs of vital substances, consisting of configured forms of energy and information.

A key variable that defines the various dynamic entities as to their "place" and use in the universe is the rate at which they process energy and information. Hence, there are varying levels of being and consciousness pervading the universe that constitute degrees of intelligence and power; these are tempered by the fact that they rely on degrees of collective coherence (i.e., the "constituents" of a system) in which Love is by any means the greatest "integrator."

A COLLAGE OF TECHNO-METAPHORS

A new and rather interesting journey is on the way, and it begins with breakthrough research on the "objective" nature of attention that will, I am "told," be subjectively linked into a collage of "techno-metaphors" that will form a new perspective on the transformative powers of the Self in its various modes of being.

It is not a theory, even though at times I think of it as a hypothesis, but even that is not sufficient to convey that these mental gymnastics are usually in search of an objective outcome, which violates the insights of the Symbol.

While the objective aspect is a definite boon to humankind, without the subject potential we would just be determined robots without freedom to think or create, which is the very essence of being human. Both are of equal importance, and without recognizing this we tend to lead lopsided lives. So, when I use such terms as hypothesis or proposal we need to keep this perspective in mind.

As I develop the book I will quote from studies and authors, using illustrations to hone in a point. In doing this I find that I need to repeat concepts in slightly different ways, as well as some of the citations, at the risk of boring the reader, to emphasize the big as well as the small pictures.

SYNCHRONOUS DANCING IN THE BRAIN

From the beginning of my journey I knew that attention is the key to understanding the unusual states that occurred to me, and if I could understand its nature it would also contribute to understanding many other things about the "mind" and its transformative potentials.

Attention is so ingrained in our lives that we rarely think of it, yet we intuitively know how valuable it is. While it's evoked non-consciously for the most part, we may have given it some thought when we need it to perform an intentional task, like threading a needle.

Passing a thread through the eye of a needle requires a moment of focused attention. The eye and the mind need to be on target, and the hand needs to be steady. Any possible distraction must be kept at bay. Less obvious is a subtle pause of breathing in an attempt to keep the body still. At a subtler level there is a complexity of biochemical and electromagnetic activities that support this rather simple act.

In technical terms, the mind-eye-hand coordination is guided by feedforward-feedback loops, which means that information is cycling back and forth between the activity and the controlling mechanism in the prefrontal cortex (i.e., forebrain) in order to meet the objective of passing a thin thread through such a tiny hole.

If the activity were more complex, say, walking along a balance beam, the scope of what attention needs to keep track of is more expansive, but the process is one and the same.

As you can see, attention is not an isolated function of the brain. Without knowing precisely what it is we can indeed say that it's holistically supported by the whole mind-body complex.

Attention is such a valued state of mind that it is sought by a mother from her child. She makes cooing sounds, gestures with her hands, and is generous with her kisses.

Teachers and business people seek to attract it, for they know that if their message doesn't get through, learning won't occur or a sale will fail.

The marketer as well as the social protester find bizarre ways to capture our attention, for they know that the mind becomes increasingly attentive as the degree of perceived novelty and strangeness increases.

Have you noticed the weird scenes on TV commercials that have little or no relevance to the product?

The magician uses it to distract us so that the novel and mysterious seem rationally unaccounted for.

The chameleon is amongst the outstanding camouflage artists of the animal kingdom. Somehow nature knew, long before humans appeared on the Earth, that the attentional process in predators can be fooled by non-evocative disguises. Warring humans carry on the same behavior of blending in with nature's colors with deceiving patterns.

Our friends see attention as a sign of respect and acceptance. When someone listens to us with undivided attention we feel socially worthy, and we value that special moment.

We dress up to draw attention, and the garb we don may be in search of a partner or perhaps to send a particular message: I am the beauty; I am the macho; I am the nobody; I am the boss; I am the scholar; I am the one who defies all of your dress codes.

To what extent a person is able to attend to something will determine how much they can learn. Most likely, their "place" in society will be linked to this ability.

Attention is absolutely necessary to succeed in playing games. One must be able to get the ball through the hoop, know where to move that chess piece, and so on.

There's no denying it, attention is a tool for survival. We use it to improve, and we use it to evolve, and even to attain "salvation."

Every spiritual system, whether religious or not, requires it of us, perhaps more than any other. One must "keep the eye on the prize," keep the mind focused on the mantra, stay with the moment and be in the "now." One must love the Lord with all thy mind, body, and soul.

There are some who aver that attention is the key to personal transformations that appear to others as magical and miraculous.

It is also the means by which we become aware of our selves, our inner attitudes, and the nature of how we are. It is the means to "know thyself."

For millennia attention has been at the forefront of human growth and development; yet until recently no one has had even an inkling of what it is. The early American psychologist William James thought of it this way:

> "Everyone knows what attention is. It is the taking possession by the mind, in clear and vivid form, of one out of what seem several simultaneously possible objects or trains of thought. Focalization, concentration, of consciousness are of

its essence. It implies withdrawal from some things in order to deal effectively with others, and is a condition which has a real opposite in the confused, dazed, scatterbrained state which in French is called distraction, and Zerstreutheit in German" (1890, 403-404).

However, this prosaically describes some of the most obvious aspects of attention, but it is not a quantifiable, replicable, scientific statement. The neuroscience of attention began in 1949 with a theory developed by the Canadian psychologist Donald Hebb.

"NEURONS THAT FIRE TOGETHER WIRE TOGETHER"

In his seminal book *The Organization of Behavior* (1949), Hebb introduced the theory of *cell assemblies* as *reverberating neuronal networks* that connect the biological function of the brain as an organ together with the higher function of the mind.

This idea had revolutionary implications about how the brain and mind were connected. It radically differed from the prevailing Pavlovian stimulus-response theory which, while still useful for understanding the nature of "conditioning" and learning, proposed that singular linear paths existed between the sensory organs and the brain.

According to Hebb, neuronal projections between various brain regions are not necessarily one-way. His conception was much more dynamic, for it brought into play widely-distributed neuronal pulsations (firings) as the *binding* mechanism for perception, learning, and memory formation he referred to as "engrams."

It incorporated the German psychiatrist Hans Berger's discovery in 1929 of brainwaves and the electromagnetic field as interconnecting agents, and also of the Gestalt school of psychology that described the assemblies as *holistic* interactions of patterned stimuli impinging on a background field of sorts, forming *gestalts*.

The theory explains how learning and Pavlovian reinforcement take place through the *repeated* associative firings of a group of neurons. These associations were held to alter the synaptic structure (the minute gap between two nerve cells through which impulses pass facilitated by a neurotransmitter) and strength between cells, creating the basis for memory formation. The theory is summarized with the phrase "neurons that fire together wire together."

EXPERIMENTAL SUPPORT FOR HEBB'S THEORY

For centuries theorizing about the mind and consciousness has been exclusive to the field of philosophy. This began to change when in 1780 the Italian anatomist Luigi Galvani discovered that frog's legs would twitch when stimulated by an electric field.

By 1924 Berger, continuing the research others had done on animals, detected electric currents in the cortex of a human brain and developed the encephalograph to record the first alpha rhythm.

The rhythm is measured in terms of Hertz (abbreviated as Hz), a measure attributed to the German physicist Heinrich Rudolf Hertz (1875-1894), indicating the frequency at which a wave, or ensemble of waves, oscillates every second.

The precursor to the modern electroencephalograph (EEG) was born. Since the time of Berger, neuropsychologists using more advanced instruments have extended the range of frequencies at both ends of the spectrum, which is broadly referred to as the "sleep-wake cycle":

Epsilon* (< .5 Hz), Delta (.5-4 Hz), Theta (4-8 Hz), Alpha (8-14 Hz), Beta (14-30 Hz), Gamma (30-100 Hz), Fast Gamma (100-200 Hz), and Ultra-Fast Gamma (200 + Hz).

[*Based on his unorthodox research with meditators and mystics, I arbitrarily added Jeffrey Thompson's (2009) epsilon parameter to the conventional EEG spectrum. It correlates with the "ground state of being" alluded to in dream yoga (see Norbu 1992)].

EXAMPLES OF WHAT THE EEG
DETECTS & SUGGESTS

An EEG is a test that detects different frequency patterns in your brain waves, stemming from the electrical activity of your brain. During the procedure, electrodes consisting of small metal discs with thin wires are pasted (with a highly conductive gel) onto your scalp. The electrodes detect tiny electrical charges that result from the activity of your brain cells. The charges are amplified by the EEG and appear as a graph on a computer screen, or as a recording that may be printed out on paper.

EPSILON

Epsilon waves are the slowest of the EEG spectrum. As mentioned, the epsilon state was discovered by separate investigations, and was dubbed epsilon by Thompson. According to him:

"Studies here at the Center for Neuroacoustic Research have shown clear and repeated evidence, in patients, of brainwave frequency patterns below the traditionally accepted lowest Delta rhythms of 0.5 Hz. This would be brainwave activity as slow as one quarter cycle per second, one frequency per 10 seconds, per one minute, or even longer. Indications of ultra-slow frequencies are evident on the EEG traces of certain patients experiencing extraordinary states of consciousness. These states seem to be associated with very high states of meditation, ecstatic states of consciousness, high-level inspiration states, spiritual insight and out-of-body experiences. Some of the higher Yogic states of suspended animation associated with deepest Delta brain states actually continue deeper into these below-Delta brainwave states, which we are calling the Epsilon State (Epsilon, since it is the next Greek letter of the alphabet after Delta). In order to explore these deeper extraordinary states of consciousness associated with Epsilon brainwave patterns, we have had to use traditional EEG equipment in unique ways and to initiate the design of specialized EEG equipment to measure frequencies this slow. Most regular EEG equipment is not set up to measure frequencies below 0.5 Hz." (1999, 1)

DELTA

Delta waves are associated with deep, dreamless sleep, and uninterrupted healing. They predominate during infancy. Increased delta is typically related to decreased awareness of the physical world. Disruption of the delta waves is found in a variety of disorders, like insomnia and attention deficit issues.

The hypothesis is put forth in this book that delta waves also connected to a vast memory field of intrinsic and cosmic information in regard to one's most primal state of being, known in the transformative literature as Essence.

THETA

Theta waves are associated with deep relaxation, drowsiness, and sleep in older children and adults. They can also be detected to predominate in young children while they are awake. Large amounts of theta activity in the brain of waking adults is rare. These waves have been experimentally correlated with the formation of memory, and this may be the reason why children are very suggestible super-learners during theta predominance.

Any kind of repetitive activity reaches a point where it down-shifts the mental system into theta. We recognize this shift when we become bored of the uniform stimuli. If we ignore the transitional impulse and allow the repetition to continue, it will amplify the context through *constructive wave interference*, leading to fractal formations, subtle variations of a particular input involved in the "thinking process" (more on this later on).

Tasks that become so automatic, like driving on the freeway, cause one to shift into theta and go on "autopilot" mode. During most of the ride one is apt to daydream while the non-conscious does the driving, according to a remembered routine. Any *out-of-pattern event* (OOPE), like an animal running across the highway, will thrust one out of autopilot and into a faster rate of awareness.

Because rhythmic iterative activity can shift the system into deep relaxation modes and enter into this non-conscious realm, spiritual development systems use it as a modality to train their pupils to enter into auspicious "alert-hypo-metabolic" states. In other words, they are attempting to *conjoin* two ordinarily "separate" states of consciousness into "one." As we shall see, this turns out to be one of the goals of transformative psychology.

Excessive theta activity correlates with poor decision making, impulsivity and slow reaction time. However, in conjunctive states of consciousness these deficits are reversed: decision making is intuitively informed and reaction time exceeds the "normal" rate by many orders of magnitude (e.g., as in creative activity, sports, and the martial arts).

ALPHA

Alpha waves are the midmost rhythms seen in normal relaxed adults without attention or concentration. They are associated with a degree of relaxed-Self-awareness. Training to remain in an alpha state can be useful for meditation and to reduce stress.

Note that being aware in other pre-alpha ranges as conjunctive states of being and consciousness is also possible, but this paradoxical coupling is rarely observed to be consistent in normal relaxed adults.

BETA

Beta waves are observed in the usual wakeful brain. It is associated with active thinking, active attention, and focus on the outside world, especially when the mind is engaged in solving analytic problems, judgment, and decision making. These waves are also emitted when one feels agitated, tense, or fearful, and in obsessive-compulsive behavior. Higher levels of beta emerge when one is in a panic state. Hence, they are related to the fight-or-flight response and in stress-adaptive syndromes.

GAMMA

Gamma waves play a particular role in cognitive processing, particularly in synchronous activity involving perception, memory formation, and the consolidation of information involved in learning. High amounts of gamma are also associated with memory retrieval, intelligence, and in sudden insights in both adults and children.

A correlation has also been found with transcendental states in highly-experienced Tibetan and Buddhist monks, displaying gamma waves in the 50-70+ Hz range.

Gamma waves have also been found to play a conjunctive role with theta waves in lucid dreaming, and the projection of thought forms and imagery to enact "paranormal" phenomena (e.g., psychokinesis, distant communication and healing).

I ASK THE PRESENCE
Q: Does the EEG frequency reflect the rate at which the mind processes information?

A: Yes, but you need to understand that information is energy configured in a certain way. The faster rates entrain the slower ones, where memory is stored in a comparative slower frequency, such as in theta, into a context encoded by the signature pattern of the faster frequency.

Q: I see, but why the faster frequency?

A: Note that when the Self is in trouble its overall physical and mental metabolic frequency increases, normally to the degree of the threat. This tells us that it is seeking to become more intelligent and resourceful. Speed is the way it does this.

Q: What determines the level of the threat?

A: The memory repertoire of the person. If they have encountered it before. If not, they'd go into problem-solving mode, that is, if time is on their side.

STATISTICAL PREDOMINANCE REIGNS

While all of the frequencies are detected in varying amounts in all of these stages, there's a statistical predominance of one group over others. All measurements are of different groups of neurons undergoing synchronous firing patterns.

ATTENTIONAL SELECTION

As Hebb had theorized, it was found that *attention occurs when a group of associative neurons pulse in synchrony—fire and wire together* (Singer 1999) in a kind of dance.

Synchrony is the simultaneous action, development, or occurrence of two or more events. The term synchrony also equates to technical terms like: being in phase, resonance, coherence, alignment, and the like. Subsequent studies found that:

> "Attentional selection involves brain processes that select and control the flow of information into the mechanisms that underlie *perception* and *consciousness*. One theory proposes that the neural activity that represents the stimuli or events to be attended to is selected through modification of its *synchrony*. Recent experimental evidence supports this theory, by showing that changes in *attentional focus increase* the synchrony

of neural firing in some neuron pairs and *decrease* it in others"
(Niebur, *et al*, 2000. Italics added).

These oscillations have multi-tasking roles that simultaneously *entrain* and *amplify* associative neurons into larger ensembles, constituting the basis for the production of attention, memory, perception, thinking, and imagination. At the same time they "deselect" (e.g., inhibit) a much larger repertoire of neuronal involvement.

The synchronous manifestations rely on the ability of certain neurons to generate *action potentials.*

ACTION POTENTIALS

In physiology, an action potential occurs when the membrane potential of a specific axon, the long threadlike part of a nerve cell along which impulses are conducted from the cell body to other cells, rapidly rises and falls, as the means to propagate signals.

Action potentials occur in several types of animal cells, called excitable cells, which include neurons, muscle cells, and endocrine cells, as well as in some plant cells.

However, the main excitable cell is the neuron, which also has the simplest mechanism for the action potential, which is also known as "nerve impulses" or "spikes," and the temporal sequence of action potentials generated by a neuron is called its "spike train." A neuron that emits an action potential is often said to "fire." In the brain, action potentials are supported by a complexity of biochemical-and-electromagnetic processes converging metabolically.

In other types of cells, their main function is to activate intracellular processes. In muscle cells, for example, an action potential is the first step in the chain of events leading to contraction.

The main players in the neuronal dance are the *pyramidal neurons.*

PYRAMIDAL NEURONS

It is important to note that neurons, with pyramidal structure, are the main input-output cells of the neuronal system responsible for synchrony.

"[P]yramidal neurons constitute the *sole* output and the largest input system of the neocortex. They form the principal targets of the axon collaterals of other pyramidal neurons, as well as of the endings of the main axons of cortico-cortical neurons" (Georgiev 2004. Italic emphasis added).

GLIA: THE SUPPORTERS & CHOREOGRAPHERS OF NEURONAL DANCING

Glial cells, sometimes called neuroglia, glia, or perineural cells, are non-neuronal cells that are the white matter of the brain. They have been estimated to outnumber neurons (the gray matter of the brain) by a 10 to 1 ratio (i.e., ~100 billion neurons times 10).

Glia were discovered in 1856 by the pathologist Rudolf Virchow (1821-1902) in his search for the "connective tissue" in the brain. He used the Greek word glia to indicate that these cells were the "glue" of the nervous system. However, it was later found that, in addition to surrounding and holding neurons in place, they also provide support and protection for neurons.

GLIA-TO-NEURON RATIOS IN OTHER ANIMALS

Subsequent research supporting the importance of glia in relation to neurons is found in the evolutionary history of their relationship. Accordingly, organisms lower in the evolutionary ladder (like jellyfish, worms) show fewer glia-to-neuron ratios than more evolved organisms, especially the vertebrates. Humans have by far the highest ratio, favoring the glia (dolphins are a contested close second).

GLIA & THEIR SUPPORTIVE ROLES

In the central nervous system (brain and spinal cord), glial cells include oligodendrocytes, ependymal cells, microglia, and astrocytes. In the peripheral nervous system (branching from the spinal cord) glial cells include Schwann cells and satellite cells.

Let's take a brief look at what these important, and hitherto misunderstood, cells do.

Oligodendrocytes

These glia form the insulation, constituting the lipid white matter around the axons of nerve cells in the central nervous system. This increases the efficiency of electrical transmission, like an insulator on an electric wire. A similar role is provided by the Schwann cells in the peripheral nervous system (the nervous system outside the brain and spinal cord).

Schwann Cells

Schwann cells form the myelin sheath on axons in the peripheral nervous system. Unlike oligodendrocytes, Schwann cells do not have multiple cellular extensions, but instead each cell wraps around a segment of axon and forms a multi-layered myelin sheath.

Other Schwann cells that do not form myelin instead engulf multiple small diameter axons into bundles. Yet another type of specialized Schwann cell encases the synaptic endings on muscle, much like astrocytes surrounding synapses in the brain. Schwann cells must perform all of the functions of astrocytes, oligodendrocytes, and microglia in the brain, as none of these glia exists outside the CNS.

As an Analog System

The orthopedic physician and researcher Robert O. Becker discovered how Schwann cells in the peripheral system are responsible in the wound and bone healing process. It was initially thought that the nervous system was the main system in conveying the intrinsic information of tissue regeneration, but his studies showed that the *alternating electrical current* (AC) of the nervous system was absent when healing occurred. Instead the *direct currents* (DC) of the Schwann cells, as an analog system, were responsible for mending broken bones and healing wounds (see Becker, et al, 1982; and 1987). This breakthrough led to new methodologies in orthopedic medicine.

The Meridian System and the Fascia

It also opened up strong possibilities for understanding the substrate of the *meridian system* of traditional Chinese medicine, including my eventual understanding of their relationship to the *fascia* (a thin sheath of fibrous tissue enclosing all muscles and other organ systems).

Ependyma

Ependyma is the thin epithelial-like lining of the ventricular system of the brain and the central canal of the spinal cord; it is involved in the production of cerebrospinal fluid.

Satellite Glial Cells

Satellite glial cells (SGCs) cover the surface of neurons in the sensory, sympathetic, and parasympathetic ganglia in the peripheral nervous system. They have been found to play a variety of roles, including control over the microenvironment of sympathetic ganglia.

They are thought to have a similar role to astrocytes in the central nervous system. They supply nutrients to the surrounding neurons and also have some structural function.

Satellite cells also act as protective, cushioning cells. Additionally, they express a variety of receptors that allow for a range of interactions with neuro-active chemicals. Many of these receptors and other ion channels have recently been implicated in health issues including chronic pain and herpes simplex.

Astrocytes

As their name suggests, astrocytes are characteristically star-shaped cells in the brain and spinal cord. They are estimated to be around 20-40% of all glia. They perform structural repair, protective, and nutrient functions to nerve tissue. While they are electrically non-excitable, they display calcium ion (Ca2+) excitability that generates waves that communicate with the neural network. Micro-cinematography has dramatically shown the extracellular broadcasting of this wave over brain tissue.

In addition to maintaining the extracellular ion balance that regulates the exchange of sodium (NA+) and potassium (K+) responsible for cellular oscillation, astrocytes are closely associated with cerebral blood vessels that regulate cerebrovascular tone by adjusting the blood supply to match local metabolic demands.

These metabolic patterns are detected by functional magnetic resonance imaging (fMRI) to generally indicate which part of the brain is being used (i.e., drawing on oxygen, glucose, and other biochemicals to fuel function) as a way to map the topography of mental function. A single astrocyte may enwrap several neuronal bodies and make contact with thousands of synapses, indirectly regulating synaptic strength and information processing.

Astrocytes which reside within the respiratory chemoreceptor areas of the brainstem's functional respiratory pH sensors have been implicated in the control of involuntary as well as voluntary breathing (Gourine, *et al*, 2010).

This has strong associations with the practice of breath control in the transformative tradition, and focus on the medulla oblongata (situated in the brainstem area) by Yogananda's Kriya form of yoga.

It is important to note that astrocytes are implicated in the generation of neural plasticity (the regenerative capacity of nerve cells), particularly in the *hippocampus*, situated in the limbic system at the center of the brain, which has association with the formation of long-term memory.

> "Neuroplasticity, also known as brain plasticity, is an umbrella term that describes lasting change [especially in dendritic connections] to the brain throughout an individual's life course. The term gained prominence in the latter half of the 20th century, when new research showed many aspects of the brain remain changeable (or 'plastic') even into adulthood. This notion contrasts with the previous scientific consensus that the brain develops during a critical period in early childhood, then remains relatively unchangeable (or 'static') afterward" (Wikipedia: Neuroplasticity. Bracket mine).

This correlates with Becker's discovery on the healing function of Shwann cells in the peripheral system.

BETWEEN THE STIMULUS & RESPONSE LIE THE ASTROCYTES

Of all the glia, it appears that the astrocytes are the most conspicuous in neural regulation, which they do by controlling the amount of extracellular potassium (K+) ions at synaptic junctions. Potassium is needed by neurons to depolarize, and if the extracellular fluid lacks the right amount depolarization cannot occur.

Since the mid-1990s, however, it was revealed that astrocytes propagate intercellular calcium-ion (Ca2+) waves over long distances in response to stimulation, and, similar to neurons, release transmitters called *"gliotransmitters"* in a calcium-ion dependent manner.

The same process also sends signals to neurons to release the excitatory neurotransmitter *glutamate*. Glutamate is an important excitatory neurotransmitter that plays the principal role in neural activation, and in assisting the neurons to form synaptic connections between each other. Therefore, glia are directly involved in the formation of synchronous cell assemblies, and much of this is attributed to the multi-tasking astrocyte.

> "Thus, a close [feedback] loop between neurons, glial networks and extracellular space could be considered as the complex frame within which *synchronous cortical oscillations arise*" (Amzica, et al, 2000, 1101. Bracket & italic emphasis mine).

When compared to neurons, the robustness of astrocytes is another example of their importance. When neurons are isolated from astrocytes and are placed in a petri dish with a supportive substrate, they fail to depolarize. On the other hand, glia function quite well by themselves.

> "An astrocyte is a self-sufficient, self-replicating cell signaling to *itself* contentedly [Indication of a positive feedback loop]. Neurons have no reason to exist except to support astrocytes. Mature neurons cannot function alone, whereas mature astrocytes have no difficulty existing without neurons" (Koob 2009, 39. Italic emphasis and bracket mine).

This does not mean that neurons are unnecessary, but that their function relies on glial involvement.

Astrocytes Process Sensory-Motor Activity

Using afferent-efferent neurons, the glia transfer energy and information to and from muscles and viscera.

"The astrocyte processes neuronal signaling and instigates the rapid action to muscles and viscera. Astrocytes are processors of sensory information and the instigators of motor action using neurons as tools" (Koob 2009, 58).

Therefore, these multi-tasking cells store and integrate information that neurons use to carry out a response, which can happen so rapidly that it acts like a reflex. Upon sending their calcium waves to signal the neuron, astrocytes pull back to not interfere with the neural-motor response. In order to do this:

"Astrocytes have been shown to shrink their cell body to let the neurons fire to each other uninhibitedly. Their end feet relax and pull up, like a child on a big wheel going down a hill. The neurons are allowed to fire rapidly without interruption or influence to create a reflex without the inhibition of thinking" (Koob 2009, 58-59).

By separating in this manner from neural activity the astrocyte isolates the thinking process, which as Koob finds is the root of thought. However, as shall be proposed, thinking involves a more intricate fractal-forming process that will be detailed further ahead.

Astrocytes Control Breathing

Since breathing is both an involuntary as well as a voluntary system it is important to note that astrocytes were found to control breathing through pH-dependent (acid-alkaline) release of ATP (adenosine trihosphate) involved in the production of energy for physiological processes (see Gourine, *et al*, 2010).

THE "NEUROGLIAL" SYSTEM

For over a century, it was believed that glia did not play any role in neurotransmission. However, we now know that they have an integral functional relationship.

"Increasing evidence suggests that glial cells are endowed with the ability to externalize their activity to the extracellular space and to neurons. Since the same activity is influenced by the extracellular ionic concentrations and the neurotransmitters released by neurons, it is suggested that neurons and glia entertain a continuous exchange of information. This behavior might have particular significance during cortical oscillations" (Amzica, et al, 2000, 1101).

Since the neuronal and glial systems are so closely integrated it would behoove us to use a more inclusive term to represent how the energy and informational exchange process of the Self is mediated.

Originally used to refer to glia, the term "neuroglia" seems more appropriate to refer to the combined system. Hence, I propose the neuroglial system (NGS). However, because the literature is so ensconced with the term "neuron" I shall continue to use it while advising the reader to realize that glia and neurons are inseparably related, though performing different yet complementary roles.

CENTRAL & PERIPHERAL NEUROGLIAL SYSTEMS

As has been described, the NGS is not exclusively occurring in the brain but throughout the entire organism. Accordingly, we need to substitute the term "central nervous system" (comprised of the brain and spinal cord) with the "central neuroglial system" (CNGS) and the "peripheral nervous system" (outside the brain and spinal cord) with the "peripheral neuroglial system" (PNGS). In the PNGS the Schwann cells do most of what the other glia do in the CNGS.

The central neuroglial system (CNGS) consists of the brain and spinal cord. It is so named because it integrates information it receives from, and coordinates and influences the activity of, all parts of the body.

THE AUTONOMIC NEUROGLIAL SYSTEM

The autonomic neuroglial (ANG) system is a division of the peripheral neuroglial (PNG) system that influences the function of internal organs.

It is regulated by the hypothalamus (a.k.a., the "instinctive center" in Fourth Way psychology) in the limbic system. The autonomic aspect is a control system that acts largely unconsciously (i.e., to the conditioned threshold of ordinary perception) and regulates bodily functions like the heart rate, digestion, respiratory rate, pupillary response, and sexual arousal. It also evokes all reflexes such as coughing, sneezing, swallowing, and vomiting.

Sympathetic and Parasympathetic Branches

The ANG system has two branches: the *sympathetic* neuroglial system and the *parasympathetic* neuroglial system. The sympathetic system is often considered the "fight-or-flight" system (or "quick response mobilizing system"), while the parasympathetic system is often considered the "rest and digest" or "feed and breed" system.

An important nerve of the parasympathetic branch is the vagus nerve, the longest nerve of the body (a.k.a. the "wandering nerve"), stemming from the 10th cranial nerve, with connections to key organs throughout the body. Of particular interest is how it affects the interface between the heart and the lungs to establish a balanced rhythm (see Porges 1995, for an excellent study on the "polyvagal" theory and its role in reducing stress).

In many cases, both of these systems have "opposite" actions where one activates a physiological response and the other inhibits it. However, in other instances this is reversed. Also, concurrent functions may occur in which both extremes are activated simultaneously, such as in sexual arousal, orgasm, and, as I propose, in higher states of consciousness.

Further ahead I shall discuss how such oppositional concurrence also plays an important role in spiritual development.

NEURONAL IMPORTANCE

While the latest research appears to render neurons as mere puppets of the glial network, they do have an important role. If they didn't, the microglia would have pruned them off the brain-scape long ago. Glia are functionally and structurally involved with neurons. For example, some of the glia connect their pods to the axon of the neuron to form myelin sheathing and influence its behavior.

While I will recognize the importance of these new findings and terms, I will also continue to use the conventional terminology because it's so widely used in the literature, and whenever the terms "neuron" or "neuro-this-or-that" arise read "neuroglia, neuroglial," etc.

Thus far, the discovery of synchrony has been a rather strong postulate about how the mental system self-organizes into cognitive moments, of which learning, memory formation, and perception are some of the most important manifestations.

However, we need to delve a bit deeper to have an idea of how synchrony manifests in the neural landscape as the means by which it separates a cognitive signal from the "noise," the asynchronous repertoire of memory encodings or constructs not selected in the FCCP, to produce a cognitive event.

BILATERAL BRAIN HEMISPHERIC FUNCTIONS

In vertebrates the brain is formed by two hemispheres, left and right, that are separated by a groove (known as the longitudinal fissure). Each of these hemispheres is organized into four areas called lobes, shown below.

In placental mammals, the hemispheres are mainly linked at the center by the *corpus callosum*, a very large bundle of nerve fibers, used to transfer information and coordinate functions between them. Smaller commissures in various parts of the brain also assist in coordinating exchange of information of more localized functions. Let's take a brief look at the function of these lobes.

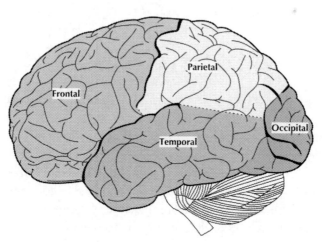

Figure 3: Lateral view of the brain.

FRONTAL LOBES

The frontal lobes, located at the front of the brain, are the largest. In humans it appears to be more developed, in relationship to body mass, than in other mammals (perhaps with the exception of whales and dolphins).

However, the psiological view doesn't place exclusive emphasis on the size of the brain and its parts, though this has importance, but more on the rate at which these regions need to process energy and information to deal with issues in the medium (i.e., solid, liquid, molecular, plasmic, etc.) in which it lives and has its being. Therefore, the spatial connectivity of the cells are more important.

The frontal lobe is the physical mode of cognition in which the process of necessity needs to be very fast in order to exert some modicum of control, given that these lobes have been found to be of great importance in executive functions of the mind, particularly with the ability to direct attention. A variety of resources to carry out these functions is available, especially the production and use of the biochemical *dopamine*.

"The frontal lobe contains most of the dopamine-delicate neurons in the cerebral cortex. The dopamine system is associated with rewards, attention, short-term memory tasks, and motivation. Dopamine tends to limit and select sensory information arriving from the thalamus to the forebrain" (Wikipedia: Brain Lobes).

PARIETAL LOBES

The parietal lobes have much to do with integrating the spatial relations the organism uses to navigate. This includes the sense of proprioception and for the sensory receptivity of touch, both of which are incorporated in the sensory motor strip colloquially known as the "homunculus" (Latin for "little man"), in which the body parts are rendered according to how much of the strip is devoted to them. The major sensory inputs (touch, temperature, and pain) from the skin relay through the thalamus to the parietal lobes. Because of its involvement in muscular orientation the parietal lobes also relate to speech and language processing.

OCCIPITAL LOBES

About 80-90% of sensory input is visual, and the occipital lobes are its physiological processing center in the mammalian brain. It incorporates many regions specialized for different visual tasks, such as visuospatial processing, color differentiation, and motion perception.

It not only plays a key role in "seeing" the outer world and one's relationship to it but also in visualizing and imagination. Hence, it has special significance in transformative teachings as part of auto-stimulatory methods to bring the various modes of being of the Self into a state of synchrony.

TEMPORAL LOBES

The temporal lobes are involved in processing sensory input into derived meanings for the appropriate retention of visual memories, language comprehension, and emotion association.

MIRROR IMAGES OF EACH OTHER

Macroscopically the hemispheres are roughly mirror images of each other, with only subtle differences from side-to-side, and even from individual to individual within a given species (an indication of subjective involvement in creative processes).

LEFT AND RIGHT BRAINS

Although much has been speculated about left and right brain function, there is considerable cooperation between the two to warrant caution in attributing absolute terms as to their function. Generally, it has been shown that the hemispheres show differences in how information is processed in terms of its scope: the right brain having greater inclusion and less specificity than the left brain, which appears to be more analytical. In some people this pattern is reversed.

Some researchers, however, find significant differences in sympathetic versus parasympathetic tone of the hemispheres:

the left being more sympathetically oriented, and the right more parasympathetically oriented (see Rossi 1986).

Size of the MC assembly would play a big role in generating a transformative force, and for this reason the analytic function is downplayed in some methodologies.

However, the most advanced insights into such transformative practices yield another revelation about the hemispheres and their relationship to the notion of "wholeness" in which both hemispheres display equilibration (see Hoffmann 2012 regarding EEG studies with subjects practicing "oneness deeksha" meditation), leading to a "neutralizing" effect and what is here referred to as the "Wholistic Response." More on this as we go along, particularly in the final chapter.

SCHISM BETWEEN SCIENCE & SPIRITUALITY BEGINS TO SOFTEN

As I searched the literature to find if spiritual systems were currently using/acknowledging this seminal information in regards to augmenting their understanding of their spiritual practices, I was surprised to see that only the Dalai Lama showed considerable interest, allowing researchers to perform EEG experiments on some of his monks.

It began in 1992, when the neuroscientist Richard Davidson got a challenge from the Dalai Lama to perform studies with his monks, which Davidson assumed would entail his current interests in why some people are more resilient than others in the face of tragedy, and, is resilience something you can gain through practice?

The Dalai Lama surprised him when he requested to know what "compassion does to the brain." At his residence in Dharamsala, India, he spoke with Davidson and said:

"You've been using the tools of modern neuroscience to study depression, and anxiety, and fear. Why can't you use those same tools to study kindness and compassion?" (Gilsinan 2015.)

THE BRAIN-HEART CONNECTION

While as far as I know, Davidson didn't quite understand the nature of the request when he began to place EEG electrodes on the scalps of the monks, who thought it was amusing since they held that he should have been probing the area of the heart, which for them was the site of love and consciousness. Yet, he found correlation with high gamma production from monks with long-term meditation experience, using the Buddhist love-and-compassion approach of mindfulness. Hence, high EEG gamma occupying large swaths of both brain hemispheres seemed to be the sole causative factor of continuous long-term mindfulness meditation. Davidson found the mental component but did not seem to grasp how it couples with the heart to form greater states of compassion and enlightenment.

While his findings are quite valuable for understanding the full picture of spiritual states in general and mystical states in particular, other research involving connecting consciously with the heart (see McCraty, 2006) not only augments the literature on gamma-evoked neural synchrony but also implicates the broad magnetic field of the heart as the broadcasting system of both fortuitous and intentional information entering the different modes of the Self (as a multifaceted sensorium) at any one time.

It is quite likely that the Dalai Lama was referring to this slow yet far-reaching pulsation of $\sim$ 1.5 Hz, extending beyond the body in a volumetric sphere of about 4 yards around, as the impersonal medium (a.k.a. an aura) by which the Self acquires and transmits information in adaptive as well as in transformative scenarios.

According to the revelations expressed in this book, the ability of individuals to entrain the different modes of the Self—physical, mental, spiritual—into states of transformative power requires, without exception, bringing these modes of being into states of collective coherence or synchrony, referred to as the Wholistic Response (WR).

Even though synchrony is a valuable component of the WR, it in and of itself is not enough to explain how information is processed by the different modes of the Self in order to explain its transformative powers, not only for acquiring knowledge but also of being and consciousness. This missing link was contained in the counterintuitive relationship of primal vortices to the nature of cognition.

COGNITIVE VORTICES
IN THE BRAIN

By the late 1970s a group of neuroscientists began to explore a compelling model developed by Walter J. Freeman III on how the mind uses vortical phase transitions between states of order (i.e., synchrony) and chaos (he expressed as "uniform disorder") to create cognitive events related to the quest for *meaning*. A synthesis of the work was subsequently presented by Freeman in his book *How Brains Make Up Their Minds* (2000).

Freeman was a neuroscientist and philosopher known for his pioneering work on how the brain generates our perception of the world. His prolific research earned him the reputation as one of the founders of computational neuroscience.

Of special interest was how he weaved certain philosophical ideas into his scientific findings. He proposed that his conclusions on how brains work agreed with the holistic philosophy of Saint Thomas Aquinas (1225-1274) on the unity of brain, body and mind, or soul.

Freeman referred to the unity as the dynamic effect of "mass action," which, while in keeping with the Hebbian hypothesis, involved how the entire population of brain cells self-organize into cognitive states.

> "This self-organized behavior, he argued, means that the images we see in our minds are the collective activity of many neurons spread over the brain, not the activity of single neurons. ... 'The brain reaches out into the environment and sees something *which is then interpreted according to its own past experience,*' Freeman said in a 1991 interview. '*First you look, then you see. The process of recognition begins from within*'" (Sanders 2016. Italics added).

The observation of "mass action" points to involvement of the "whole brain" in every goal-directed cognitive moment as it interfaces with the world through perception, and subjectively explores variations of a "theme" in search for "meaning."

Whether we are aware of it or not, all cognitive and behavioral events are motivated by an aggregation of inherent and/or acquired memory constructs that seek to generate some form of perceptual stability from the hot-wet-chaotic environment of the brain and the constantly changing outer environment.

LIKE THE VORTEX OF A HURRICANE

In his studies of laboratory animals (particularly on the olfactory bulb in rabbits) and humans he noticed a vortical pattern of this global process. According to Freeman (2009),

> "These patterns resemble satellite video images of the vortex of a hurricane."

Obviously, something is being mediated by vortical spin that appears to have relevance to a fundamental process in nature, since spirals and vortices are seen throughout the universe. Let's take a closer look at how they might work in the brain.

VORTICES INVOLVED IN THE ATTENTIONAL PROCESS

Freeman's topological patterns of the cognitive process showed the formation of two vortices (he also referred to as "cones"): one that gyrates towards the apex, producing an "implosive" effect, and the other reversing the gyration, producing an "explosive" (more like an "unwinding") effect. See Figure 4 for a Symbol-inspired illustration.

We need to keep in mind that the cognitive cycle at the human level is a fractal equivalent of the one used by the universe at large. According to this rationale, of which more will be explained as we go along, everything is fashioned from the PUF as waves and vortical condensations, occurring in self-referential spin states on different scales of existence.

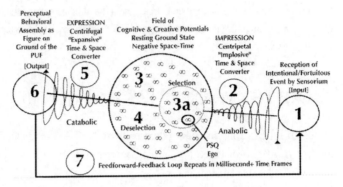

Figure 4: Interpretation of Freeman's vortices involved in cognition.

In this regard, the uniform field of the PUF becomes modified into waves that turn into all sorts of vortical waves, of which various kinds of vortices play certain roles.

Before going into the nature of the these waves let's briefly parse the above model to have a general idea of how it functions as a 1-to-7 cognitive cycle.

THE 1-TO-7 COGNITIVE CYCLE

As we zoom in on the cognitive cycle we realize that the formation of attentional assemblies is just one, though essential, aspect of it. It is also responsible for identifying inputs, creating and updating memory constructs, thinking, imagination, creating neural assemblies, and generating perception. Let's use the numbers that appear on the model as points of reference.

(1): INPUT

Everything that we perceive in the inner and/or outer worlds is formed and mediated by various kinds of energy waves, especially: sonic, biochemical, pressure, and electromagnetic radiations that impinge on a variety of receptors throughout the various modes of the Self, which I refer to as the *sensorium* (1).

Prior to interacting with the sensorium these waves behave in a Hertzian (Hz) manner, such that they radiate linearly from point A to multiple points, like a radio transmitter connecting with multiple receivers, which in the sensorium are the existing memory constructs (MCs) already poised to receive a pattern of information with which they can resonate.

As the Hz waves get close to the sensorium they begin to curl up and form scalar vortices (see Meyl 1996), creating of themselves the intricate patterns that constitute the self-referential nature of the MCs (indicated by the infinity symbol: ∞).

Hence, Hz waves carrying information are transformed into vortical-MCs that populate the field of cognitive and creative potentials (FCCP). Note that each MC is a mini-cognitive system in its own right. Therefore, each MC is a unit or complex of intelligence.

First We Look, Then We See

Strangely enough, research consistently reveals that we don't experience the input upon arrival to the sensorium. According to Freeman, "First we look, and then we see," and the "seeing" part of the process occurs after the stimulus has been selected at 3a in the FCCP and amplified to manifest at 6 in the cycle (see 6 below for more information). Accordingly, we may also think of the sensorium as a *field* poised by its varied repertoire to create meaning.

Intentionality

The input stage is where we can insert an idea or image that has a certain level of uniqueness in terms of the combinatorial possibilities existing in the FCCP. This stage highlights the subjective potentials suggested by the Symbol, and it instantiates the nature of our freedom.

(2) IMPRESSION: THE COGNITIVE CRUNCH

The ingoing vortex (centripetal, implosive) converts the incoming data as inputs from the sensorium towards its apex in a kind of "crunch." In doing so, it is acting as a space-and-time converter, simply because by changing its scale (space) it is also altering its time. In this case it is shifting towards negative space-time; to be discussed further ahead. However, the change in scale does not alter the context.

The information is organized by the structure of the *ingoing vortex* into their spectral order so that it can resonantly interact with the appropriate *memory constructs* in the FCCP in order to form the selection at 3a.

Psychophysiological Status Quo

The selection is influenced by the *psychophysiological status quo* (PSQ with affinity to the ego), a powerful MC that conditions the mental system to respond in habitual ways in its "quest" for stability. Its multiple streams of information are interpreted and edited in and near the limbic system of at the center of the brain.

(3) FIELD OF COGNITIVE & CREATIVE POTENTIALS

The field of cognitive and creative potentials (FCCP) has similarities to the PUF, the most fundamental field of universe, in the sense that it is its fractal equivalent (a.k.a. a microcosm). In this rather technical regard we can conceive of human life to be fashioned in the "image of the Creator."

Like the PUF, the FCCP is considered to consist of an unknown "substance" (e.g., "dark matter") occupying all of the space of the Self in its various modes of being—physical, mental, spiritual—in which a uniform condition prevails (similar to the concept of a pervading "super symmetry" in physics). Like the field of a magnet, it envelops these modes of being as interfacing *auras*.

Pure Potential

In the resting undisturbed state the field, or regions thereof, instantiates *pure potential*, from which all "things" and "non-things" are made from its wave modifications. The FCCP is similarly fashioned. To keep things simple, I shall also refer to both with the term "field" to express this cosmic equivalence.

Field Sensitivity

When the field is in a state of equilibrium it becomes very sensitive to stimuli or how it is modified. At this stage its acts as a sensorium. So, the field and the sensorium are essentially the same thing but serving two aspects of the cognitive process: as a receptor system and as a conserver of information.

Ingoing Vortex Forms in Low Pressure Zones

Upon being stimulated, the field develops low pressure zones or *nodes* that act as a "sink holes" through which the input energy, typically in the form of sonic or Hertzian electromagnetic (HEM) waves, are drawn.

Upon traversing the rather small sink hole, relative to the volume of the field, the linear sonic or HEM waves curl up, together with the information they carry, into an ingoing vortex. This is not any different in principle to how a tub full of water (high-pressure zone) squeezes down the drain hole as a whirlpool.

Energy and Information are Conserved

In the vortical state the energy and information carried by these sonic or HEM waves are conserved by dint of their self-referential-spin dynamics. Hence, we have the wherewithal for *memory* to exist.

Vortices Become HEM Waves & Vice Versa

When vortices are by some means properly stimulated (based on threshold parameters of their spin bonds) they revert to being transverse sonic or HEM waves and transmit the information they have stored from one place to another.

Note that HEM or sonic waves and vortices are created from the modifications of the singular "substance" of the field and are constantly converting into each other (see Wolff 2008 for a similar concept in physics). Hence, the singularity at the "crunch" affects the field and the field in turn affects the singularity.

All "Objects" are Forms of Memory

As we move along, it is important to keep in mind that the fundamental constructs of the universe (such as particles and atoms) are, according to this revelation, essentially vortical phenomena iterating by their self-referential spin a particular identity.

Hence, all the elementals (with affinity to alphabets) of quantum physics, chemistry, and biology are actually memory constructs (MCs) in signature vortical configurations, forming the basis of greater complexities (the so-called "building blocks" of the universe).

Since vortices exist on all scales of existence we conclude that memory is real and universally distributed throughout the space medium.

No Two Objects or Events are Identical

Generally, a field of any kind is made up of a collection of similar "objects," each of which has a degree of frequency and contextual *signature* such that no other object or event is exactly like it. That is why a hound can trace our molecular print and find us in a crowd.

This is not any different in principle to a human population of individuals that share certain attributes in common, yet are unique in their own right (this includes so-called "identical twins," who are very similar but not identical. The hound dog can prove it).

Please Don't Disturb Me or I'll See You

In its equilibrated resting state the field becomes conscious of any disturbance. Hence, the "sensorium" is a very general term for an array of receptor fields distributed throughout the organism. It not only involves cells but also molecules and subtler energy states working in concert or individually.

While neurons and glia are highly sensitive receptor systems, all cells in the organism are receptors of pressure, biochemical, and electromagnetic stimuli. They acquire specialized roles by their frequency/resonance thresholds. When cells, for example, are in a *balanced-resting* state they are said to be in "capacitance."

Motivation and Arousal

The resting state can be viewed as the baseline of a system in particular or the Self in general. It serves as an indicator of *balance* between opposing forces. Anytime its dips below or rises above these parameters it causes the system in question to become *aware* of the differential. It is then "motivated" to correct the imbalance, given that the equilibrated state is where the system "wants to be."

This is the basis of "homeostasis" for the Self in its physical mode of being. However, all the modes of the Self have a layer of "embodiment" in gradients appropriate to the rate at which they process energy and information.

The balanced state is where the system stores its energy and information, which it uses to accomplish certain objectives by discharging to perform a certain kind of work. It then goes through a stage of recharging and recouping its potentials. This is true for all the modes of being of the Self.

Energy entering the sensorium is yet to be interpreted as to its nature and relationship to balancing the needs of the Self in its various modes of being, of which the conventional terms of "body, mind, and spirit" are used with their technical meaning of *frequency*—rate at which they process energy and information.

Having a subtler makeup than its physical aspect, the field envelops the grosser modes of the Self with an auric envelope, similar to the field of a magnet. Keep in mind that all MCs are both object and field phenomena, which in the transformative literature has been expressed as "worlds within worlds."

(3a)-(4) SELECTION-DESELECTION

At the non-conscious level the input enters as an unknown context to the system, which it must interpret with its associations conserved as memory in the field before it can respond to it.

It's important to note that the selection-deselection process occurs either from a fortuitous or intentional input. In the former mode the cycle runs on autopilot while in the latter one a more conscious activity takes place by dint that the system accelerates to compensate for the OOPE.

Anabolic Like

Because it is constructing a percept from a repertoire of possibilities it has anabolic attributes, similar to tissue building.

The Leftover Is Huge

The fact that a deselected portion of the FCCP remains by default once a selection is made may not seem so crucially important. It allows for a sculpting effect to take place and "prevent everything from happening at once" (Einstein).

However, according to the principles of *transformative psiology* (the new science of being and consciousness being developed in this book) it is a key variable in the transformative process, given that the left-over energy would be needed to generate the Wholistic Response, a state in which much larger portions of the energy configurations (MCs) are employed to generate a higher state of being and consciousness.

Tapping Potential Energy

It is key because the quantity of how much of the Self is included at any point in time determines the power to enact a qualitative transformation from physical, to mental, to spiritual, defined by their energy and information processing rates. This obviously requires a concerted amount of energy, which is potentially available in the FCCP (involving local as well as nonlocal access).

Because each MC is a conserved "configuration of energy" the deselected amount of energy is greater than what has been selected. It therefore follows that under ordinary conditions of awareness there's an enormous amount of energy that's not being used in the subliminal realms of the FCCP. How then can it be tapped?

Conventional forms of thinking would attribute the acquisition and store of energy to the digestion and metabolism of solid foods and liquids, but this only accounts for what the physical mode of the Self needs to sustain its physiology via mitochondrial function (involving ATP and oxidative chemistry).

Given that there are some amounts of subtle energies in such foods (e.g., prana, chi, od, etc.) the amount is not sufficient in magnitude to effectively produce the transformative level of synchrony.

From a psiological perspective information mediated by the EM spectrum is also considered as being "food." As we shall more clearly see in subsequent chapters, this subtle energy is trapped in the enfolded vortical configurations that constitute the MCs. It obeys the principle that HEM waves transmit energy and information while vortices hold them in place.

Gradients of Fractal Energy as Subtle Energies

Fractal dynamics simply describe the mechanism with which MCs are fashioned in diminishing scale throughout the FCCP as a filigree of self-similar harmonics.

The subtleties of this ingoing progression constitute the energetic nature that those in the transformative tradition refer to as: subtle energies, prana, chi, orgone, od, manna, and so forth (see Müller in Waser 2004 for an in depth look at the mathematics and science of fractal extension).

Currently, the term has taken on the theoretical attributes of "zero-point energy" by enthusiasts interpreting the thoughts and achievements of the genius Nikola Tesla, and the like. Within this group we find the German engineer Konstantin Meyl, who interprets the fractal extensions in terms of vortical scalar waves (1996).

The Food of Spirits and Angels

Because each gradient level of energy has a specific value (a.k.a. the eigenfrequency in physics, representing the natural resonant frequencies of a system), it has use in "feeding" the cognitive and creative operations of all "higher being bodies" (e.g., spirits, angels) that process energy and information at faster "octane" rates within their signature frame of reference. The signature degree and amount of subtle energy processed by a system attracts other systems (entities) to one's sphere of existence by the principle of "need" and/or resonance.

The vibratory signature and state attracts other entities by the principle of resonance. In spiritist circles this is the basis of why "good" and "bad" spirits are attracted to one's being.

(5) EXPRESSION

The term "expression" means that the compressed information at 3a is amplified (also, magnified, expanded) into positive time-space by the eversion of the ingoing vortex (e.g., like turning a sock inside out).

It now transitions towards the larger-slower legs of the vortex. Nevertheless, it is not a full blown manifestation within the conditioned threshold of the observing Self.

Dimensional Scale Transitions

As the outgoing vortex expands the scale of the selected information, it is also progressively changing its dimensional format from 1D, to 2D, 3D, 4D, and so on. Here, dimension means scale, such that there's always three/four dimensions to every event that only varies in scale. So, a point has 3-4 dimensions too small to be detected by the human eye or instruments.

(6) OUTPUT

It is at this phase that what we "looked" at (i.e., thought of, imagined, visualized, sensed, etc.), becomes a perceptual and behavioral reality for our conditioned level of awareness. This is true for all modes of being of the Self, though occurring in faster operational rates.

This time-delay before awareness is not something we experience nor are apt to "logically" deduce to exist. To the unsuspecting observer, stimulus and response don't appear separate but seem to concur, as it should in order to produce a "seamless" percept and sense of reactivity. Nevertheless, this brief time-delay is important to the Self in order to know what the stimulus means and how to properly respond to it.

200-500 Milliseconds of Reaction Time

It is now a well-known fact that after being stimulated 200-500 ms go by before we are aware of the event. Because it is happening so fast we are typically not privy to the non-conscious aspects of the process. However, a "readiness potential" (RP), like lightning preceding thunder, ushers in the full event.

"One discovery related to the awareness of intention is the readiness potential (RP). Kornhuber and Deecke (1965. NIB) were interested in analyzing brain activity preceding voluntary movements. They asked subjects to perform self-paced finger movements while recording the subjects' electroencephalogram (EEG). By averaging many segments of EEG activity in the time window preceding each finger movement, they discovered a distinct electrophysiological potential recorded at the vertex of the scalp reliably preceding self-paced movements. They named this event-related potential the Bereitschafts potential (BP, German), also known subsequently as the readiness potential (RP)" (Scholarpedia: Proposed/Awareness of intention).

When analyzed with the EEG they found that:

"The RP is a slow, progressively negative electrophysiological potential whose onset occurs *approximately 500 ms before an initiated action.* The exact onset time of the RP depends on the nature of the task being performed, but *it consistently precedes conscious awareness.* It is surprising to find that awareness of a seemingly voluntary action might actually occur after unconscious preparation. Kornhuber and Deecke's demonstration provides evidence that voluntary actions are preceded and perhaps initiated by *unconscious* neural activity" (Ibid. Italics added).

Phenomenal Consciousness

At this stage, however, we may wonder how it is that we become conscious of phenomena being amplified. While the synchrony of pulsing assemblies (choreographed memory constructs) at 6 would suggest the mechanism of how this occurs, I am "informed" that this stage organizes the constructs, but that it is the CONTRAST of the created MCs to the isotropic field that is the key factor on all scales of existence.

Gestalt

In other words, it is the isotropic "field" that experiences its modifications. In this regard, we can think of the field as a sensitive "screen." The contrast between these two attributes is the basis of a *gestalt*, in which the field as "ground" becomes aware of a "figure" in its midst.

Absorption

In taking this to its "logical" ramifications, we must also conclude that the MCs, as self-similar fractals of the whole, also have a field (as fractal equivalence) and therefore experience themselves in contrast to the field in which they are embedded. However, there are exceptions.

For example, if the vibratory rates of the field and the MC in question are close, it's quite likely that such a high degree of being in-phase with one another will blur the sense of "separation." This vibratory closeness would have affinity to the notion of "absorption."

This often happens when we are in the midst of "great beauty" where the harmony of the situation envelops our being in the filigree of its pulsations.

Catabolic Like

Because there's an expansive aspect to this phase it displays the attributes of catabolic metabolism, building a construct on the way out as it dissipates.

Local As Well As Nonlocal

I propose that negative-energy states are a fundamental attribute of the FCCP, which consists of vortical soliton waves as the ideal information-conserving receptacles. In this regard, the FCCP is an enormous memory system that encompasses local as well as nonlocal cosmic interactions. One could say that it is the "informational library" of the non-conscious domain that forms what we conceive of as "space."

Scope of Phenomenal Consciousness

Reviewing and analyzing the neuropsychological research, the neuroscientist Lucia Melloni and her associates came to the conclusion that most of the studies on neural synchronization have been done on spatially restricted neural assemblies, without distinguishing between local and global coordination. They concluded that the determining factor in the formation of a conscious gamma-evoked assembly is the degree to which it is distributed throughout the brain.

> "The neural signature of *unconscious perception* would be *local coordination* of neural activity and propagation along sensory processing pathways, whereas *conscious perception* would require *global coordination* of widely distributed neural activity by long-distance synchronization" (Melloni, *et al*, 2007, 2858. Italics added).

This is an important distinction. Synchrony per se is, therefore, not the sole prerequisite for perception and phenomenal consciousness (although smaller more-localized assemblies partake in subliminal cognition in faster temporal frameworks). The size and scope of the neural distribution is the key factor. Spatial and temporal aspects are involved. In this regard we can see how a spatial quantity creates qualitative effects by a greater connectivity of the dots, so to speak.

Now, if we follow the complementary research, we have a parameter of sorts that instantiates a quantitative factor in understanding the notion of consciousness in general, and that of "holiness" to mean involving the "whole" of the Self in regards to spiritual development.

EEG Frequency and its Entraining Effects

Hence, the size and scope, the *spatial* aspect in the neuro-glial population of the assembly are important variables affected by the faster processing rate, the *temporal* variable (i.e., frequency).

It seems obvious that the amount of memory to be integrated into phenomenal contexts would be directly proportional to the processing rate and its entraining effects on the memory repertoire of the field of cognitive and creative potentials (FCCP).

Therefore, larger assemblies encompassing larger amounts of memory would require a faster processing rate to "bind" them together, which in turn would depend on the degree to which the mind is focused (Weiss, *et al*, 2003).

Speed Has Greater Binding Power

By expanding Melloni's discovery, we can postulate as a general rule that when the rate of mental processing increases it expands by default the number of memory constructs it is able to entrain and have access to. In other words, it has greater "binding" power (a.k.a. unity, yoking, re-legion, etc.).

Itzhak Bentov in his delightful and informative book *Stalking the Wild Pendulum* gives us an example of this rhythm entraining effect.

> "When electronic circuits are built so that they contain oscillations (of electronic type common in radio and TV circuits) that happen to oscillate in frequencies that are close enough to one another, there will be a tendency of the oscillations to get locked in step with each other and oscillate at the frequency of one of them. Usually, *it is the fastest oscillator that will force the slower ones to operate at its pace*" (1981, 37-38. Italics added).

Extends Beyond the Body into the Cosmos

On its way out it merges with the ultra-slow epsilon wave that extends beyond the body into the cosmic potentials of the FCCP.

While we are not aware of it, it is part of a F-F loop that informs the cosmos as well as the microcosmos of the individual to maintain the universal cosmic order. It is in this regard that we and the universe are bound in a dynamic co-creative relationship.

Therefore on its return to 1, it iterates the initial percept, which is also affected by cosmic inputs from the FCCP that are refined by the down-scaling of the ingoing vortex as the process repeats.

Intuitions

This cosmic aspect of the cycle connects us to the universe as a whole and brings in information the PSQ tends to filter out. At times the filtering does not work and information is unpacked in the outgoing vortex as an intuition, a feeling, a hunch. In this rather mystical regard the universe is constantly having a secret conversation with us, one that can be drowned out by environmental noise.

(7) THE QUANTUM MOVIE OF OUR LIVES

The F-F loop ties the whole process together into a recurring personal as well as a cosmic cycle. (Note: I will also use the term "feedback" to also indicate the F-F loop.) This 1-to-7 looping repeats itself in millisecond and faster time frames, producing a movie-like series of frames of information that we experience as a *standing wave* (e.g., the Star of the Symbol) as a modification of the isotropic field, forming a gestalt of phenomenal perception.

As mentioned, typically about 200-500 ms of cognitive processing precedes a conscious event. The speed of the cycling increases beyond this rate in the ultra-subtle mental and spiritual functions of the Self.

Simultaneous Tripartite Modes of the Self

All these levels are occurring simultaneously in the Self, but in differing ratios. Ordinarily the physical aspect predominates, instantiating a motivational pattern in which the physiology rules the mind and spirit in the satisfaction of its needs. The transformative psiological goal seeks to change the ratio so that the spiritual rules the mental and physical in the satisfaction of its needs.

Though we typically only see a portion of it, this "computational" process of personal and cosmic significance is going on all the time on multiple levels of existence, given our individual and composite fractal equivalents.

Iteration of the "Future"

It is at this juncture where "what goes around, comes around." However, every return is not identical to what went on before. What we habitually or intentionally enact repeats itself as behavior and events in our lives, with nuances of difference.

We can augment this differential by thinking of ourselves in creative and compassionate ways that can lead to a new way of life. We are back to the opportunities given by the input stage.

Again: the input stage is where we can insert an idea or image that has a certain level of uniqueness in terms of the combinatorial possibilities existing in the FCCP. In other words, this is where the subtleties of a personal and unique idea can be amplified by the cycle into perceptual and/or behavior realities. Change begins at the subtlest levels of being and scales down to the grosser modes of being.

The "Butterfly Effect"

Chaos theory shows how subtle initial conditions may affect the recursive equation (i.e., in which the output is enfolded with the input), with the example of the "butterfly effect", stating that the flutter of a butterfly's wings in Brazil may cause a cascade of events many miles elsewhere. Note that there are positives and negatives to this scenario.

Because of the amplifying effects of positive feedback loops (see THE MIND IS A CONTROL SYSTEM for details) a novel input at 1, however small, may very well instantiate the "butterfly effect" and become a full-blown reality if the loop locks into a hyper-repetitive cycle. Small things are important!

The Benefits of Strangeness

The notion of uniqueness is relevant to the history of individuals, written in the repertoire of the FCCP. Any desire to make changes in one's Self produces by contrast to this history (especially the interpretation mediated by the PSQ) a sense of strangeness that by default may bring forth levels of fear and angst and all that pertains to them.

While this turn of events poses to dishearten the inexperienced "transformer" it does bring benefits by dint that the Self in its slower modes of being and consciousness will accelerate the rate at which energy and information are processed to meet the "challenge."

Accelerating the Cycle Through Focus

Every nuance in the cognitive cycle is interconnected. For example, strangeness or any out-of-pattern event (OOPE) evokes attention, and the degree to which we attend to it (i.e., attention span) causes the cycle to accelerate by the fact that it is producing a positive F-F loop.

Step-Down Transformer

Note that the outgoing vortex is transforming the point attractor (the crunch at 3a) in negative space-time into slower time as it "expands" the scale of the information towards positive time-space. Hence, it has the role of a *step-down transformer* that satisfies the needs of the PSQ to interact with the slower frames of reference of the "outer" world.

It expands the "lines" of information into a resolute format that can be repeated by the synchronous assembly, which is the 1-to-7 process cycling in synchrony (a.k.a. a "steady state"). The outgoing vortex performs this deceleration effect by bucking the effect of the ingoing vortex with negative feedback (more on this further ahead).

Attention Span

The ability to sustain a percept X number of times through the cycle determines attention span. Interestingly, the degree of focus and duration of attention span is what accelerates the 1-to-7 cycle. However, we need to keep in mind that the cycle is the complementary activity of contracting and expanding flows of vortices.

Rate of Time Altered by Hypnosis

Studies done by the psychiatrists Linn Cooper and Milton Erickson (1959/2006), using orally induced hypnotic suggestion and a metronome as a measure of time, showed that they were able to alter the perceived rate of time by the subject while in trance.

The context suggested to the subject that the rhythm of the metronome was of a much slower time unit where each beat or second became extended.

> "The initial experiments involved the use of a metronome which was striking at a constant rate of one stroke per second. The suggestion was given to the hypnotized subject that the metronome was being gradually slowed down, and the reports indicate that the subject actually experienced a marked slowing in the rhythm. In other words, the seeming duration of the intervals reportedly became greatly prolonged" (2006, 2).

During that altered time frame the subject was able to perform certain prescribed imaginary tasks involving dress design in minutes that normally would have taken hours. From this one would assume that the mind had been entrained by hypnotic suggestion to increase its rate to process information. Hence, the millisecond-time delay was reduced.

With this piece of information we can see how "context" provided by the hypnotist (or in auto-suggestive moments by the person to its Self) modifies how the mental system is affected to operate at faster or slower rates.

While the faster rates do what the slower rates do, for them to be effective they need to be amplified.

This suggests that if attention can become conditioned to function at certain frequency thresholds, its range of sensitivity can be changed by autosuggestion involving a temporal context. It would be, for example, a matter of changing a perceptual habit, the certain threshold of experiencing time driven by external cues, such as the position of the sun or with clocks.

The Maharishi's "alert-hypo-metabolic state" implicates the conjunction of high gamma and theta bands, as the conjunction of "intent" with "memory potential," respectively. In such instances, the higher frequency will program or bypass the time keeper to operate with a different temporal frame of reference. The cognitive cycle (CC) could be said to operate at, say, CC^2, CC^4, and so on.

Altering the Metabolic Rate & Fractal Time

From the foregoing we can deduce that the changing of processing time to achieve almost any goal, albeit a physical, mental, or spiritual one, can be attained by altering the metabolic rate. We find the essence of this technique in the age old methods of shamanism throughout the history of all cultures.

This directly affects the rate of the cognitive cycle, which tells us that the rate at which the cycle is modified defines its level of existence, as assessed from the logarithmic baseline of physiological interactions.

MORPHOLOGY OF
WAVES & VORTICES

"Dark energy is the term that scientists have given to the mysterious 'something' deemed responsible for the accelerating expansion of the universe. However, unlike gravity, which pulls things together, physicists and cosmologists still can't explain what dark energy really is or how it does what it does, despite the fact that it theoretically makes up a substantial part of everything." - Pedro Ferreira

"If you want to find the secrets of the universe, think in terms of energy, frequency and vibration." - Nikola Tesla

"In the wave lies the secret of creation." - Walter Russell

The cognitive model that has been proposed thus far is based on the notion that the universe and its fractal expressions, occurring in all scales of existence, are fundamentally cognitive.

The cognition, which instantiates thinking, imagination, memory formation and retrieval, intelligence, and perception, is mediated by a self-referential process involving two kinds of waves, namely: Hertzian electromagnetic (HEM), and vortices.

These HEM waves and vortices are how a medium (i.e., the PUF, and the FCCP as its fractal equivalent is modified. (I shall also refer to both as the "field.")

NO ONE ACTUALLY KNOWS THE
TRUE NATURE OF THE FIELD

No one actually knows of what the field consists. It can only be subjectively inferred from quantum and other viable theories, all of which have reliable swaths of objectivity, and subjectivity as well. Recall that subjectivity is a legitimate and valuable state of mind that is the basis of our "free will."

In this chapter more will be explained about the nature of the field, and the waves and vortices that are created in its midst.

THE INTUITED FIELD & HOW IT IS MODIFIED IS THE ONLY REALITY YOU'LL EVER NEED

The revealed idea that everything—e.g., physical, mental, and spiritual—emerges from the fundamental nature of a field is in keeping with the following statement by Albert Einstein.

> "We may therefore regard matter as being constituted by the regions of space in which the field is extremely intense. ....
> There is no place in this new kind of physics both for the field and matter, for the field is the only reality" (Quoted in Capek, 1961, 319).

In the resting/ground state the isotropic (uniform) PUF/FCCP is motionless. Though we cannot fully know the nature of what is not moving (since science can only detect comparative movement) we can assume that it has the attributes of a scalar volume suggestive of massless bosons—subatomic particles, that have zero or integral spin and follows the statistical description given by the Indian physicist Satyendra Nath Bose and Einstein. It was referred to as a Bose-Einstein Condensate (BEC), and was hailed by the physics community as a "new phase of matter."

Below, we (the Presence and I) interpret an experiment that intimates how the field (PUF/FCCP) is likely to be modified into boson-like states that progressively become more concentrated.

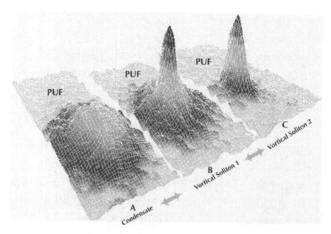

Figure 5: Psiological interpretation of the development of a
Bose-Einstein concentrate.

Note that where the PUF is indicated there would be an isotropic undisturbed "smoothness." In A there is a mild bulge where a low pressure zone has formed, forming a node or sink hole towards which the field is attracted. In B a vortical soliton forms, as the field pushes and increases the pressure through the opening. In C the vortical soliton becomes more condensed.

Bosons form part of the standard model in physics of elementary particles and are known as "gauge bosons." Here, however, a boson is used as an example of how particles are actually highly compressed waves that create a "sense" of materiality. In this regard there is no "wave-particle duality," but a morphological continuity in which the nature of particles is based on how densely the "substance" of the field is modified into various kinds of vortices.

Therefore, all kinds of particles/objects are simply rolled-up waves, not as ultra-tiny balls independently flitting through space (see Wolff 2008, for a clear and scientific exposition, based on the Schrödinger wave equations, on the absurdity of independent particles).

We can think of the field to consist of no-spin states or too smooth to create any kind of converging interference effect. This would be the "ground state."

Again, the PUF is the ground in which the FCCP is embedded. The PUF is deemed to be isotropic, such that when it is anywhere "measured" (see Wolff 2008) it remains the same. In that highly equilibrated state it is *pluripotent* in the sense that it can, by its condensational aspect, self organize into any phenomenal "thing," of which the Self in its various modes of being is included.

Hence, the condensational attributes are what a system perceives as phenomena that are made conscious by how they contrast to the smoothness of the field. Such phenomena also have the fundamental attributes of a system with in-and-out movements that facilitate inputs, throughputs, and outputs.

In its primal resting state the PUF constitutes, not energy as we know it, but potential energy (PE) that by its condensations fabricates kinetic energy (KE), "objects" that can transmit and store energy and information by dint of their configured undulations and iterative vorticity where input and output are enfolded.

Let's begin to understand this hypothesis with the undulatory nature of Hertzian electromagnetic (HEM) waves and how they shape-shift into vortices and back again.

HERTZIAN EM WAVES MORPH INTO VORTICES

Waves in the organism are measured as Hertzian electromagnetic oscillations in which the electric field and the magnetic field alternate 90 degrees to one another as they propagate. Hertzian waves are transversal, which means they vibrate at right angles to the direction of their propagation.

According to the German professor of engineering Konstantin Meyl (1996), when these transverse waves encounter the "near field" of antenna (albeit of any kind of sensor or bio-receiver) they curl up into vortical scalar waves, and when they are emitted they unfurl in the "far field" back into HEM transverse waves. Here's an illustration to help explain the process:

MACRO TENDING
Scale suggested by the Hubble Telescope

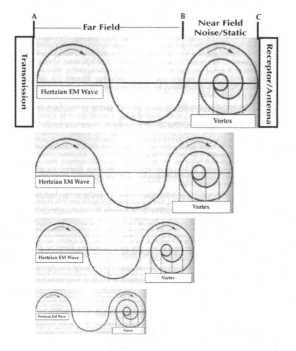

MICRO TENDING
Scale suggested by the Plank Constant

Figure 6: Curling and Unfurling of HEM Waves & Vortices
towards opposing scales.

Hence, depending on its proximity to the antenna/sensorium, waves transform between these two geometries.

If we consider that the dendritic tree of each neuron/glia also functions as antennae, then the HEM waves flowing through the organism (including the heart wave) will, on encountering them, curl up as scalar standing waves around them into vortices carrying energy and information, which the cell absorbs and stores in its interior (in its watery matrix: see Montagnier 2010).

Since there are many dendrites for each neuron, each would converge into the pyramidal cell and need to be synthesized into a cognitive vortex. Then it is sent through the axon, and retransmitted to other neurons as information; it would then need further re-synthesis to culminate in the larger vortical formation that is created in the output assembly.

This synthesizing role would seem to apply to the geometry of the pyramidal neurons, appearing to have golden ratio attributes. Therefore, their form and function are complementarily related.

Meyl has deduced from his research and experiments with Tesla transmitters and receivers that the synthesized information would be sent through the axon in the form of vortical rings or tori (tiny donut-like formations).

Since vortices are also able to form balls, it is not inconceivable that spherical soliton constructs would spin and "flatten" (by dint that fast-turning spheres tend to flatten) into discs that collectively move through the axon as a pressure wave, just like a sound wave transmits a signal by a profile of varying pressure.

This soliton alternative agrees with the theory of "soliton hypothesis" developed by Thomas Heimburg and Andrew D. Jackson (2005). The model is proposed as an alternative to the Hodgkin-Huxley (HH) model in which action potentials are based on alternating ion shifts across membranes.

The soliton hypothesis claims to explain how action potentials are initiated and conducted along axons, based on how the thermodynamic attributes of the axon membranes facilitate signal transmission in the form of solitons that travel as density pulsations (i.e., push-pull, expand-contract, dynamics) through a medium. During this process the membrane thermodynamically transitions between relative "liquid" and "solid" states.

In the liquid state it is more conducive to soliton movement, while the solid state is not. Hence, the soliton is amplitude modulated like light pulses moving through a poorly-conducting medium (i.e., dialectic) like fiber optic cable, which has greater concentrative effects (see Meyl 2012b).

Because the lipid membranes surrounding the neuron are poorly conducting, they minimize the dissipation of energy as the soliton moves through the axon. Hence, it is a very efficient system for preventing energy losses and conserving information.

Therefore, the HH electromagnetically based model, would not be a stark contradiction to the soliton solution. Both would be implicated, given that HEM waves convert into scalar vortical waves in the near field, and vice versa in the far field.

The EEG picks up the Hertzian waves in the far field, that is, off the scalp. Electrodes contacting the depolarizing cell would distort the vorticity and read as an electromagnetic wave. That may be one reason why scalar waves have not been detected, since in the near field they manifest as "noise."

Scalar vortical waves have not been detected since they are the far-field emissions of Hertzian waves detected by the EEG off the scalp. When the electrodes directly contact the cell, as was done with Freeman's topological studies with animals, vortices form on the phase-space maps. Hence, being in contact with the cellular-near-field would bring up the vortical standing waves while in the cellular-far-field, the Hertzian EEG would be present.

The Hertzian component would describe digitized frequency-modulated (FM) pulsations, while the longitudinal waves would implicate a homogenous analog field that would be modified by high and low pressure zones, as do amplitude-modulated (AM) sound waves in the medium of air.

Therefore, I do not see the two models as being contradictory, but as having a complementary relationship, based on Meyl's findings. The Hertzian pathway appears to foster phenomenal consciousness with greater specificity (i.e., favoring locality) while the non-Hertzian one provides the basis for a more generalized connectivity (i.e., favoring non-locality), perhaps within the substrate of the FCCP, which has been attributed to the fascia, amongst other supporting systems, and the meridian system (see Bai 2011).

This would allow for frequency modulation taking place in the brain to somewhat "isolate" itself from all the metabolic complexities occurring throughout the rest of the organism during the waking states. Only when the organism sleeps does the digital mind separate from the senses (leaving the hearing to some degree), and reconnect through resonance via the analog 1 Hz EEG frequency bridge.

However, particularly during moments of "great danger" or extreme "novelty" does the psychophysiology find it beneficial to bypass the PSQ from the editing threshold of the limbic system in order to analogically connect to the Self as a whole.

During these moments the soliton pathway would become quite valuable, for it would entrain the entire mind-body complex into a spiritual state of power: it would produce the higher being body, composed of a harmonious population of solitons, I call Psilotons.

As is well-known, HEM waves can be modulated to transport information across the volume of the field. As they travel they create high pressure regions that are apt to encounter low-pressure zones in the field and be attracted to their "vacuums," so to speak. The sensorium is like an antenna and its low-pressure attributes are what attract HEM signals to its surface. This generates the ingoing vortex in the FCCP, which needs a more analytical description.

ANATOMY OF THE VORTEX

Though not commenting on Freeman's observations, professor Meyl, whose specialty is the physics developed by Nikola Tesla and the nature of vortices as longitudinal waves (i.e., scalar waves), revealed that vortices are *frequency converters*.

> "With this property the vortex proves to be a converter of frequency: the vortex transforms the frequency of the causing [electromagnetic] wave into an *even spectrum*, that starts at low frequencies and stretches to very high frequencies" (Meyl 2012b, 8. Bracket & italics added).

Figure 7 below illustrates this principle with the EEG. As you can see, by changing the frequency, changes are also being concurrently made to the temporal dimension of the space-time equation. The movement may spiral towards the "blue shift" or the "red shift".

Not only is the ingoing blue-shifted vortex changing the parameters of space and time towards the micro-realm of negative space-time, it is also doing other interesting things that beg our attention.

VORTEX ENERGY THEORY OF DAVID ASH

In line with the "as above so below" principle, we see the wave and the vortex as elemental forms to instantiate fundamental processes on all scales of existence.

In his book Vortex of Energy: A Scientific Theory (2012) David Ash has presented a revised vortical theory of energy based on the works of William Thompson and the vortex metaphysics of the ancient seers of India.

Ash's theory works around the belief by Thompson that our world is formed of two fundamental forms: the wave and the vortex as modifications of the aether into electrons and atoms.

He begins by reformulating his idea by swapping the word "aether" for "energy" and "atom" for "subatomic particle." This allows Ash to translate the axiom of his "new vortex theory" into a single statement:

"Energy, as particles of motion flowing in waves or spinning in vortices, forms everything" (33).

Based on this axiom, Thompson's theory and those of the seers of India are brought into the parlance of modern physics. Accordingly:

1. The vortex is a dynamic three-dimensional spiral that would explain the 3D nature of light to form the 3D of matter and space, including the corpuscular shape of subatomic particles.
2. Vortical spin can account for the inertia of mass, and oppositional spin can explain the negative and positive charge of subatomic particles and the forces they create.
3. In the spherical vortex there are only two ways in which energy flows: in or out of the center. When the vortex unravels, its mass is transformed into energy. Different compressions of the vortex would explain the amount of energy that it releases.

These postulates resolve some of the biggest problems in physics. Ash concludes that:

"The major contribution of the vortex model is it shows how spin sets up *static inertia*. The vortex reveals spin as the cause of the apparent stasis that is characteristic of *mass*. The power of the vortex idea is that in a single stroke it can account for *potential energy,* nuclear energy, and the inertia of mass all of which, according to Richard Feynman, cannot be explained satisfactorily in current physics" (34. Italics added).

THE CREATION OF FRACTALS

Note that it is progressively diminishing the *scale* while not significantly altering the information. In chaos theory this is known as *scale invariance*, which is the attribute of *fractals*, a curved or geometric figure, each part of which has *similarity* to the whole (though not necessarily being identical). So, fractals have holographic-like properties on all diminishing scales of existence.

FRACTALS AS MEMORY CONSTRUCTS

What is interesting about fractals is that their information, like holograms, can be up-scaled to form the progenitor object or state to which they are similar. This attribute lends itself beautifully to the notion of forming memory constructs (MCs), and recalling them.

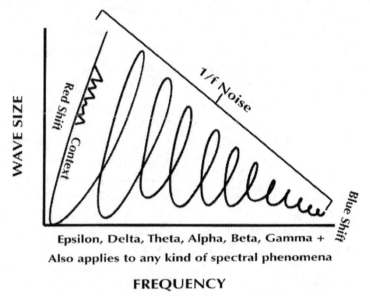

Figure 7: EEG Cognitive Vortex (not to scale) exemplifying contractive and expansive transitions of scale. Also applies to any kind of spectrum.

COMPRESSES TO A SUPER-DENSE POINT

The centripetal process compresses the energy and information to a super-dense nonlinear point (a.k.a. "point attractor"). The squeeze allows the information to somewhat escape its linear grips, and flow more freely throughout the field and interact with other MCs.

At this point the information is subject to some morphing, due to its quasi-similar fractal formations.

The interaction evokes a pattern of information and, at the same time, updates the memory in the FCCP while creating a selection at 3a, and a deselection at 4 by default. At this point, the conditioned aspect of the mind as a psychophysiological habit interacts with the selection at 3a to influence the output at 5.

EXPLOSIONS, IMPLOSIONS, & SYNTHESIS

As the ingoing vortex moves towards its apex, it acquires a high degree of density and takes on a nonlinear attribute, leading to an *implosion*, characterized by fusion (see Donavan 2012, for a detailed and finely illustrated presentation of this process based on the golden ratio).

Accordingly, the outgoing vortex would have fission (explosive) effects while the ingoing vortex would display fusion, which is the fundamental nature of *synthesis* (i.e., integration, cognitive binding).

However, it is labeled as having a fragmenting catabolic effect by dint that at its implosive stage it fractals (fragments) to access associative MCs in the FCCP that will form the assembly of the outgoing vortex.

As it reduces toward gamma+ it transitions towards the shorter wave sizes of the sonic or EM spectrum (with affinity to the "blue shift") and in the opposite direction it unwinds towards the longer wave sizes (with affinity to the "red shift").

THE MULTITASKING VORTEX

In Figure 7 we glimpsed a model of how vortices, particularly the conical form, compress and decompress energy and information as the key dynamic that transforms the nature of space and time for the information being processed.

Fractals burst from this compression like a Fourth of July sparkler, or how a plant emerges, in a different temporal framework, from a "seed." Information is updated when they are "broadcasted" throughout the field. Hence, the cycling is perpetually interfacing with itself in recursive F-F loops.

SPACE & TIME OF VORTICAL INVERSIONS

The ingoing vortex converts information from the positive time-space realm that forms the sphere of our phenomenal universe, to scale-invariant compressions in negative space-time of the inner kingdoms (although there's relativity of scale).

The outgoing vortex converts the selected information back into positive time-space. The parity of space and time can be arbitrarily represented with a number line:

Neg. Space-time...-3- -2, -1 - n0w - +1, +2, +3...Pos. Time-space

According to this hypothesis, information constructs are modified by the in-and-out going vortices, which are identified as "time and space converters."

The ingoing vortex is transforming the space and time of the information into smaller configurations. In a relative sense, it is shifting the scale to the smaller numbers.

The outgoing vortex is transforming the space and time of the compressed selection created by the ingoing vortex at 3a into larger configurations. In a relative sense, it is shifting the scale to the larger numbers.

This reciprocal time and space scaling is sustained by the entire cognitive cycle in millisecond+ time frames. By the speed of the F-F loop the two aspects merge into the experience of NOW.

Ratios of Space to Time

Accordingly, the ingoing vortex transforms information existing in positive time-space, where the ratio of time predominates, and "things" are subject to decay, to one in negative space-time, where the ratio of space predominates, where "things" are not subject to the ravages of time. It transitions the information from an electro-magnetic ratio to a magneto-electric one.

The former produces for us the temporal objects and events of the universe while the latter provides the wherewithal for memory to endure. When electro-magnetic predominance wanes, our intrinsic information is conserved in the magneto-electric domain.

When the dynamical spin of the cognitive cycle is in an equilibrated steady state, the impression being cycled produces a standing wave, signified by the zero in the center of the number line.

Electro-Magnetic/Magneto-Electric

Positive time-space would have the properties of being more electrical, with a weak magnetic component. This interesting proposal was made by the Los Alamos Scientist John V. Milewsky to argue for the parity of light (in Echel 2018). Hence, it would be electro-magnetic whereas in negative space-time the magnetic properties would predominate and would be labeled as being magneto-electric.

Hence, MCs are encoded in negative space-time by the weak (conserved) electro-magnetic signature (EMS), and from there it would be resonantly evoked by the ingoing vortex with a selection at 3a.

It is then equilibrated as a Hertzian wave (where the electro and magnetic components are equal) and, finally, converted to the electro-magnetic format by the outgoing vortex and converted into positive time-space.

Direction of Spin Creates Charges

If the vortex spins in a counterclockwise manner it produces a negative charge. If it spins in a clockwise direction it produces a positive charge.

Fuse to Form Toroidal Soliton Waves

At the cusp where magneto-electric and electro-magnetic waves interact, they create more-enduring toroidal soliton (TS) waves that form the MCs in the FCCP.

The TS releases its information to the outgoing vortex into positive time-space, where serial time prevails and "prevents," as Einstein averred, "everything from happening at once."

SPIRAL WAVES INFLUENCE THE SPATIAL COHERENCE OF THE NEOCORTEX

The neuroscientist Xiaoying Huang and associates have found that spiral waves play important cognitive roles in the neocortex of the brain.

> "Although spiral waves are ubiquitous features of nature and have been observed in many biological systems, their existence and potential function in mammalian cerebral

cortex remain uncertain. Using voltage-sensitive dye imaging, we found that *spiral waves occur frequently in the neocortex* in vivo, both during pharmacologically induced oscillations and during sleep-like states. *While their life span is limited, spiral waves can modify ongoing cortical activity by influencing oscillation frequencies and spatial coherence and by reducing amplitude in the area surrounding the spiral phase singularity*" (Huang, et al, 2010, 978. Italics added).

VORTICES SENSITIVES SEE IN THE BODY

Studies with sensitives (seers) also report that they are able to see the psycho-information field as consisting of spiral and vortical waveforms, which they construe to be thoughts.

One of the earlier studies of "higher sense perception" was done by the neuropsychiatrist Shafica Karagulla. In her book *Breakthrough to Creativity: Your Higher Sense Perception* (1969), she discovers that there are people leading ordinary lives who are endowed with the gift of seeing dimensions of human reality not available to ordinary perception.

She reports that not all of her subjects were equal in sensitivity or in the ability to acquire the same information. (However, as the psychologist William James noted, it takes the discovery of one white crow to create a paradigm shift.)

One outstanding subject, for example, is able to see "...the physical organs of the body and any pathology or disturbance of function" (124), even though she has had no medical training whatsoever; yet is able to translate what she sees in terms that can be easily translated into medical jargon. However, for Karagulla it is the other things that this subject sees that fascinate her, and which is also of strong interest to me.

> "She observes a "vital" or energy body or field which sub-stands the dense physical body, interpenetrating it like a sparkling web of light beams. This web of light frequencies is in constant movement and apparently looks somewhat like the lines of light on a television screen when a picture is not in focus [signifying a nonlinear wave interference pattern]. This energy body extends in and through the dense physical body and for an inch or two beyond the body and is a replica of the physical body" (124, Italics and bracket added).

Another quote from the same passage supports my notion that the vibratory speed of this subtle body exchanges information at a much faster rate with a comparable field, which we all have. The "future" seems to be already formulated in the subtle vibratory network of our "invisible" reality. Consequently, awareness of it can predict what is to happen in the gross body, which in this case has to do with a person's health.

> "She insists that any disturbance in the physical structure itself is preceded and later accompanied by disturbances in this energy body or field. Within this energy body or pattern of frequencies she observes eight major vortices of force, and many smaller vortices. As she describes it, energy moves in and out of these vortices, which look like spiral cones. *Seven of these major vortices are directly related to the different glands of the body.* She describes them as also being related to any pathology in the physical body in their general area. The spiral cones of energy that make up the vortices may be fast or slow, rhythmical or jerky. She sometimes sees breaks in the energy pattern. Each major vortex as she describes it more minutely is made up of a number of lesser spiral cones of energy [vortical fractals?] and each major vortex differs in the number of these spiral cones" (124. Italics and inquiring brackets added).

From this we can see that the larger vortices are major energy systems and the smaller ones constitute the subsystems (i.e., fractals of the larger vortex). As is well known in physics, vortices may converge and enter into phase, forming larger ones of greater amplitude.

PHYLOGENESIS OF BRAINWAVES

Darwinian evolution does not contradict the holistic postulates being developed thus far. Therefore, we can see that the brainwave potentials humans display can be attributed to the phylogenetic complexity of human evolution.

This suggests that the growth of the human neocortex is the adaptive result of dealing with challenges mainly coming from the outer environment. What distinguishes the holistic postulates is the contributing attribute of subjectivity as inseparable from objective outcomes, which are both reductive and emergent, and manifest in terms of ratios to one another. These ratios are not mentioned in Darwin's writings.

Therefore, research that shows brainwaves to have a phylogenetic basis is relevant to the holistic "history" in the development of being and consciousness. Reductive determinism does not exclusively answer basic questions about how an organism evolves from a determined framework of precursors. It instantiates an intelligent and creative process based on immutable predetermined instants. Lacking the subjective component, it is totally absurd.

On the other hand, human evolution (whether from primates or not) has been tending towards greater levels of subjective self-determination. This does not preclude subjective freedom from more "primitive" levels in the animal kingdom but points to a predominance of a species-wide process, whereas with humans there is an individual subjective potential that cannot be ignored.

THE TRIUNE BRAIN HYPOTHESIS

In keeping with this rationale, we can agree with the "triune brain" hypothesis developed in 1990 by American physician and neuroscientist Paul MacLean (1913—2007), which proposes that the human brain is in reality three brains in one, acquired through evolution (not necessarily the result of accidental effects) in the following order: the *reptilian complex*, the *limbic system*, and the *neocortex*.

Using McClean's concept, the neuroscientist Denis Schutter proposes that there is a signature EEG frequency pattern that is expressed by this phylogenetic history. He proposes that

> "More complex networks operate using *higher frequency ranges*. In line, slow wave activity corresponding to delta (1-3 Hz) and theta (4-7 Hz) frequency range stems from the 'evolutionary older' subcortical structures (including the brainstem and septo-hippocampal complex)" (2012, 47. Italics added).

Research involving electric stimulation of these brain areas lends some support to this view.

> "Electric stimulation of the brain-stem ascending reticular activating system (ARAS) elicits 1-4 Hz (delta) cortical

responses, while electric stimulation of limbic areas evoke distinct 7 Hz (theta) activity. Fast wave activity in the alpha (8-12 Hz) and beta (13-30 Hz) frequency range, on the other hand, is suggested to find its origin in thalamo-cortical and cortico-cortical circuits respectively" (Ibid, 47.).

Therefore, the dominant low frequencies reflect neural systems that are more "ancient" in the evolutionary ladder while higher ones correlate with increasing development and complexity of the neocortex.

As a general rule, we can say they are more "instinctive" to avoid the connotation that they are of compromised intelligence as the term "primitive" may suggest. One may consider it as mental processing occurring in different space and time scales.

As we have delineated how the vortex "has the ability to transform energy-and-information from one scale to another without distorting the information," we see how these larger wave forms also contribute to perceiving larger scale formats that are integrated with smaller ones through cognitive fractal crunching. This allows for perceiving differentials of scale via integrated-scale-invariant formats.

INGOING VORTEX INTEGRATES HISTORY OF THE SELF WITH OOPES

Note that each band of the vortex carries information riding on the backs of the EEG spectrum. Therefore, the stimulus entering the sensorium is influenced by the history of the Self, going "back" to its cosmic as well as its more current times. It sweeps the FCCP from the large epsilon waves down to the gamma+ ones.

As has been proposed by a number of neuroscientists, each band of the EEG carries information based on the phylogenetic evolutionary history of the psycho-organism (see Schutter 2012). This history is conserved in the FCCP, where all memory constructs present a repertoire of combinatorial possibilities (the subjective potential) integrating the past and the future into the NOW.

According to this rationale, epsilon would carry the most fundamental cosmic precursors of the Self. Delta would enfold these with the intrinsic information of instinctive and movement functions. Theta would enfold the delta stream into a composite, manifesting in the editing function of the limbic system with regard to expectations.

Alpha further refines the theta synthesis while beta helps to isolate aspects of perception made "relevant" by the limbic editor. The gamma+ frequencies modulate areas of the neocortex in greater attentional and creative refinements. In a very "real" way, each cognitive event implicates the whole universe.

All this takes place when the attentional system instantiates a still moment in the form of a node.

THE ATTENTIONAL VACUUM: WHEN NOTHING ATTRACTS EVERYTHING

Hence, it is not inconceivable that the attentional process is instantiated by low-pressure states to create the ingoing vortex (*vrittis* in Sanskrit). This probably begins when the Self is confronted by an OOPE in which two opposing potentials come face-to-face and cancel out and leave behind a node, a "still zone" or "vacuum" (see Müller in Waser 2004 for scientific details).

In that instant, for example, there is a concomitant and equally potent possibility of a "yes" and a "no," constituting the wherewithal of all decision-making events.

It is a well-known psychological fact that OOPEs arouse the mental system to operate at faster rates (move from resting states towards gamma+), attestation that the mind accelerates to become more intelligent, perceptive, and responsive in the midst of physical, mental, and spiritual contradictions.

The degree to which the mental system becomes focused (regardless of the scale of what it is attending to) equates quite literally to the extent of its shift towards the higher EEG frequencies to resolve the contextual conflict. It does this by accelerating the percept at 6 with the F-F loop at 7.

THE RELEVANCE OF CONTRADICTION

However, it is the contextual conflict that produces a low-pressure node as the result, as we shall see, of two equally strong possibilities clashing in the field of perception.

To have an idea of what effects such a clash would produce we'd need to interpret the scenario from a technical perspective, using the scientific principles of *wave interference*, of which more will be expressed in Chapter THE MIND AS A CONTROL SYSTEM. Briefly, because the mind is quintessentially a wave system it is subject to the effects of *constructive* as well as *destructive* interference.

INTERFERENCE

As we shall more clearly see as we go along, the scientific principle of wave interference sets the "rules" of how waves communicate with one another. This is accomplished with two kinds of interference: *constructive* and *destructive*.

It uses these two modalities to sculpt out decisions and perceptions based on how much (the computational factor) the constituents cohere and rise in amplitude. More details on this in the Chapter THE MIND IS A CONTROL SYSTEM.

CONSTRUCTIVE INTERFERENCE

Constructive interference occurs when two or more simple or complex wave contexts share considerable similarities and therefore can merge in phase with one another. Their peaks and troughs merge like fingers in a glove. When this occurs, the Self experiences a rise in amplitude and a sense of satisfaction.

DESTRUCTIVE INTERFERENCE

Conversely, destructive interference occurs when two or more wave contexts are not to any significant degree similar. Their peaks and troughs outrightly clash and produce low amplitude effects and a sense of dissatisfaction.

These low amplitude effects are technically low-pressure effects of simple or complex patterns of wave interference. The low-amplitude-and-pressure differential from the resting state (the line separating the peaks and troughs of an oscillating wave) of the Self indicates that something is "not right," and an ingoing-cognitive vortex develops in that "nodal space" to resolve the contradiction.

We need to keep in mind that all these interference effects result from auto-modifications of the field (PUF, FCCP). I use the term "auto" to emphasize the self-referentiality of the cognitive process.

EMERGENCE OF THE PSILOTON

Most vortices modify the field in a certain manner and eventually dissipate, returning the energy back to the enormity of the overpowering field, not any different in principle to how a whirlpool in the ocean dissipates into its vastness. This dynamic typically happens with the duration of short-term memory (a.k.a. "working memory").

However, if the iteration of the percept goes on to repeat in the 1-to-7 cycle for an "extended" period of time, the vortex will nonlinearly morph via positive feedback into a more robust and enduring kind of pattern known as the *toroidal soliton* (TS), sporting a spheroidal shape with an axial aperture through which an energy stream flows through it and out back over its surface to the inflow.

Now, the information processed by the vortices will acquire the attributes of long-term memory, physically related to operations occurring with the hippocampi in the vicinity of the *limbic system* at the center of the brain.

The TS is a hybrid of a torus and a soliton. What makes this critter so interesting for psychology is that it instantiates in an integrated manner the entire cognitive cycle in the form of a self-referential energy-and-informational sphere.

We may quite assuredly assume that this is what seers detect as an *aura,* and the superconducting quantum interference device (SQUID) detects as the broad magnetic field of the heart wave (see McCraty 2006) extending beyond the body about 3-4 yards, similar to a TS.

What makes the TS so compelling is that it can fractal into self-similar fragments of itself, such that the small can be incorporated into the large in a seamless manner. To accomplish this the TS needs to step up its in-out cycling speed and begin to bifurcate (branch out).

This fractal effect can be likened to breaking a magnet into tiny fragments that retain the same polarity and magnetic field but on smaller scales. It needs to be taken into account that the smaller magnets do not structurally mimic the larger one, but are functionally the same to the progenitor magnet in a scale-invariant manner.

THE PSILOTON

The 1-to-7 cognitive cycle intimated by the TS would more appropriately be labeled and described as a *Psiloton*, depicted in Fig. 8 below as a 2D crosscut plus a 4+D simulation.

The 1-to-7 numbers used in the model in figure 4 are superposed throughout its volume, and spatial coordinates have been added to indicate how the sphere spins and the energy flows.

The flow enters its top ("North" pole) and spirals through its axis in a counterclockwise direction, converting the information into negative space time. It inverts at the center where another Psiloton is produced, at that point of "criticality" where an implosion takes place at the blue shift, and exiting at the bottom ("South" pole) in a clockwise direction towards the red shift and positive time-space, in a continuous recursive F-F cycle at 7 occurring in millisecond+ time frames.

This is not any different in principle to the flow of the field around a magnet (see Davis 1974, showing, however, the in-and-out flow converging at the center like a figure 8 though not violating the principle of self-referentiality).

Each implosive phase of the Psiloton cycle converts the information carried by the vortex into an array of fractals that create the wherewithal for the selection-deselection effect at 3a and 4. These fractals have similarity, though are not identical, instantiating the thinking process. They restructure into a compressed fractal Psiloton that is taken up by the outgoing vortex which then amplifies and unpacks it into a state of perception and/or behavior at 5-6.

Fractal MCs are distributed throughout the Psiloton volume to form continuous fractal zones—FZ1, FZ2, FZ3, FZ4,...FZ0—that define the depth and scope of the FCCP, based on degree of subtlety and spin, that is, the rate at which they process energy and information. Hence, every aspect of the Psiloton is populated by sub-Psiloton structures that are producing cognitive moments that contribute to the overall effect at 5-6.

Because information is initially being processed at the subtlest of fractal levels, the zones closer to FZ0 are privy to information before it is affected by the PSQ occupying influence in the denser zone closer to FZ4...upon being amplified.

However, there are always remnants filtering through the PSQ barrier that can be picked up as intuitions, etc., when the observer is able to separate the loudness of the noise from these subtle signals.

Each Psiloton is Unique

In keeping with Freeman's observations, the 1D Psiloton dots positioned throughout its volume are "uniformly disordered" by dint that they each share objective similarities with internal signature, not any different in principle to the common hexagonal frames of snowflakes framing unique crystalline patterns, indicating that nature is "programmed" to always include objective with subjective potentials.

FCCP is a Configuration of the PUF

We need to keep in mind that the FCCP is populated by Psiloton constructs, which are essentially modifications of the PUF that it experiences as contrasting phenomena to its isotropic nature. We also need to keep in mind that the PUF is not the Creator per se but an aspect of its unfathomable being.

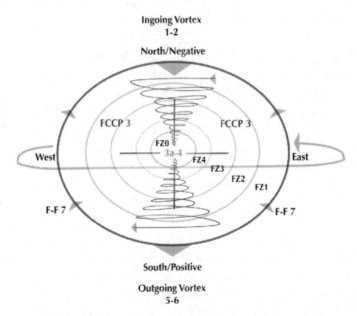

Figure 8: Anatomy of the toroidal-soliton attributes of the Psiloton, that more integrally defines the entire 1-to-7 cognitive cycle as a singular self-referential dynamic. Author.

Returning to the Source

Should Psilotons for one reason or another lose their spin, their form-and-functional context would be destroyed and the configured energy would dissolve back into the PUF. In this regard, the fundamental precursor of energy is eternal, the context temporal, and since we are rooted in this primal state, we are quintessentially indestructible. In merging back into the PUF we enter the mysterious field of pure potential.

It is here assumed that contextual erasure occurs as a result of destructive interference (e.g., when the peaks and troughs of similar waves of two or more systems clash).

In the cosmic eschatology of transformative systems this is known by various names: e.g., "pralaya" in Indian philosophy, and the "End of days" in some forms of Christianity. It does, however, occur on all scales of fractal existence.

A SENTIENT FIELD AROUND ALL THINGS

According to the 16th century Swedish philosopher, scientist, and mystic Emanuel Swedenborg, and numerous other seers, the aura is the subtle sphere or ovoid that surrounds all things in the universe and contains their signature history. This would include the magnetic sphere of the Earth and all celestial bodies.

> "It was also perceived that a sphere flows forth, not only from angels and spirits but also from each and all things that appear in the spiritual world—from trees and their fruits, from shrubs and their flowers, from herbs, and from grasses, even from the soils and from their very particles. From which it was patent that both in the case of things living and things dead this is a universal law. That each thing is encompassed by something like that which is within it, and that is continuously exhaled from it. It is known from the observation of many learned men, that it is the same in the natural world—that is, that there is a wave of effluvia constantly flowing forth out of man, also out of every animal, likewise out of every tree, fruit, flower, and even out of metal and stone. This the natural world derives from the spiritual and the spiritual from the Divine" (Divine Love and Wisdom 2009a, No. 293).

According to what has been revealed to me, the identity of these varied forms in the natural and spiritual world are sustained by the coupling of the inflow with the outflow.

THE TORUS, THE BASIS OF SELF-REFERENTIALITY

The toroidal and solitonic features of the Psiloton combine their attributes to make for an ideal cognitive system.

When we look at the donut-like structure of a torus enfolded upon itself like a smoke ring, it almost goes without saying that it visually displays the principle of iterative self-referentiality in nature.

If indeed cognition is based on the nature of waves and their interactions, then the torus as a wave system displays how the cognitive process interacts with itself to create emergent psiological effects by enfolding the "past" and the "future" into the present moment. The enfolding involves cross-frequency couplings that emerge from the interference of two circulations, one moving around the tubular ring and the other around the circumference. When the torus morphs into a soliton it adds these dynamic attributes to its stabilizing effects.

THE SOLITON: THE BASIS OF STABILITY

The soliton is a type of vortical energy construct well known to physics which is found in a variety of environments in the form of a stable, self-organizing wave that maintains its energy and shape in a variety of challenging conditions. It has been defined as a "vortical standing wave." It is the only kind of torsion wave that, as far as I know, can account for conserving memory as configured forms of energy.

Energy and information cannot be stored by radiant "straight" lines, only the repetitive enfolding effects of vortices can do that. The infinity loop in the Symbol does not literally describe how the looping works, but instantiates the notion of dynamic equilibration in which the input and the output are linked in a recursive interfering process to create the standing waves of phenomenal reality.

These fascinating solitonic bundles of energy have long been observed in a variety of mediums: air, in quantum space, plasma, liquid and solid crystals, and in magnetic and other domain structures. They have also been implicated to be the basis of biological systems and their means of cognition (e.g., see Davia/Carpenter 2005; also Petoukhov, 1999).

The soliton phenomenon was "first" observed, in 1834 by John Scott Russell, while he was riding a horse along a narrow and shallow canal in Scotland. It appeared as a "rounded smooth well-defined heap of water" propagating "without change of form or diminution of speed." In order to study it he reproduced the phenomenon in a wave tank and named it the "wave of translation;" it was subsequently called a "solitary wave" or "soliton" (see Wikipedia: Soliton).

While typical soliton shapes are spheroid and ovoid (like the "cosmic egg" I envisioned), as shown below, they can take on a variety of shapes, as seen, for example, in the formation of elongated "rogue waves" and tsunamis that travel hundreds of miles across oceans and onto shores without attenuating.

What renders such unusual stability to the soliton is its ability to equilibrate dispersive and nonlinear forces. Without this balance, objects at all scales and levels of existence would either dissolve or collapse (as has been shown with certain stars). It may very well be the basis of "inertia."

Figure 9: Typical stable soliton shapes.

So, one can see that to be a relatively enduring phenomenal construct, including memory, requires a balance of these oppositional yet complementary forces (i.e., energy flow and resistance must equilibrate according to Ohm's law. More on this to follow).

LIGHT SPONTANEOUSLY TIES INTO KNOTS

From the foregoing, we may have asked: how is information (i.e., the configuration of energy) stored electromagnetically in negative space-time? To answer this question we'd need to refer to Lord Kelvin (William Thompson) and the spin offs of his brilliant theories.

In the 1860s, Kelvin became interested in the structure of atoms when he observed smoke rings, and proposed that atoms were shaped like vortices spiraling around each other in the aether. This was similar to the way knots loop and twist to form patterns.

TOROIDAL VORTEX

Vortex rings manifest in turbulent flows of liquids and gases, but are rarely noticed unless the motion of the fluid is revealed by suspended particles—as in the case of smoke rings, which can be produced by smokers, in the firing of cannons, and in mushroom clouds. However:

> "A vortex ring, also called a toroidal vortex (donut shape), is a torus-shaped vortex in a fluid or gas; that is, a region where the fluid mostly spins around an imaginary axis line that forms a *closed loop* [self-referential]. The dominant flow in a vortex ring is said to be toroidal, more precisely poloidal. A vortex ring usually tends to move in a direction that is perpendicular to the plane of the ring and such that the inner edge of the ring moves faster forward than the outer edge. Within a stationary body of fluid, a vortex ring can travel for relatively long distances, carrying the spinning fluid with it". (Wikipedia: Vortex Ring. Italics and bracket added).

If vorticity, as Lord Kelvin proclaimed, is indeed the robust structure of electrons and atoms, then light can be structured in an umpteen number of simple to very complex self-referential convolutions.

Recently scientists have discovered that this is precisely what occurs when light traps itself in soliton formats, which I postulate to be how energy (e.g., the life force) and information are trapped and conserved as a MC.

Anton Desyatnikov (and associates) describe a theory using simulations of spinning optical vortices to show that their propagations spontaneously excite knotted and linked optical vortices into soliton configurations, as in Figure 10.

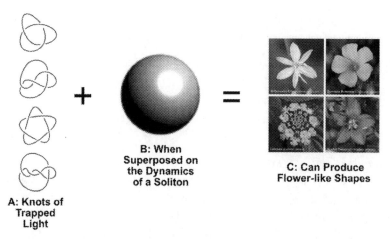

A: Knots of Trapped Light

B: When Superposed on the Dynamics of a Soliton

C: Can Produce Flower-like Shapes

Figure 10: Illustrating the concept of light trapped in vortex loops and superposed on the self-referential dynamics of a soliton, yielding flower-like patterns.

These interesting symmetries occur quite spontaneously when:

"The nonlinear phase of the self-trapped light beam breaks the wave front into a sequence of optical vortex loops around the soliton, which, through the soliton's orbital angular momentum and spatial twist, tangle on propagation to form links and knots. We anticipate similar spontaneous knot topology to be a universal feature of waves whose phase front is twisted and nonlinearly modulated, including superfluids and trapped matter waves" (Desyatnikov, et al, 2012, 1).

We can also see how subtle energy patterns can be formulated quite similarly into fractal symmetries with considerable individuality in the FCCP. A compelling example of this possibility is found in the research performed by the Japanese scientist Masaru Emoto (1943-2014), and described in his book *The Hidden Messages in Water* (2004).

Briefly, Emoto showed with compelling photographs that when the messages (written on the container or thought-projected onto the sample) were positive (e.g., love) the crystals, generated by freezing the samples, and viewed under a dark-field microscope were geometrically beautiful and symmetrical, while when negative messages (e.g., hate) were used the crystalline effects were distorted and ugly.

As will be discussed further ahead, Emoto's work has close affinity to that of Luc Montagnier's electronic transmission of a DNA sample into

125

water, which is of immense importance considering that the organism consists of 70% water, of which 85% is in the brain.

THE PSILOTON INTIMATES THE
BASIS OF ALL SYSTEMS

In the toroidal vortex we can see the fundamental self-referential dynamics of a system. It has the wherewithal for inputs, throughputs, and outputs that are recycled in a continuous manner.

We can see its variations exemplified in the digestive system, the respiratory system, and the cardiovascular system of organic life on Earth. It is also essential to the nervous and glial systems in the processing (metabolizing) of information that supplies the Self with the means to make decisions based on what has been previously learned.

THE NATURE OF UNIVERSAL RESPIRATION

The reciprocal activity between potential energy (PE) of the field and kinetic energy (KE) of its Psilotonic modifications creates the oscillation of universal "respiration," known in ancient spiritual systems as the "breath of God." Its psiological implications can be found in the Greek word "psyche."

> "The basic meaning of the Greek word *psyche* was 'life' in the sense of 'breath', formed from the verb *psycho* ('to blow'). Derived meanings included 'spirit', 'soul', 'ghost', and ultimately 'self' in the sense of 'conscious personality' or 'psyche'. The association of 'spirit' and 'breath' is not unique to Greek or western cultures..... The linkages between 'spirit' and 'breath' were formed independently by ancient peoples who at the time did not have any real contact with one another" (Wikipedia: Psyche).

Time & Timelessness

The breath not only animates the universe and all that pertains to it but also brings with it the wherewithal for "time" to manifest. It is the "soul," and the "anima" and "animus" of Jungian psychology.

Note that in this regard the PUF/FCCP is not yet imbued by time as we know it, but by the potential of spin as the basis of all kinds of materiality. Hence, with each respiration the universe on all scales of existence is shifting between objective time and subjective timelessness in ultra-second time frames.

According to the Turkish engineer Hakan Egne, this integral concept brings together the currently disparate classical and quantum physics by the interference effects of interactive geometries. In this regard he states:

> "A particle obeying classical physics at super atomic level is a field obeying modern physics at subatomic levels. The quote from Einstein (mentioned above) describes such particles. In other words when the super fluid medium (space) takes an orderly form at a certain region (condenses) and flourishes a circulation, it forms matter (Egne 2012b, 1. Parentheses added).

Matter is formed by how the field is vortically modified in such a way that it acquires areas of enduring robustness. What endows these areas with such endurance is based on how the waves circularly "repeat" their signature identities by giving back and taking from the field in a steady and reliable manner.

In other words, it is the vorticity of the wave in which its output and input are linked in the form of a *torus*, a donut-like structure with the fluidity of a smoke ring traveling in the field.

Note that in one direction its circular spin "wants" to escape the toroidal flow but is held back by the concurrent flow in the opposite direction (perhaps due to a Coanda effect).

What's so compelling about the dynamics of the TS is that it provides the wherewithal for energy to be configured and conserved by its self-referential circulation.

THE MIND PIXEL

In this regard, Psilotonic spin at its most fundamental level would more comprehensively have the attributes of a "mind pixel" envisioned by the scientists Hupping Hu and Maxin Wu:

> "We postulate that consciousness is intrinsically connected to quantum mechanical spin since said spin is embedded

in the microscopic structure of spacetime and may be more fundamental than spacetime itself. Thus, we theorize that consciousness emerges quantum mechanically from the collective dynamics of "protopsychic" spins under the influence of spacetime dynamics. That is, spin is the 'pixel' of mind. The unity of mind is achieved by quantum entanglement of the mind-pixels" (Hu, et al, 2004).

In electronics a pixel is a minute area of illumination on a display screen, one of many from which an image is composed like a pointillist painting.

The key difference with the TS and the pixel is that unlike the locked-in pixel on a computer screen, it is the minutest processing mechanism of cognition that can combine with other Psilotons to create larger collective ensembles of "minding."

Signature Information Micro-extends Towards Hyperspace of Negative Space-Time

While each fractal appears to have standard Euclidean geometry, its Hausdorff dimension allows it to micro-extend infinitely inward without increasing the size of the progenitor polygon. In this sense, the negative progression micro-extends towards *hyperspace*, which I interpret to be the realm of negative space-time. Note that the tesseract is an ingressive continuation of the cube, so that within C smaller 3D fractal cubes will form until they appear to the naked eye or magnifying instruments as a 1D point, or totally "disappear." This is what I mean by dimensional crunching.

THE PSYCHO-PHYSICAL LANGUAGE OF WAVES

"Religion, mysticism and magic all spring from the same basic 'feeling' about the universe: a sudden feeling of meaning, which human beings sometimes 'pick up' accidentally, as your radio might pick up some unknown station. Poets feel that we are cut off from meaning by a thick lead wall, and that sometimes for no reason we can understand, the wall seems to vanish and we are suddenly overwhelmed with a sense of the infinite interestingness of things." - Colin Wilson

There's a mysterious relationship between the physical organism and language that is of special interest to psychosomatic medicine, to those who are curious about "placebo/nocebo" effects, and to those who employ hypnosis to enact changes of the Self. It can be summarized as: the physiology affects the psychology and vice versa.

According to the F-F loop of iterative causation, habits of body and mind are constantly enfolding and unfurling with OOPEs. This self-referential process generates the contexts with which we experience our meaning of the world from previous and present moments of learning.

Learning translates into the acquisition of memory, which at its most fundamental level equates to the formation and interactivity of waves and vortices that interfere with one another to create the subjective and objective nature of reality.

WAVE INTERFERENCE

We live and have our existence in a universe consisting of various wave potentials with infinite configurations.

Information travels through the psycho-organism in the form of biochemicals and configured sonic and/or electromagnetic waves that obey the principle in physics mentioned earlier as *interference*.

This dynamic not only attests to what goes on inside us but also how we communicate with others using our various potentials to make signals with movement.

Here I will use this interactive principle of interference to explore the *contextual* meanings these wave configurations produce. First let's take a brief look at how waves are considered and measured in our technological times.

COMPUTER ANALYSIS OF WAVES

Neuroscientists have been able to study the individual and collective oscillations of brain cells by analyzing specimens and subjects using computers programmed with *Fourier Transform Analysis*. Transform analysis was invented in 1822 by the French mathematician Joseph Fourier (1768–1830), when he realized that some functions could be written as an infinite sum of harmonics, based on the principle of interference.

This is, for example, the basis of analyzing the EEG and the magnetoencephalograph (MEG). It is also the basis of other bioamplifiers such as the magnetic resonance imaging (MRI) machine.

A Wave is a Wave is a Wave

If you've seen waves occurring in water you've seen the basic activity of waves and vortices in all sorts of media and on all scales of existence. The same wave activity takes place in a cell phone as well as in the nervous system, and other cells.

Amplitude

The most practical example for understanding electromagnetic waves is to visualize the ocean's waves. Note that there are all sizes of waves, which indicates their varying *amplitudes*. The larger the wave the greater is its amplitude. These larger waves are made of smaller waves that are in *phase* with one another, so they add up. In this case the baseline measure of amplitude (a.k.a. the baseline) is the shore.

Frequency

Some waves move towards the shore more quickly than others. The rate at which they move is called frequency. For example, in the EEG, epsilon is a large very slow moving wave while gamma is a small very fast moving set of waves; the froth at the edge of wave trains are white water fractals.

SOURCE OF NUMBER IN THE UNIVERSE

From the foregoing we can glean an ontological "truth": wave interactions express mathematical rules of addition, subtraction, division, and multiplication. Waves are intimately related to numbers and numbers are intimately related to waves. It is well known how waves naturally divide into harmonics such that the fundamental wave as 1 divides into 2, which then divides into 4, then 8, and so on.

Waves, in whatever kind of medium, will generate these mathematical calculations (computational theory). They may be ever so simple or utterly complex.

All the waves are carrying and processing information in fractal scales in a scale-invariant manner. Nevertheless, the fractal harmonics, although subtler and thereby weaker, are processing the information at faster rates. It is in this sense that they are attributed with being subliminal or non-conscious, while the larger waves enter into the conditioned threshold of consciousness.

Self-organization

We can glean from this example that there is a self-similar fractal progression of linearities. Nonlinear progressions are more complex, yet are mathematically endowed. Both provide the wherewithal for a system to "self-organize" (the principle behind "emergence" and "holism").

The Secret of the Cosmic Egg

If we consider a sphere or ovoid shape to be the fundamental then it will broadcast waves from its outer surface and generate fractal harmonics from its inner surface. Hence, structure and function share an inseparable relationship. Note that the fractal harmonics will evolve inwardly towards infinity (-∞) without enlarging the fundamental.

Again, the Symbol's aura equates symbolically to eggs (incubators) of all kinds, seeds, womb, ova, bandages, musical instruments, cooking vessels, the skull, and so on. The ingressive harmonic oscillations that take place in these varied containers have objective and subjective potentials that produce the infinite variety of things we discern in the universe.

Without this variety the universe would collapse into itself into one ginormous mass.

The notion of unique fractal properties emerging from the geometry of the container implicates sacred geometry as a serious science in transformative psychology, especially in the design of sacred symbols, art, and architecture.

We Live in a Sea of Frequencies

Comparatively speaking, our brains operate on waves that are very low frequency (from < .5 cycles per second) to on average about 50 cycles per second. This seems to be a threshold that our psychophysiology can handle to deal with practical things.

Of course, the non-conscious domain operates at very fast speeds, but even these do not compare to the electromagnetic interactions that take place almost to the speed of light (186,000 miles per second), depending on the density of the media.

The Resting Frequency

All waves systems have a *natural resonant frequency* (a.k.a., "eigenfrequency" in physics) when they are at *rest* (i.e., not static but "standing in place"). When measuring brainwaves, scientists seek to access the resting frequency to form a baseline from which other brainwaves deviate in an up or down manner.

In transformative psychology the resting frequency would be considered as one's Essence parameter.

Structural Synchrony, Signature, and Snowflakes

By now you may have wondered how a synchronous assembly can display incoherent thoughts, feelings, and sensations, given that the constituents of the assembly would presumably need to be in-phase.

This would-be conundrum can be resolved with the effect of *entrainment*, in which a faster frequency (in the beta and gamma ranges) can bring two or more disparate frequencies that, from a mathematical and physiological perspective, cannot enter into a synchronous relationship to pulse in sync (see Pletzer 2010) and "should," therefore, not appear in our consciousness.

If, however, we consider that MCs share a common geometric framework, like the hexagonal pattern of snowflakes, yet contain a signature pattern within it, the outer common shape would provide the wherewithal for synchrony to take place and yet convey an incongruent message. Then entraining effect would then allow for paradoxes and OOPEs of all sorts.

THE LANGUAGE OF WAVE INTERFERENCE

The principle of *wave interference* governs how waves aggregate, dissociate, and integrate to form cognitive and contextual outcomes.

> "Interference is a phenomenon in which two waves superpose to form a resultant wave of greater or lower amplitude. … [Its] effects can be observed with all types of waves, for example, light, radio, acoustic, surface, surface water waves, or matter waves" (Wikipedia: Interference).

Waves interfere with one another in two basic ways: *constructive*, and *destructive*. We need to consider this in terms of linear waves and vortices.

CONSTRUCTIVE INTERFERENCE

Constructive interference occurs with waves as well as vortices that share a similar geometry and frequency, and are turning in the same direction. This allows them to attract and merge their "lines" of force, and they are said to be in *phase* or *synchrony*. This graphic illustrates the constructive effect:

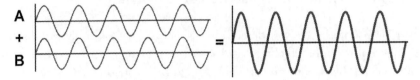

Figure 11: Constructive Interference.

Upon two 20 Hz waves merging their values *add,* and result in a greater amplitude and consequent power, which endows it with some *psychic mass* on other waves in its area of influence.

When vortical wave systems (indicated by the </> signs on the Symbol) merge (>> or <<), their combined energy rises in amplitude, and a larger vortex forms. The same can be said of *vortical soliton* autokinetic (AK) waves.

Consonance: "Yes"

Consonant effects are based on constructive interference, and they have compelling psychological implications for describing how the mental system "accepts," "absorbs," "remembers," "incorporates," "assimilates," "receives," and the like. We can simply say that this is its "yes."

It could be signaling "yes" to personality or Essence, or to both. If the personality is not in sync with Essence, agreeing with it will leave a deep sense of discontent, even though the results of the agreement may be "successful."

Positive Emotions are Consonant Vibes

Emotions are vibratory interactions occurring in the non-conscious realm. These interactions are emanations that converge from thoughts, imagination, and the repertoire of existing memory constructs in the FCCP.

When these vibes enter into contextual coherence they produce a positive state by how much they rise in amplitude.

We could generally use the terminology in physics to define positive amplitude as the maximum extent of a vibration or oscillation, measured from the resting position of instinctive, movement, and intellectual equilibrium.

Temporal Entrainment

The observation of how two oscillating systems become synchronized was made by the Dutch physicist Christiaan Huygens (1629-1695), who invented the spring-driven pendulum clock, amongst numerous other major contributions.

In 1666 he noticed that two clocks mounted next to each other on a wall had shifted from asynchrony to synchronous oscillations. The two clocks synchronized with their pendulums consistently swinging in opposite directions, 180° out of phase, but in-phase states can also result. In his report of the incident to the Royal Society he referred to it as "an odd kind of sympathy." The concept is now known as *entrainment*.

Nature Always Seeks Minimal Energy Expenditure

Entrainment occurs when small amounts of energy are transferred between two systems that are out of phase in such a way as to produce negative feedback, through which they eventually assume a more stable phase relationship, and the amount of expended energy gradually reduces to zero.

In other words, the two systems enter into a resonant relationship that requires a minimal amount of energy to sustain. It appears that entrainment and synchrony are nature's way to conserve energy. So in this very general way the universe appears to have a design or purpose built into its wave configurations.

Resonance

Resonance is created when one frequency "adjusts" by increasing or reducing in order to come to rest with another frequency. Therefore, nature is always seeking efficient low energy-spending relationships.

Accordingly, each frequency is changed by the presence of the other frequency until a form of union is established, thereby facilitating the flow of energy and information that allows for a reciprocal exchange.

Essence is an Inherent Signature Unity

Consequently, we cannot blame nature for all of our contradictions, for it wants us at the fundamental level of energy and informational exchange to be harmoniously interconnected.

It is important to know that Essence operates at this natural level, but the PSQ interrupts this smoothness with inhibitory blockages that keep us locked into the conditioned "default mode network" (DMN).

That is why in Scripture we are advised to "become as a 'child' again" (i.e., rebirth).

Harmonic Oscillators: the Bridge to the Modern World

Huygens' seminal observations contributed to the concept of resonance in physics and the resonant coupling of harmonic oscillators, which also gives rise to sympathetic vibrations. The concept became instrumental in all kinds of wireless transmission, which as we well know, transformed the time-space nature of mass communication by expanding its scope and speeding it up. Every time we use a cell phone we can thank Huygens for his discovery.

Therefore, when two or more systems, similar in frequency and structure, are gyrating in the same direction they will *attract* and form phase relationships that are deemed by nature to be most efficient. By their similarity of movement they are able to exchange energy and information.

Bliss, Euphoria, Ananda

In equating amplitude with emotions we can see why wave mediated states that are in phase with one's Essence and/or PSQ will produce positive emotions, and why those that are not in phase with them produce negative ones.

Translating this into transformative psychology, we can understand why the different nerve plexus, known as chakras or centers, are able to make decisions about what is going on in their particular region of influence.

That's why, for example, some people speak of a "gut feeling." Then again, others may speak of their heart being "torn" with grief.

Conversely, when the polarities of these centers are properly balanced via feedback, the whole mind-body system aligns into a neutral state, which draws on a subtle energy source from within the organism and/or the surround and produces the feeling of bliss (a.k.a., euphoria, oneness, nirvana, etc.).

Many, if not all, of these in-phase effects are amplified by a dopamine-delivering system in the frontal lobe area near the limbic system of the brain called the *nucleus accumbens*, which has also been dubbed the "pleasure" and "reward" center.

The Cascading Effect

This faster wave construct recruits other waves to oscillate at its pace, which then has a cascading effect on other memory clusters that become the basis of a larger assembly. Large cascading effects may be experienced as a general feeling of cortical fullness and pleasant electrical discharges in the head.

DESTRUCTIVE INTERFERENCE

Waves that interact in a destructive manner share the same geometry and frequency, but their peaks and troughs are not aligned (180° out of phase) and on interacting have a subtractive-effect and cancel out. This graphic illustrates the destructive effect:

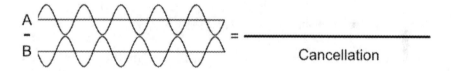

Figure 12: Destructive Interference.

CANCELLATION DOES NOT MEAN EXTINCTION

However, because of the law of the *conservation of energy and matter*, a fundamental law in physics, cancellation does not mean extinction of the energy, but only the configured form of the information occurring in the medium of space (a.k.a., the "field"). In other words, it loses its contextual information, given that information is how energy is configured on the legs of these vortices.

Dissonance: "No"

Recall that waves that display different frequencies are subject to the subtractive effects of destructive interference. Two or more waves bucking their peaks or gyrations as they approach one another would produce a cancelling effect. They are said to be totally out of "phase," forming a dissonant state. We can also, in a very general way, say that this is the system's "no." However, to avoid contextual extinction, the MCs in question, existing as fractal units of intelligence, would avoid coming into contact with obliterating foes.

Another Kind of Duplicity

This is why contradictions rarely interact in the same psychic space. This creates by default a contextually based division other than a vibratory border to separate from the non-conscious domain. Hence it more explicitly involves *personality* ruled by the PSQ.

It could be signaling "no" to personality or Essence, or to both. If Essence is rejected while agreeing with a project motivated by personality, it will leave a deep sense of discontent, even if the results of the agreement are "successful."

Negative Emotions are Dissonant Vibes

Emotions are vibratory interactions occurring in the non-conscious realm. These interactions are emanations that converge from thoughts, imagination, and the repertoire of existing memory constructs (FCCP).

When these vibes enter into out-of-phase states they produce negative feelings by how much they lower in amplitude below the resting position of instinctive, movement, and intellectual equilibrium.

BEAT FREQUENCIES

An interesting thing happens when the waves/vortices are somewhat similar but are not the same in frequency and structure. In such cases they still undergo a combination of constructive and destructive interference, producing a kind of hybrid wave. This is technically known as a *beat frequency* or *heterodyning*, like so:

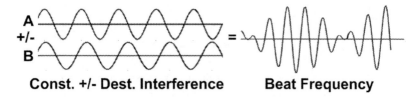

Const. +/- Dest. Interference   **Beat Frequency**

Figure 13: Beat Frequency.

As you can see, the size of the wave varies between increases and decreases in power (amplitude). Because of this complexity, beat frequencies don't appear to carry the same emotional impact as constructive states or those of total cancellation, since most minds rarely acquire high degrees of *contextual coherence*, in which the memory constructs are in a high level of being in or out of phase.

Beat frequencies can be viewed as interactions between "positive" and "negative" states that yield a value which is experienced by the degree to which it rises above or below the resting position of instinctive, movement, and intellectual equilibrium (i.e., the system in its relative state of rest; a.k.a. "Essence" in transformative psychology).

According to physics and psiological revelation, waves are the most fundamental constructs in the universe. As they interact they produce more complex objects, that of themselves form the basis of organic life as memories with particular identities, iterated in Psiloton formats as enduring standing waves. These complex organisms are sustained by a molecular substrate of modifiable codes (bio-alphabet) with which numerous organisms are fashioned.

Therefore DNA, as the most established bio-habit, interfaces with the mental system of thoughts, emotions, and sensations in a "causative" and auto-affected manner. Prior to the 1990s there was very little scientifically known of this obvious co-creative interface. Soon after, this all began to change.

DNA: THE MOLECULAR PHASE OF SYNTAX

We now enter into the fascinating research by the Russian scientist Peter Gariaev and his associates (1994) revealing that DNA/RNA functions just like the arrangement of words and phrases to create well-formed sentences in a language. According to their findings:

The evolution of biosystems has created genetic 'texts', similar to natural context dependent texts in human languages, shaping the text of these speech-like patterns.

That the chromosome apparatus acts simultaneously both as a source and receiver (similar to radio waves) of these genetic texts, respectively decoding and encoding them, and

That the chromosome continuum of multicellular organisms is analogous to a static-dynamical multiplex time-space holographic grating, which comprises the space-time of an organism in a convoluted form.

The process is based on a quantum holographic principle that mimics the basis of computer design by dint that it carries a copy of itself.

Hence, it replicates its information in a self-referential manner by altering the sequences of its invariant nucleic-acid alphabet, consisting of A (Adenine), G (Guanine), C (Cytocine), and U (Uracil), much like the arrangement of words and phrases to create well-formed sentences in a language: syntax.

According to the Symbol, these nucleic acid units are the objective potentials that the subjective aspect of the instinctive mind uses in a combinatorial manner to create the physical mode of the Self. The letters are "invariant" because they do not easily change (i.e., are locked as finite potentials) regardless of scale, except in relationship to one another (i.e., subjective potentials). They form the letters of a bio-alphabet by their signature vibrations, and can be arranged to combine in certain sequences (e.g., genetic engineering, and epigenetics), that also acquire signature addresses.

All this is mediated by "biophotons" (inner light: see Popp 1998), acoustics (i.e., sonic waves), and the electromagnetic field, which interfaces the organism with the "outer" cosmos via the magnetic field of the toroidal heart wave that surrounds us like a huge donut.

The information is held in the electromagnetic field and is configured by the "acoustic field," based on complex wave interference, to construct the organism out of the available material. We can envision what's going on by how the system is talking to itself resonantly-and-dissonantly in the sculpting of its physical reality.

It is important to note that this multidimensional, auto-circular process cannot happen with the simple seriality of language, but must include mechanisms that correspond with the basic features of quantum communication/information transfer known as nonlocal *quantum teleportation*, which consists of two inseparable signal processes: one classical, one quantum.

> "The latter is instantaneous transmission from X to Y (unlimited in principle as to distance), but which cannot be used without the other, which is transmission from X to Y by conventional means at the speed of light or lower. In the case of DNA, therefore, it is the existence of the genetic text of the organism itself which constitutes the classical signal process of quantum teleportation, able to facilitate the quantum mechanical signal processes of both the copying of the DNA as its own blueprint, and of the construction of the organism (for which DNA is the blueprint) in a massively parallel way by the means of quantum teleportation" (Gariaev 1994, 16).

The quantum wave process not only emerges into the classical structures of the organism but is also the DNA basis for linguistic syntax to occur. Even though the objective potential of language is already established at birth, it must be subjectively informed and cultivated by the social and physical environments.

Because the classical and quantum aspects of language are energetically related by dint that information can be mediated and modulated vortically and conserved by soliton waves, we can see why thoughts and speech can translate to the deepest levels of our being and form the basis of psychophysiological enactments.

THE GENERATION OF CONTEXT

Each time the ingoing vortex converges towards its apex it selects/deselects (at 3a and 4) a certain context—e.g., an object, thought, image of an event. On the divergent out-spin it creates an assembly (at 5), having conscious and/or non-conscious potentials, depending on its scope and complexity. The most obvious conscious assembly is what we experience in the surround. With each reciprocal gyration, it draws from and updates existing memory constructs (∞) in the FCCP in millisecond time frames to incorporate the subtle-to-extreme nuances of our constantly changing reality.

Contextual configurations are, for the most part, fortuitously evoked by the environment. They can also be intentionally evoked as novel events, mainly from existing memory constructs subjectively arranged in new ways.

It is now well known that the physiology and psychology interact and affect each other based on contextual frames of reference (see Lipton 2005).

PHENOMENAL NATURE OF CONTEXT

All contexts, being energy configured with signature patterns, produce vibrations (e.g., sounds, forms, etc.) that obey the principle of interference, which are organized and synthesized by ingoing vortical transitions that accept inputs "as is."

POWER OF CONTEXT

Inherent and acquired contexts are always involved. For example hunger creates the context of needing to be fed, which associates with all that is related in the FCCP to that need. This relationship forms a memory construct (MC). Such non-conscious primal needs color the development of acquired beliefs and attitudes in an abstract manner. For example, the mental notion of "feeding" a habit pattern, which technically points to recurring F-F loops that energize it. This has implications for all kinds of habits.

PSYCHIC MASS

The coined term *psychic mass* pertains to MCs that have acquired (i.e., intentionally and/or through associative evolution) the ability to entrain other MCs by dint of their *contextual coherence*, which is the degree to which its ideational constituents are syntactically in sync.

THE CONTEXTUAL POWER OF IMAGINATION

The notion that imagination and visual perception are phenomenally similar has been noted at least since the time of the Greek philosophers. Plato, for instance, describes mental imagery by using the metaphor of a mental artist painting pictures in the soul. This correlates with the estimate that 90% of our sensory processing is visual.

"Studies involving functional magnetic resonance imaging (fMRI) reveal that, like any other perceptual moment, the memory constituents of visual imagery, or 'seeing with the mind's eye,' are composites retrieved from long-term memory" (Ganis, *et al*, 2004, 226).

THE PROCESS IS IMPERSONAL

Whenever I dwell on a particular context, attention as a neutral psycho-accelerant will amplify its positive or negative or neutral aspects (via the F-F loop of circular causation). The process subserves all of one's beliefs, regardless of what they are. Hence, amplification of negative-or-positive contexts undergoing constructive interference are subject to the amplifying effect.

BELIEF: A SELF REFERENTIAL CONTEXT

The ability to modify and control the context of thoughts and emotions into a belief system is perhaps the most powerful human attribute in transformative psiology.

Recall Selye's great discovery: how one interprets the situation determines whether it will come across as "stress" or "eustress" (i.e., good stress). Of course, the interpretation is more effective if it is imbued with contextual coherence.

CONTEXT AS IMAGERY

Because larger neuroglial assemblies have greater transformative power the use of imagery is a key asset.

This is due to the fact that imagery always involves the incorporation and integration of more MCs into cognitive ensembles than thoughts; however, thought plus imagery is also quite expansive.

AROUSAL, RATE OF PROCESSING INFORMATION, & LEVELS OF UNDERSTANDING

Gurdjieff's transformative system introduces for the first time a unique way of viewing the processing capabilities of the different centers.

Ouspensky defined these rates based on the speed of the intellectual center, which he designated as X. The movement-instinctive center functions at 30,000 faster than X. The optimal speed of the emotional center, the sex center, and the higher emotional center is 30,000 times that of the movement-instinctive center, while the higher mental center (the symbolic "brain" of the logos) exceeds them by the same amount.

Conventional forms of psychology have not embraced this idea. However, the spectral frequencies of the EEG strongly suggest that the psycho-organism uses faster frequencies to more intelligently bind and process information.

This is supported by the fact that when the system is confronted by different kinds of contradictions, as well as certain levels of complexity, it becomes aroused, which automatically accelerates the EEG to higher levels, depending on the degree of challenge. Each higher-faster level of processing may lead to greater degrees of understanding.

THE VIBRATORY RELATIVITY OF MEANING

Wave dynamics do not of themselves produce meaning, but express how it is modulated contextually according to an inherent (i.e., instinctive) and/or acquired value system. Therefore, the same wave format may express oppositional messages in differing cultures or situations. Hence, the frame of reference rules over the context.

For example, a thief may successfully steal money from a bank, and experience the amplified emotion of "success" from the synchrony of constructive interference. In this case the goal is in concert with the event, which creates the positive feedback loop that is behind the amplification of the feeling of "success."

The same wave format may also produce the same emotional effect to an altruist who gives money to a project that would improve the lives of poor people.

Nevertheless, context is very important because it forms the nucleus of the belief system, which has effects on the degree to which we are integral, to what is allowed to enter into consciousness, and to what is rejected or repressed.

However, the inherent quest for overall coherence is what can bring the Self into a state of peace with itself and the world. This is why Self acceptance, as one is, is so valuable in the development of coherence.

If the belief system produces divisiveness, such as the realization that one is (however rationalized) selfishly stealing from others, it will interfere with the transformative quest to bring the entire mind-body complex into an energetic synchrony. Divisiveness always has a disempowering effect to the system. This integral objective can only be supported by a *nonjudgmental* emotional state modulated by understanding.

With this bit of knowledge we can deduce all the psychological defense mechanisms (rationalizing, projecting, cognitive dissonance, etc.) that occur in considering the mind as a wave system.

We can tag incompatible memory constructs as being out of phase with one another; this is why they create all the suffering that we experience when they occupy the same psychic space-time in the information field. Such incompatibilities produce low energy states due to their beat frequencies.

Even if we are not fully aware of them, they produce a sense of angst in the non-conscious realm, quite likely due to the subtractive-destructive result which "steps down" energy.

This is how we experience energy when we are anxious or angry. Even though these negative states are experienced as a surge of energy, the adrenalized boost is (in the final analysis) dissipated through the movement center or some kind of psychosomatic effect. Hence, it is an energy loss.

Feelings of happiness, joy, elation, and even bliss are due to when the physical and mental constructs are in synchrony, in harmonic phase with one another.

LEVELS OF CREATIVITY & UNDERSTANDING

Each higher level of being and consciousness of the Self processes energy and information at faster rates, forming more refined interference patterns (in keeping with experiments in Cymatics: see Jenny 2000). If we, for example, double the rate at which these faster processes take place, we'd have a fractal continuum of octaves, in the Gurdjieffian sense.

It therefore follows that each higher level will be characterized as having a much greater degree of freedom in regards to creativity and understanding. This mental acceleration would provide greater access to subtler levels of the FCCP, pertaining to individuals and other systems, and allow for acts of creation (in all of its aspects) by a higher mind to appear magical and miraculous to a lower one.

For example, my first readings of a variety of holy books gleaned a rather literal interpretation of the material; upon acquiring a higher level of being and consciousness, I saw the same content in terms of myths and allegories brimming with multiple levels of meaning. In my current state, I glean a more inclusive objective-subjective meaning mediated with "pure" symbols related to a metaphysical science, which I find challenging to express in this book. In this regard, processing information "quicker" implicates spiritual states of being.

There is an incident related by Swedenborg in his *Arcana Coelestia* (AC 1758/2009b), in which he tells of his encounters with spirits and angels and how they absorb (inflow) information.

> "Since I've been in the company of spirits and angels constantly for nine years now, I have very carefully observed what *inflow* [impression] is like. When I thought, I could see solid concepts of thought as though they were surrounded by a kind of *wave*; and I noticed that this wave was nothing other than the kinds of things with the matter in my memory and that, in this way, spirits could see the full thought. But nothing reaches 'normal' human sensation except what is the middle and *seems to be solid* [comparatively speaking]. I have compared this surrounding wave to spiritual wings by which the object of *thought is lifted out* of the *memory* [by outgoing vortex?]. This is what brings it to our attention" (AC No. 6200. Italics and commentary/inquiry brackets added).

While in these spiritual journeys, Swedenborg further observes that angels and spirits process information at much faster rates than what is possible for humans to do in their physical mode of being. He states:

> "I was able to determine that there was a great deal of associative matter in that surrounding wave substance from the fact that the spirits in a subtler sphere [cosmic egg/aura] knew from it all I had ever known about the subject, drawing out and absorbing in this way everything proper to a person. ….. For example, when I was thinking about someone I knew, then his image appeared in the middle as he looked when he was named in human presence; but all around, like something flowing in waves, was everything I had known and thought about him from boyhood. So that the whole man as he existed in my thoughts and affections, was instantly visible among the spirits. Again, when I thought about a particular city, then the spirits knew instantly from the surrounding sphere of waves everything I had seen and knew. The same held true for matters of knowledge. (Ibid. Italics and bracket added).

One could say that there's a hierarchy of perception that separates one realm from another. We can interact with these realms of subtlety by speeding up the rate of mental operation. In the following passage from *Divine Love And Wisdom* (DLW 2009a) this becomes more evident.

> "In regards to this matter I have conversed much with the angels. They said that they have a clear perception of it [regarding the sun of the spiritual world] in their own spiritual light, but that they cannot easily present it to man, in his natural light, owing to the difference of the two kinds of light and the consequent difference of thought" (DLW No. 291).

Boredom

When the system is subject to "perceived" repeated *uniform stimuli,* it experiences boredom and is lulled towards the hypo-metabolic frequency ranges of sleep—delta and epsilon.

The conditioned either/or mind keeps us dithering between possibilities. It is constantly struggling with: "Is it this? Is it that? It can't be both!" If it is overwhelmed by the indecision it will shift towards sleep like a possum playing dead.

Functional Dissociation

Dissociation is the process by which the mental system uses interference to create specificity of thoughts, feelings, sensations, and actions by "separating" (through attentional focusing) memory constructs from one another into synchronous clusters.

This naturally occurs as the system progressively transitions from gamma to epsilon, the ultimate *resting state*.

In this ground state of the mind no two memory constructs are able to enter into a phase relationship. Unable to combine into assemblies, the mental system is void of "phenomenal" consciousness, showing no contrast to the PUF, forming the basis of dreamless sleep.

They are separated by vibrations characterized by the irrational numbers of the golden ratio series (Pletzer 2000; Weiss, et al, 2003). According to Freeman (2000) they are "uniformly disordered," like the Brownian motion of particles suspended in a fluid, which satisfies the meaning of *entropy*, the thermodynamic condition in which a system can do no work from a kinetic point of view.

The FCCP provides the memory repertoire from which the mind progressively becomes "made up" as it transitions towards gamma while performing the orderly work of *negentropy*, as coherent "meaningful" contexts.

Contextual Dissociation

Contexts are formed by energy configured in a variety of ways by the summation of constructive and destructive waves. They are never identical, but in many cases, they are close enough in similarity that they can combine into constructive clusters of meaning.

Contexts that don't agree with one another tend to inhibit one another in the psychic space of the observer. Their wave attributes do not fully match the waves of those with whom they "disagree".

By repeating a "rejecting" percept in the 1-to-7 cognitive cycle, larger clusters of MCs can be dissociated by iterative entrainment. Since contradictory perceptions tend to hog our attention, this over-dwell on them generates powerful negative emotions, which the Self seeks to excrete via the mechanism of projection.

Inhibition

Inhibition is normally involved in all assembly forming processes as *negative feedback* mediated by inhibitory neurotransmitters. It has a sculpting effect on memory constructs in the FCCP, allowing certain contexts to come forth and others to remain in the background in the 3a and 3b selection/deselection process. It is, therefore, not seen as a negative emotional state by psiology.

Suppression

Suppression involves memory constructs—contextualized by thoughts, feelings, and sensations—that, for some reason or other, deny the existence of other ones that are contextually unacceptable, using destructive interference and/or through signature vibrations.

Since the Psiloton self-referential flow sweeps through the FCCP, to maintain alertness and conceptual as well as behavioral potentials, energy must be used to separate the suppressed MCs from the selection at 3a. Hence, this amount of energy is lost for transformative contexts.

Blockages

It may also involve the psychosomatic prevention of biochemical and/or electrical flow that supports the function of an object of suppression (in the transformative tradition it is referred to as "blockages").

Multiple Personalities

Suppression is the basis of dissociative identity disorder (DID, a.k.a., multiple personality disorder). Each personality has a unique PSQ that evokes signature psychosomatic effects that keeps them apart. While the non-conscious is always aware of them, the limited conscious threshold may only have inklings of their existence and may be able to cue them to manifest behaviorally. They may appear as entities "possessing" the Self.

Intentional Phase Mediated Transitions

Dissociation also may occur as a *phase transition*. In this case it is not a suppressive act, but one in which the attentional process can sustainedly observe a memory construct, regardless of its complexity, and transform it through *positive feedback* into another contextual level of order, using a desired state of being contained in another memory construct to entrain it.

The context is how the observer translates memory constructs being stimulated by an internal and/or external event. This allows the observer to transform the input.

For example, the Hungarian endocrinologist Hans Selye (1907-1982) showed that stress can be transformed into "eustress" (good stress) by how the subject interprets the initial impression (Szabo 2012). This is often the basis for coping creativity.

Neutral Dissociations

Phase mediated dissociation is also the key process in the transitioning of the Self into its various modes of being (i.e., physical, mental, spiritual): a "higher being body." According to this method, the mode being transformed is neither rejected or accepted but simply neutrally observed and, therefore, amplified through positive feedback. Therefore, it doesn't inhibit, suppress, or in any way deny its existence.

As far as I know, this kind of transformation is not known in the orthodox psychological literature. However, it forms a key concept behind the methodologies of the transformative tradition.

In Scripture we find its implication in terms like the "quickening of the spirit" and the "watching" of the "kingdom within," the contextualized population of memory constructs in the non-conscious mind (see Sanford 1987, and O'Connor 1971, on the "inner population").

PHYSIOLOGICAL ASPECTS OF THE PSQ

While the whole neuroglial network is generally involved in the arousal process, three systems have specific relevance to shift it from sleep to relative wakefulness: 1) the locus coeruleus (a.k.a., the reticular activating system), situated in the brain stem, 2) the limbic system, consisting of a number of subsystems, situated below the cortex at the center of the brain (six are key), and 3) the frontal lobes. I refer to them as the "arousal and editing system," depicted in Figure 14 as a lateral view.

While the locus coeruleus has a general arousing effect via the thalamic pathway, the other structures in the limbic zone have a more selective editing role on how the system will be aroused and respond to stimuli. These factors contribute to the *contextual* and *expectational* nature of the assembly.

Context processing involves a brain region called the hippocampus, and its connections to two other regions called the prefrontal cortex and the amygdala.

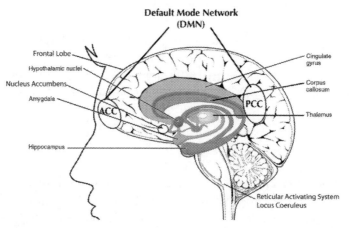

Figure 14: The Limbic System (lines and sites are approximate).

Locus Coeruleus

The locus coeruleus is a small nucleus located in the pons, the part of the brainstem that links the medulla oblongata (the lowest part of the brainstem, containing control centers for the heart and lungs), and the thalamus; it is the main source of noradrenaline in the forebrain. Together with other nuclei located in the upper anterior part of the brain stem, it belongs to what used to be described as the 'ascending reticular activating system', an area critical for arousal and wakefulness.

Locus coeruleus neurons have extremely wide projections and are innervated by only a few brain-stem nuclei and forebrain areas. The activity of locus coeruleus neurons varies not only with arousal but also with specific cognitive processes, resulting in concerted release of noradrenaline in multiple target areas, and very complex effects depending upon local parameters. This key neuromodulatory system is currently thought to be critical for numerous functions including response to stress, attention, emotion, motivation, decision making, learning and memory.

Thalamus

The thalamus is a large mass of gray matter situated above the hypothalamus and in the center of the limbic system. It is a multi-functional hub for integrating subcortical sensory-motor information (with the exception of the olfactory system) and relaying it to the associated primary cortical areas for the regulation of consciousness, sleep, and alertness.

Therefore, it is a key "gate" through which the level of consciousness is regulated, based on how OOPEs are interpreted.

In the transformative tradition it may be the "higher emotional center" that works closely with the *pineal gland*, also situated in the limbic control center.

Hippocampi

The hippocampus (named for its resemblance to the seahorse) is a major component of the brains of humans and other vertebrates. There are two hippocampi: one in the limbic region of each hemisphere. It plays an important role in the consolidation of information from short-term memory to long-term memory, memory consolidation during sleep, and is involved in spatial memory that enables navigation.

However, in the time of memory research of Lashley, the memory function of the hippocampi (the hippocampus in the two hemispheres of the brain) was not known. Their involvement in memory function was discovered from studies done on the patient Henry Molaison who in 1953, at the age of 27, lost his hippocampi as a result of surgery to alleviate an intractable epilepsy. Soon after the surgery the consequences became disturbingly clear.

He couldn't form new memories. Though he recognized his parents he could not remember the doctors and nurses that cared for him in the hospital, and he couldn't find his way to the bathroom, even after going several times.

As tragic as it was, Molaison's case (1953-2008) contributed enormously to understanding the role of the hippocampi. Though not conclusive, it revealed their importance in memory formation and retrieval, but it didn't point to where in the organism memory was "stored."

Amygdalae Remember OOPEs

The amygdalae are two almond-shaped structures located within the limbic system of the two brain hemispheres. Research reveals their primary role in the processing of highly-charged emotions (mainly fear but also joy), memory, and decision-making. Connected to other limbic structures, especially the hippocampi, the amygdalae are sensitive to out-of-pattern events (OOPEs).

An OOPE is by its very nature a combination of opposing impressions or energy patterns that evoke attention. OOPEs are characterized by their strangeness, particularly how they associate in some way to highly charged emotional events which may be discomforting or threatening to the observing system, or may encompass non-threatening forms of novelty, such as are presented by the various arts.

Obviously, recognizing the nature of an OOPE requires a comparison with the ensemble of memories that contribute to one's signature PSQ, which can vary from person to person.

OOPEs are edited by two structures embedded in the underside of both hemispheres of the brain, known as the amygdalae (Figure 14). They are responsible for remembering (or encoding) and identifying OOPEs and sending emotional signals to the neocortex to "watch out," "figure out," or "let it go."

Research indicates that during fear conditioning, sensory stimuli reach areas of the amygdalae where they form associations with memories of the stimuli. For example, when the amygdalae have been excised in laboratory rats they have no fear of walking in front of a cat (see Austin, 2000). You know how that will end up.

> "Up to now, we have been observing how learning takes place in adult rats. These adults have learned to fear through the process of conditioning. But the amygdala had been primed long before. It was already genetically programmed to help generate primal fear. A normal rat innately fears a cat. Seeing a cat, it freezes. However, rats lose this instinctive fearful behavior after they have had lesions of the amygdala. They will even climb up on the back of a sleeping cat after you have rendered the cat harmless with a hypnotic drug" (Austin 2000, 177).

From this we may assume that some "fearless" or "naive" individuals may, for some reason, lack proper functioning of these structures, or perhaps by alteration of the stimulus-response connection. This will reflect as a personality trait.

While there may be some innate fears, such as instinctive thresholds like body heat and pain, most of them must be learned by the unsuspecting child in conjunction with his physical and social environments. Quite often, we acquire an irrational set of fears that will not by their nature hurt us in any way except by how we think they will.

The association between stimuli and the aversive events they predict is thought to be reinforced by an iterative process known as long-term potentiation (LTP), a sustained enhancement (hours to days) of rapidly repeated signaling that may reach 100 Hz +, between affected neurons that may contribute to learning and long-term memory.

However, analysis by Shors, et al, (1997), of LTP indicates that it is not directly related to learning and long-term memory but is implicated in strengthening of neuronal connectivity that contributes to learning and long-term memory. In this more convincing interpretation, LTP is more apt to be involved in the amplifying effects of the arousal and attentional system that triggers an assembly.

Regardless of the precise nature of LTP we can safely surmise that rapidly occurring iterative processes are behind the triggering and magnification of cell assemblies and their signature connectivity. The highly focused mind is in a hyper-iterative state.

As Hebb had proposed, the connectivity is not one-to-one but involves the uniqueness of the neural configuration occurring throughout the neural population. The notion that learning and memory are a configuration of the multiple inputs that occur during a learning moment is supported by the psychological concept of habituation.

While the amygdalae are involved in every F-F cognitive loop, it is possible to redefine the context of their memory constructs as motivational forms of "eustress" (Selye), or as parts of "bisociative" states (Koestler), involving the acceptance of paradox, juxtaposition, etc.

In some people there is the tendency to exaggerate OOPEs to the point where they become overwhelming. The exaggeration is the result of dwelling on negative emotions and amplifying them into vicious positive feedback cycles. In this regard it could be said that their amygdala was hijacked.

"Amygdala hijack" is a term coined by Daniel Goleman in his 1996 book *Emotional Intelligence: Why It Can Matter More Than IQ*. Drawing on the work of Joseph E. LeDoux, Goleman uses the term to describe emotional responses from people which are immediate and overwhelming, and out of measure with the actual stimulus because it has triggered a much more significant emotional threat.

Cingulate Cortex

The cingulate cortex, a part of the limbic system, is involved with learning and memory formation. The combination of these two functions makes the cingulate gyrus highly influential in linking behavioral outcomes to motivation (e.g. a certain action induced a positive emotional response, which results in learning).

Nucleus Accumbens: Source of Pleasure & Addictions

On focusing on the center of the brow (i.e., frontal lobe), one establishes a connection to the executive area of the brain and, at the same time, has the potential to connect to the pleasure center located in the *nucleus accumbens*, a *dopamine-driven* set of neurons that acts as a reward system to reinforce behavior by priming it to a specific context.

In the language of wave interference, it would be triggered when the perceived input matches (i.e., is in phase with) a goal: the confluence indicates "Yes." Or, when the perceived input produces a counterintuitive insight, such as when previously disparate phenomena are seen in a complementary relationship: the "Aha!" response.

If sufficiently activated it produces synchronous activity related to pleasure, and can be the source of addictions. Its connection to the executive functions can be viewed as providing the motivational wherewithal for reinforcing certain behaviors.

The Dopamine Reward System

Dopamine plays an important role in the regulation of reward and movement. As part of the reward pathway, dopamine is manufactured in nerve cell bodies located within the ventral tegmental area (VTA) and is released in the nucleus accumbens and the prefrontal cortex. Its motor functions are linked to a separate pathway; lack of it in this pathway has been linked to Parkinson's disease. The relationship between the two pathways suggests a powerful connection between positive emotions and will (the ability for volitional action).

> "An important part of this motivational link is related to the prolonged iterative effects of a learning process called long term potentiation (LTP). Excitatory synapses on dopamine neurons in the VTA can undergo both long-term potentiation and depression" (Wikipedia: Long-Term Potentiation).

Joy, Euphoria

We feel good because the dopamine-driven activity is a powerful source of entrainment, producing synchrony in the particular assembly. If strongly activated it can generate broadly distributed coherent states that can range from joy to euphoria (translates as *ananda* in yogic lore). Though not anatomically identified, in the Fourth Way it is has affinity to the *higher emotional center.*

The Default Mode Network (DMN)

Cognitive functions arise from the orchestrated activation and cooperation of a wide variety of networks in the FCCP. Studies have found that there are two main hubs in the brain that organize these networks: *intrinsic* (a.k.a. the default mode network, DMN), and *extrinsic* modalities of perception.

The extrinsic portion of the brain is activated when individuals are focused on external tasks, like playing ping pong or replacing a light bulb. On the other hand, the intrinsic or DMN turns on when people reflect on matters that involve themselves (autobiographically) and their personal emotions. The DMN has been shown to be active when a person is not focused on the outside world and the brain is at wakeful rest (e.g., in alpha), such as during daydreaming and mind-wandering.

The DMN is commonly found in certain brain regions known as the *thalamus,* but is more pronounced in the *anterior and posterior cingulate cortex* (ACC, which includes the medial prefrontal cortex, and PCC) and in the medial temporal lobes (see Figure 14).

However, we need not dwell too much on these technical terms, but note the role of DMN in regards to psychophysiological function and the extent of its connectivity during the resting and attentive states.

THE MIND IS A CONTROL SYSTEM

In addition to what has thus far been presented, we need to also take into consideration that mental systems as well as machines are basically complex "control systems," mediated by F-F loops (Wiener 1989).

MENTAL SYSTEMS MUST OPERATE FASTER THAN THAT WHICH IS BEING CONTROLLED

A common example of this F-F principle is found in driving a car along a highway. The awareness of the driver must, of necessity, operate faster than the speed of the car in order to keep it on the road and to avoid unexpected events.

In this scenario, the physical nature of the road affects how the driver will react, and how the driver responds is an ongoing reciprocal process. So, it is not inconceivable to assume from a psiological perspective that the physiology affects the psychology and vice versa.

It therefore follows that the mental system, while affected by the physical one, controls the physiology of the body in order to maintain a dynamic, yet complex, equilibrium, called "homeostasis."

MUST EXCEED SPEED OF BIOCHEMICAL PROCESSES

In order to do this it needs to exceed the speed of bioelectrical processes, which have been estimated by Herbert Fröhlich (in Smith 1998) to take place at the electromagnetic-resonant-frequency of biomembranes, at around 100 GHz (1 Giga Hz is = to 1 000 000 000 Hz, or, to put it in more familiar terms, 1 million cycles per second times 100), involving about 50 trillion cells that constitute the physical organism.

SCOPE OF TASK IS ENORMOUS

Because the scope of the task is so enormous, the cognitive system regulates it into a separate frequency range (which is what we call the "subconscious") and creates another much slower frequency to attend to out-of-pattern events (OOPEs) occurring in the outer environment that may directly affect the critical biochemistry to maintain homeostasis.

This familiar range of vigilance we refer to as our "consciousness," but it does not, as you can see, have the scope, depth, and speed the non-conscious (all subliminal processes) uses to process energy and information. The threshold in which we normally function is considerably smaller.

Of necessity our conditioned consciousness must function at a much lower rate to perceive the slower movement of objects and events in the outer environment. This schism produces the "duality" transformative teachings allude to in their development practices.

However, there are moments when OOPEs overwhelm our limited-conscious threshold and the mental system finds itself needing a faster intelligence to deal with the situation at hand, and it quite logically opts to relegate the issue to the non-conscious domain. In those moments, the Self in the physical mode of being begins to process information at the 100 GHz range or faster, if it is to control the organism to do its bidding.

These are the fortuitous moments in which people perform "miraculously." If they are not totally overwhelmed, they move faster, perceive more, and think quicker. More importantly, however, is that they have crossed the threshold of the "normal" and have entered the "spiritual" zones of the "paranormal" by dint of being accelerated into another level of being and consciousness.

More precisely, the people caught up in such scenarios have entered the processing speeds of the fractal domains and are able, by dint of the 1-to-7 cycling, to amplify them into behavior at 6. The dire nature of the events is what thrusts the Self into arrested attention and its consequent accelerative effects of Psiloton processing.

In that zone the inner world speeds up and has the vibratory wherewithal to resonantly connect with other concurrent fields of information, expanding the memory repertoire to glean vital information from the "collective unconscious" with which to control the cellular network.

In many cases this set of dynamics extends control over physiological events, for example, as is reported as "placebo" (i.e., positive outcomes) or "nocebo" (i.e., negative outcomes) effects found in the medical literature (e.g., see Maltz 1969).

While many have experienced the placebo-or-nocebo effects and fortuitous-mental-accelerative shocks of critical OOPEs and their associative "paranormal" experiences, the ability to engender them intentionally through some form of mental suggestion or mind control is the common experience of "endowed" or highly trained "seers."

Be that as it may, we ought not forget that such amplification effects both positive as well as negative outcomes.

OOPES CAUSE MIND TO ACCELERATE

It is a psychological fact, when interpreted properly, that when the mental system is programmed by suggestion and/or is challenged by OOPEs it immediately increases the EEG frequency to deal with it. In other words, it usually doesn't opt to go to sleep (but may do so as a defense option, like a possum) but experiences an upsurge of energy and mental alertness.

While hyper vigilance is also known to have dire effects on the organism, the point being emphasized here is how the organism needs to increase the rate at which it can process information as a means to solve seemingly intractable problems.

Not only must it increase its processing speed by shifting the EEG towards gamma+, but must, at the same time, access the non-conscious domain that has relevance to the biochemical processes of basic instinctive processes at the cellular level by incorporating in the vibratory pattern frequencies tending towards the large slow waves of epsilon.

To do this it appears to couple the two ends of the spectrum, the epsilon coupled to the gamma+, so to speak, into a F-F loop. From this perspective the mental system is cross-coupling its entire frequency spectrum in a holistic manner.

What we glean from the literature in neuroscience is that memory in the slower frequencies is being created, updated, and amplified in millisecond time frames into the form-and-function of phenomena we perceive.

THE F-F LOOP OF HOLISTIC CAUSATION

Like any other wave phenomena, vortices interact using F-F processes (which will also be designated as feedback) based on the principle of interference.

Interference is a phenomenon in which two or more waves interact to form a resultant wave(s) of greater, lower, or the same amplitude. These interactions determine the systems' "yes," consisting of high amplitude, "no," consisting of low amplitude, or "maybe," a value somewhere in the middle.

Feedback is the means by which the mental system uses the effects of wave interference to generate conscious and/or non-conscious events in order to make decisions based on an inherent and/or a desired goal.

Decisions are made by how the memory constructs in the field of cognitive and creative potentials (FCCP) resonantly merge or dissonantly diverge from one another, due to vortical spin. When associative vortices spin in phase they produce a collective result described as a synchronous assembly.

Feedback is the process by which a system exchanges energy and information with itself and/or another system or subsystem in order to attain a habitual (i.e., expected) or novel outcome.

This occurs as a wave or energetic phenomena, since solid objects (e.g., cells) are too slow and cumbersome to in-and-of-themselves enact a proper exchange that can oscillate to generate a standing wave of perception. Information is carried as energy configurations mediated as waves on the bands of vortices that occur within and/or emanate from these cells.

It is important to note that all energy and information exchanges require an alternating "F-F" process, "give-and-take," "transmit-and-receive," relationship.

The principle is exemplified by two tuning forks of the same frequency. When the example is used, the explanation rarely includes that the oscillatory exchange is actually an alternating activity.

When one is struck in close enough proximity to the other one, they begin to send and receive vibrations in an alternating manner. We don't see this concatenation because it takes place at the speed of sound (343 meters per second; 1,125 ft/s). Without this F-F process an energy-and-informational exchange cannot take place.

At the human level, for example, a conversation occurs when two people alternately talk and listen. So this rule applies to micro as well as macro levels of "behavior."

At the quantum level, for example, one particle has to be "up" and the other "down" for energy exchange to occur. According to the astrophysicist Milo Wolff:

> "Energy exchange is the source of information. ... *Everything we measure in Nature requires an energy exchange that tells us something has happened.* This is a requirement to find truth. Experience tells us that acquisition of knowledge of any kind occurs only with an energy transfer. *Natural law describes energy exchanges. Storage of information, whether in a computer disk or in our brain, always requires an energy transfer. Energy is required to move a needle, to magnetize a tape, to stimulate a neuron.* There are no exceptions. Thus finding the energy transfer mechanism between particles is part and parcel of understanding the electron and the natural laws. You cannot accept any statement about the measurement of a natural event unless you verify the energy exchange that allowed it." (Wolff 2008, 136. Italics added).

The mind-body system is constantly interacting with the environment through F-F (i.e., output-input) mechanisms that help it maintain its current identity or modify it to adapt to challenging conditions in the outer as well as its inner worlds. It mainly sustains its identity by repeating what worked before, constituting its repertoire of habits and expectations in the FCCP.

In this regard, all activity of the psycho-organism is purposeful, and the F-F mechanism is what endows the system to recur, somewhat as is, or to adapt by altering the environment and/or itself. Most of its adaptive possibilities rely on its subjective abilities to expand consciousness and to create new combinations of what it knows or has learned.

This is the means by which a system steers its behavior in pursuit of a habitual-or-novel goal. As Freeman suggests, all mind-body behavior is purposeful, a concept not embraced by many scientists.

Habitual activity "seeks" to repeat itself in the life of the psycho-organism. It produces a complex, which I have referred to as the PSQ, which is a more inclusive term that supplants the common notion of the "ego."

Feedback processes can be quite simple and very complex. For example, the entire mind-body complex is "held" together by a plethora of F-F loops that modify the "fields" in which they exist. To help understand what they entail we need a simple basic model to help explain the 1-to-7 cycle.

A holistic metabolic feedback process could be sparsely modeled with 6 variables—*negative* and *positive feedback, out-of-pattern events* (OOPEs), *expectations/habits*, existing *attractors or memory constructs* (∞) in a memory pool, referred to as a 1D *Field of cognitive & creative potentials* (FCCPs).

All interact to produce a standing *vortical soliton* wave, the 6th variable at the center, indicated by a NOW, a 3D *vortex*, which enfolds and dimensionally reduces back into the field as a holographic 1D memory construct. It cannot be overstated that the process, as Freeman's research supports, repeats in millisecond (thousandth of a second), or faster time frames.

It incorporates 4 vortical movements (2 inward, "implosive," and 2 outward, "explosive") mediated by two kinds of feedback, that, in order to get a closer look at the cognitive cycle at 7, could be modeled thusly:

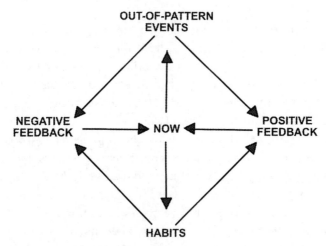

Figure 15: The Feedforward-Feedback Loop of Holistic Causation.

Like with the Symbol, information is simultaneously moving towards and away from the NOW center at the Star.

As you can "see" this multi-causal loop is not a stimulus-response effect, but the convergence of multiple effects that cannot be reduced to the constituent parts. The neuroscientist Wolf Singer, one of the discoverers of the attentional assembly-forming process, shares a similar insight:

> "It has also become clear that the brain is by no means a stimulus driven system. Rather, *it is self-active, permanently generating highly structured, high dimensional spatio-temporal activity patterns*. These patterns are far from being random but seem to reflect the specificities of the functional architecture that is determined by the genes, modified by experience throughout post-natal development and further shaped by learning. These self-generated activity patterns in turn seem to serve as priors with which incoming sensory signals are compared. Perception is now understood as an active, reconstructive process in which self-generated expectancies are compared with incoming sensory signals. The development of methods that allow simultaneous registration of the activity of large numbers of spatially distributed neurons revealed a mind-boggling complexity of interaction dynamics that eludes the capacity of conventional analytical tools and because of its non-linearity challenges hypotheses derived from intuition. These new and fascinating insights impose revision of concepts and unravel explanatory gaps that were not visible a few decades ago" (Singer 2013, 2. Italics added).

The above model is simply introduced as a means to have an appreciation of the 1-to-7 mind-boggling process that repeats itself in millisecond time frames that generate a morphing-movie-like depiction of "objective reality."

What's even more mind-boggling is that this sense of "substance" is fashioned from the mysterious space medium we, for want of a more descriptive term, refer to as "primal field" of sorts, which I have dubbed the PUF (Pluripotent Unified Field), the "substance" from which all is fashioned.

> "The wave structure of matter, is very simple - in fact it could hardly be simpler being composed of only two binary waves that are formed of a single substance, the wave medium. Traveling in opposite directions, the two waves are mirror images of each other. Many ancient philosophers, who carefully observed nature and deduced some of its properties, had predicted that nature is based on a single substance" (Wolff 2008, 46).

The F-F dynamic appears to be the means by which the mind-body-spirit wave complex sustains its signature identity. Let's take a closer look at each of these variables.

TWO KINDS OF FEEDBACK

From my readings in cybernetics, particularly the works of Norbert Wiener (1954), I learned that there are two kinds of feedback: *positive* and *negative* (not necessarily having to do with ethics or social values; yet, applicable).

IN OR OUT OF PHASE

The F-F mechanism uses the psychophysical language of waves as its frame of reference. Hence, it compares the behaviors of the Self in any or all of its modes of being to an inherent and/or selected goal. In simple terms, it is always generating agreement, disagreement, or neutrality in regards to meeting it. At a more fundamental level the system is assessing phase relationships. When examined closely with these generic terms we would find that we are constantly in this F-F mode, regardless of how simple or complex the nature of the activity.

POSITIVE FEEDBACK

Positive feedback occurs when a *similar* portion or all of the output of a system (in this case, selected from the memory field) is returned as input in a *self-referential cycle*. For example, when two signals are similar in frequency (temporal aspect) and structure (spatial aspect) they add linearly (e.g., $2 + 2 = 4$) by constructive interference. This repetitive activity causes a system to increase the strength of the output (a.k.a. "gain" in electronics).

Positive feedback occurs when the fed-back signal is in phase with the input signal. Under certain gain conditions, positive feedback reinforces the input signal to the point where the output of the device oscillates between its maximum and minimum possible states. Hence, it has, at a certain stage in the process, a catabolic-like energetic effect.

A common technical example of a positive feedback loop occurs when someone is talking into a microphone that is too close to the speaker. The input at the microphone is amplified (a positive feedback process in itself) and sent as output to the speaker, which is picked up by the microphone and re-amplified in a recurring cycle, producing an accelerative *vortex* that manifests as a screeching sound.

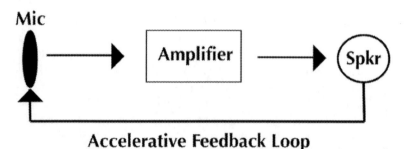

Accelerative Feedback Loop

Figure 16: Dynamics of a Positive Feedback Loop.

Note that the key to positive feedback cycles is the recurring effect of *similarity* of motion and/or context, which is explained scientifically by the additive-amplifying effects of *constructive wave interference*, cyclicly enfolded. Here, the energy and informational exchange is characterized by a recurring "agreement."

In human terms, two people would be pumping each other up by a conversation in which both parties are "yessing" each other. In psychological jargon they are *reinforcing* a particular belief and/or behavior.

At the individual level it is the technology that produces "contextual coherence." While it has a sustaining effect, its redundancy can lead to a kind of friendly mental shake up, or chaos when the iteration increases its cyclic velocity beyond a certain threshold.

Neuroscientists have discovered this iterative phenomenon occurring in the brain. They've dubbed it as "long-term potentiation" involved in the arousal of attention.

Reinforcement

Repetition of a percept (i.e., the 1-to-7 loop) according to Pavlovian conditioning, produces a habit and reinforces it. What actually happens is that it increases the number of memory constructs and their temporal associations, thereby forming larger assemblies when cued.

Long-term Potentiation

If you've ever heard one of your favorite songs repeating inside of your head it's quite likely that you have encountered the effects of long-term potentiation (LTP), which is actually the iterative activity of a positive feedback cycle with an amplifying effect.

> "Specifically, we propose that LTP is the neural equivalent of an arousal or attention device and that it acts by increasing the *gain* [positive feedback] of neural representations of environmental stimuli. If one assumes that an environmental cue is represented in the brain as a synaptic response or pattern of responses, then the induction of an LTP-like phenomenon would magnify that response(s), allowing for more efficacious detection of stimuli in general. Such an increase in gain (and consequent perceptual awareness) could then modify learning by increasing the likelihood that contingent relationships between stimuli are recognized. *Thus, we are proposing that LTP is not a mechanism for memory storage and retrieval but that it does play an incidental role in memory formation*" (Shors, et al, 1997, 609. Italics and bracket added).

Chaos theory informs us that if such a cycle exceeds a certain limit it will go into chaos. Now, though this sounds a bit extreme there is an aspect of chaos that endows the mental system with the ability to think, that is, explore different aspects of a particular theme.

The Positive Feedback Loop of Patanjali

Interestingly, in *The Yoga Philosophy of Patanjali* we find the following commentary by Samkhya scholar and practitioner Swami Hariharananda Aranya (1983) about similarity in the meditative process of the one-pointed mind that strongly suggests Patanjali knew about constructive wave interference.

> "The mind which is pointed to one direction only, i.e., holds on to one thing only, is called a one-pointed mind. Patanjali has defined it later as a mind wherein, on the *fading away of one thought, the same thought arises again in succession.* In other words, when one thought vanishes from the mind and the next that

arises is *similar* and there is a continuity of such successive states, then the mind is one-pointed" (1983, 4. Italics added).

Repetition is the recurrence of a similar object (i.e., context) in the mind, and because the wave signatures are "similar" they result in a constructive wave, which increases additively in amplitude (power).

Attachment

Knowledge of interference tells one how the mental system interfaces with the inner as well as the outer environments through F-F loops. It tells what the system likes, dislikes, or is indifferent to, whether occurring consciously (using larger assemblies) or non-consciously (using smaller assemblies in fractal time). It also goes towards explaining how the perceptual field is formulated by standing waves (see Lehar 2000).

In transformative psychology the principle of constructive interference helps to define the notion of *attachment*, which in psychology has affinity to the notion of *absorption*, and the phase transitions of neuroglial populations in the expansion of *consciousness*. It also shines a new light on the nature of the iterative rituals (e.g., mantra, prayer, movement, ceremony, etc.) that prevail throughout the transformative tradition.

So, it's a valuable tool that can enrich spiritual psychology's somewhat prosaic expressions with scientific ones. Both are valuable. The former creates a context, which the latter processes into transformative enactments.

NEGATIVE FEEDBACK

Negative feedback occurs when the result of a process influences the operation of the process itself in such a way as to reduce changes. It contributes to the regulating effects of a system to maintain stability, accuracy, and responsiveness by reducing the effects of fluctuations in a timely manner. Hence, it has a anabolic-like effect on the system.

In the stream analogy, negative feedback is the aggregation of all of what contributes to prevent the attributes of the laminar stream from changing.

Like positive feedback it can be conceived in a variety of ways. Generally speaking, it is anything that blocks, impedes, resists, dampens, reduces, prevents, restricts, and so on, to sustain or block a particular function or operation.

In electronics negative feedback occurs when the fed-back output signal has a relative phase of 180° with respect to the input signal (upside down).

This situation is sometimes referred to as being out of phase, but that term also is used to indicate other phase separations, as in "90° out of phase." It employs destructive interference that reduces the amplitude of the signal to a desired output. Negative feedback can be used to correct output errors or to desensitize a system to unwanted fluctuations.

In organic life it is what sustains the metabolic functions of the organism, involving all of the "norms" that ensure its survival in a changing environment.

In the automobile negative feedback would be the brake, or any force that slows it down. In the cruise control system of a car it is what maintains the selected speed by regulating the amount of fuel.

In a moral code it urges what not to do (i.e., the "no," the "don'ts") that clamp down on liberal behavior. This establishes its PSQ.

Whatever jives with the vibratory value of the PSQ is accepted, which manifests as a rise in amplitude we experience as a feeling, which occurs when two or more wave systems are in phase.

In order to be effective, negative feedback requires a frame of reference or value (e.g., the value selected on a thermostat) against which it can "measure" differentials.

From a medical perspective it is found in the reference ranges of biochemicals and rates of operation of the different organ systems based on the biological clock situated in the hypothalamus.

In a more generalized way it is found in the vibratory signature of its natural resting state, known as *resonance*, which I construe to be the vibratory rate of one's *Essence* (a.k.a., the "child within"), the signature rate of the Self's phenomenal existence.

Negative feedback seeks to maintain a system's identity, so that each metabolic moment of tissue breakdown (i.e., catabolism) and tissue reconstruction (i.e., anabolism) is faithfully repeated. It is responsible for maintaining the PSQ, mediated by the limbic system and/or to a strong memory construct (i.e., belief).

Information (i.e., configured energy) that does not resonate with the PSQ and does not pose a critical threat is vented psychosomatically through the musculature and psychological defense mechanisms, like the expression and projection of negative emotions.

Knowledge of interference tells one how the mental system interfaces with the inner as well as the outer environments through feedback loops. It tells what the system likes, dislikes, or is indifferent to, whether occurring consciously (using larger assemblies) or non-consciously (using smaller assemblies in fractal time). It also goes towards explaining how the perceptual field is formulated by standing waves (see Lehar 2000).

Negative Attachment

In transformative psychology the principle of destructive interference helps to define the notion of *negative attachment*, which in psychology has affinity to the notion of maintaining the PSQ, and whatever appears to not jive with what it has rejected by reducing its influence on the entire system.

RATIO OF NEGATIVE TO POSITIVE FEEDBACK

In general, negative and positive feedback are needed to co-create a "manifestation," and more specifically, to form a synchronous assembly.

Positive feedback entrains a resonant context while negative feedback "holds" it in place. The attentional process is sustained by a complementary balance between negative and positive feedback (Attention: NF=PF). We experience the enduring equivalence as recurring moments of "stillness." The extent to which we can sustain the balance constitutes our attention span.

Positive feedback produces the energy while negative feedback is what contains it. The rate at which a system oscillates is determined by a combination of negative and positive feedback, forming a balanced ratio, where a cancellation effect occurs forming an "attracting" node.

Further ahead, we shall postulate that this point of equilibrium is based on the golden ratio (1.618...). Positive feedback tends towards accelerative activity while negative feedback counteracts it. The density or viscosity of the medium in which the waves/vortices oscillate (e.g., intracellular and extracellular fluid) is also a contributing factor of negative feedback, since it may "slow" down accelerative effects.

For example, it is easier to stir a cup of tea than a cup of molasses. The uniformity of the medium is also a contributing factor. There are many variables with this dynamic.

As a general rule, when the ratio of negative feedback to positive feedback is low, the system may lack enough counter force (e.g., grounding) to prevent it from spiraling out of control, like when someone explodes with anger.

Conversely, when the ratio of negative feedback exceeds that of positive feedback one may feel compelled to, for example, not speak one's mind. Liberalism and conformism are the two extremes of this psychological dynamic. Ideally, the two are in a complementary relationship.

However, the two are needed in some ratio of relative balance to oscillate in sync and form a conscious assembly.

This is biochemically mediated by inhibitory and excitatory neurotransmitters. The attentional process requires a blend of positive to negative feedback to sustain itself. This dualistic complementarity is found throughout nature.

OUT-OF-PATTERN EVENTS

Out-of-pattern events (OOPEs) are those that are not immediately assimilable by the PSQ, and they come into two basic categories: unexpected and intended (like subjecting oneself to new experiences). Both may contribute to the development of the Self.

Intent means purpose. It implicates the freedom to choose an unconditioned response to incoming stimuli. It is this creative aspect that modifies the FCCP by exploring combinatorial possibilities in relationship to a goal, which itself is subject to morphing via the imagination.

It is important to note that there are two streams of information constantly entering the mind-body complex: the conscious and the non-conscious (estimated by Szegedy-Maszak, 2005, to be to be 5% and 95%, respectively!). The former takes place within a threshold conditioned by the limbic editor while the latter is continuous, and is sensitive to conditions that are above-and-below the conscious threshold.

Will

As a key part of this holistic dynamism, will is based on an iterated desire to attain a certain novel objective. The objective is iterated in the mind by the 1-to-7 cycle and is reinforced by behavior into a habit, a memory construct that is encoded in the FCCP with a certain "gravitas" or "psychic mass" by dint of how it is associatively constructed. This means that it elicits a relatively large assembly, primarily in the intellectual and movement centers.

To generate a new habit pattern, negative feedback is used to isolate the intent (this happens automatically by the context one desires at the input phase, 1) and iterates through thought, imaginations, and/or behavior, of the desired goal until it becomes a reinforced memory construct in the FCCP. This is required for all kinds of "transformations." Memory precedes behavior.

One way to reinforce a desired memory construct is to expand its context through associative imagination, that is, by connecting it to other constructs (ideas, objects, etc.) to generate a broader pattern, hence a larger assembly.

This is the technique used since antiquity, known as mnemonics: the study and development of systems for improving and assisting the memory. Being in the NOW is also a powerful memory-enhancing method because it includes one's Self in the larger contextual surround of the moment.

GOAL DIRECTED BEHAVIOR

To determine if the goal and behavior match, the system uses the F-F loop (based on the principle of wave interference) to compare the input to the output in the 1-to-7 cycle.

It "knows" that the goal is not met when the characteristic waves of the output of the system do not match (i.e., sum up) those of the input. Hence, there's a degree of dissonance (i.e., lack of being completely in phase). This is the system's "no."

If not in phase, the mental system is prompted to begin a learning process of trial-and-error in order to attain the match. Depending on the level of motivation and other challenging factors, the feedback process will continue until a match takes place and the match produces a higher amplitude, which the psycho-organism feels as "satisfaction."

It is in this manner that feedback steers the system towards the goal. The observer knows whether it has occurred by dint of constructive and/ or destructive wave interference.

Should the wave signatures of the goal and the input match, they enter into a state of constructive interference, which the observer experiences emotionally as a positive rise in amplitude.

In that case, the learning cycle is apt to pause or stop. Now the subjective nonlinearity of the learning process produces a fairly stable linear, more objective, result.

Also, keep in mind that the number of times the feedback loop cycles depends on the degree of "novelty" and/or "complexity" of the desired event.

OUR TWO BASIC FRAMES OF REFERENCE

By now you may have surmised that the mental system is "designed" to respond to contradictions, which are, quintessentially, anything that does not jive (is out of phase) with our two basic frames of reference:

1. Essence: it encompasses potentials we are born with, and is typically active until about 6 years of age. It is referred to as the "child within" by many transformative systems, and it contains the potentials of the "true Self," and its vibratory signature.
2. Personality: is normally active with a strong predominance of Essence until the psychophysiological status quo (PSQ) or social frame of reference begins to develop and overpower it.

Essence can be thought of as the unity of a signature pattern we acquire during gestation in the womb. It is referred to in Scripture as "destiny," given by Higher Intelligence (HI). It represents the basic "material" that can be reconditioned to foster spiritual development because it is more "unified" than personality.

Notice how an infant responds to stimuli with all of its being. For example, when it cries or shows pleasure, its entire body participates in the response (i.e., the Wholistic Response). It has one mind-body language, the intrinsic information of which has been postulated to be conserved in the continuity of the fascia.

Gestation is a delicate period because it can be positively or negatively affected by exogenous factors, which include the mental and physical state maintained by the mother's emotional state and nurturance and the immediate environment.

While the two must find compromise, the degree to which these two systems agree (i.e., resonate) with each other determines the core feelings people have of themselves.

The mental system compares impressions entering the system to its signature and acquired vibratory frames of reference to assess them.

All sorts of psychological problems can be traced to the extent these two systems don't agree with one another. Neurosis and psychosis tend to develop when, for one reason or another, there's a huge discrepancy between the two. When properly balanced the Self acquires the state of "individuation" put forth by Jung.

Since one's signature essence is strongly established (a.k.a. in the transformative tradition as "destiny" or "fate"), the adjustment is rightly placed on altering the more labile personality into a supportive role. Therefore, in transformative psychology being true (i.e., faithful and accepting) to one's essential Self is the measure of health (i.e., being whole), and the source of "holiness." Shakespeare expressed it perfectly:

> "This above all: to thine own self be true, And it must follow,
> as the night the day, Thou canst not then be false to any man"
> (Hamlet Act 1, scene 3).

When Essence is properly nourished it produces a strong sense of Presence (a.k.a. the "fullness of God" and the "Holy Spirit" in Scripture), simply because energy and information more readily flow uninterruptedly through the energy channels.

People intuitively know what their Essence is. It is reflected in desires related to what one "loves" or would "love" to be doing (i.e., being) if given the right circumstances. However, it can be suggested by one's "body type" (see: e.g., Berman 1926; Friedlander 1993). Jung's typology would also be instrumental in assessing Essence in order to make a comparison to the person's current circumstances and life style (see Fudjack/Dinkelaker 1999b).

These two basic frames of reference always directly and indirectly effect how the mental system determines what it accepts, is neutral about, or rejects.

CONSCIENCE: ESSENCE'S SIGNAL OF "TRUTH"

Essence is that part of us that precedes personality in human development. Since it can be thought of as an inherent "habit" of nature (i.e., the materializations of the "life force"), it has influence on the development of the PSQ, conventionally referred to as the "ego," providing a kind of "referential style."

As a person develops, the PSQ takes on, due to growing socialization and peer influence, a more controlling role, likely to contradict the needs and desires of Essence, which, according to research (e.g., Bai 2011; Becker 1987), has a strong affinity to the meridian energy network mediated by the continuity of the fascia and its vast watery element (about 70%) throughout the organism.

In this developmental regard, Essence and personality can have degrees of coherence. I have postulated that when there's a "large" discrepancy between the PSQ and Essence, psychophysiological disorders are likely to crop up.

Over time, as this discrepancy grows, one will begin to feel a hidden source of anxiety, more accurately, as a kind of angst, signifying a feeling of deep unfocused fear or dread, about oneself and the human condition in general.

An interesting aspect of Essence is its relationship to the non-conscious domain where "raw" perception registers objects and events in their unedited state. Hence, it harbors in the FCCP the "suppressed" unvarnished history of every individual.

Whenever the PSQ edits what it perceives to satisfy group-think (i.e., "other directed" beliefs), Essence feels the contradiction (i.e., dissonance) to its nature. When compared to other trivial aspects of the personality Essence "knows" it exists, and moreover, wants to blossom, but finds itself in a conceptual prison. It came forth to me as the vision of the multi-colored bird encircled by a stone wall and stymied by ignorance that it could "fly," rise above, its predicament.

It has instinctive, emotional, and intellectual needs that are suppressed by personality's acquired "psychic mass." Whenever it encounters signs for its expression, Essence responds with feelings of unrequited desire. This occurs more poignantly when personality denies it has this deep propensity for a signature existence at the core of its being.

The P300

This recognition appears to have a strong affinity to a particular EEG signal, known as P300, that neuroscientists have discovered to subconsciously register in the brain that indicates a subject has experienced a particular event, regardless of whether it has been denied.

For example, if a subject is exposed to a series of random names and occasionally his own name, a P300 response is evoked whenever his name is called. Because the response is happening at the non-conscious level the subject cannot control its expression whenever familiar objects or events are perceived. So, if he denies that his "real" name is so-and-so the P300 response will prove him to be lying.

After the P300 was shown (as early as 1988) that it could be used to identify college students who were lying about having stolen something, the method was refined to be used as a lie detector by the neuroscientist Lawrence Farwell to detect lying in criminal cases (see Wen 2001). Although the P300 is reliable in 87.8% of cases, it has been accepted as a useful "lie detector" with some success in a number of court cases.

Pursuit of Wholeness is a Divine Mandate

The postulate being explored here is that the discrepancy between the non-conscious mind that forms the mental system of our Essential nature manifests not only as "deep conscience" when such discrepancies arise, but also as an indication that the moral mandate by Higher Mind (a.k.a., God, the Absolute, the Source, the Divine, etc.) exacts of us a state of total harmony. Note that all our suffering stems from lack of personal and social coherence due to this dividing factor.

Therefore, the Biblical commandment "Thou shalt not bear false witness against they neighbor," could be extended to include "nor against thyself." The P300 is watching!

Moreover, the degree to which we acquire this harmony in ourselves and with others marks the level of our personal evolution. Whenever we deviate from this mandate we "suffer" the effects of asynchrony.

Again: studies (in Kopell 2009) using EEG and MEG to detect neural synchrony suggest that several clinical conditions, such as schizophrenia, autism, and Alzheimer's disease are associated with the inability to sustain synchrony and coordination of distributed brain processes, probably due to psychosomatic and/or structural effects.

While there are many intricacies about this response it points to the need for harmonizing Essence with personality, not only to avert psychological aberrations but also as the basis for "unity." This means, as Ouspensky taught, that if we "lie" to ourselves in personality, it cannot be hidden, for Essence is privy to the discrepancy. Lying, or "the bearing of false witness," divides us, and is one of the most pernicious contributors to psycho-fragmentation.

However, as we shall explain in the final chapter, the transformative meaning of unity is not what is usually thought of in conventional terms.

THE BODY DOESN'T LIE

A variation of this theme is also found in the book *Your Body Doesn't Lie: Unlock the Power of Your Natural Energy!* (1979) written by the psychiatrist John Diamond. He advocates a finger method of using muscle testing to determine how the body will respond to questions and exposure to a variety of substances. By asking the body (i.e., the the Self in the physical mode of being) if the question or substance resonates with its essential nature, a person can have an initial line of communication to find out the nature of Essence and what harmonizes with its signature wave configuration.

UNIVERSAL RATIONALE FOR ESSENCE'S SIGNATURE

To ensure that the similar "objects" of the universe don't clump together into a ginormous mass, Universal Intelligence (mediated by the self-created Pluripotent Unified Field) employs a dynamic process that produces "chaos" to keep them dissonantly separate to some degree. This process is the antithesis to resonance, or phase locking, which is what connects similar objects together.

By increasing the rate at which a system turns its output to its input (i.e., its self-referential cycling), it reaches a point of period doubling (based on the Feigenbaum numbers: see Gleick 1987) that brings the system into a state of chaos.

In this state it generates fractals, self-similar fragments of itself. While having similarity (i.e., uniformity of information) they are not identical, and therefore have a slight dissonance with one another, which allows for enough disconnection amongst them to, if given enough energy, interact nonlinearly in novel ways. Otherwise we'd not be able to think or create.

Note that chaos is simply described as a state in which mental or physical objects are not linearly, or in psychological terms, not meaningfully, connected. However, they have, while not apparent, nonlinear relationships that contribute to an "implicate order" (see Bohm 1980).

Our Signature is as Unique as a Snowflake

Like what has been observed of snowflakes, no two people are identical, even identical twins (see Williams, 1979, on our "biochemical individuality"). A dog's sensitivity to the molecular print can always pick out the "right" twin (who, by the way, has different finger prints). Though we all share similarity (analogous to the hexagonal frame of the snowflake), there's a uniqueness (seen in the crystallized pattern within the snowflake) to each person.

It is referred to as signature, and it is quite evident, for example, in the fingerprint, the voiceprint, and the molecular print a hound uses to find one in a crowd.

The immune system (particularly, the education mediated by the thymus gland) uses this molecular signature to detect what is not harmonious (i.e., contradictory) to the organism.

Everything in nature has this uniqueness. Even the tiniest particles (which, according to Wolff 2008, are actually scalar wave soliton constructs) have it, which is known as an eigenfrequency (the unique natural resonant frequency of a system), indicating its particular position and vibration in space. Like everything else, the particles have a high degree of similarity, but are not identical, as is commonly thought.

From this phenomenon we can discern that nature would not be able to produce the enormous variety that exists in the universe without employing a process that generates signature vibrations. To do so it relies on the principle of wave interference in order to nonlinearly create difference and combinatorial potentials.

Hence, there's an entraining or driving frequency carrying a signature wave configuration used to resonantly connect with memory encodings in the non-conscious (simply, the vast repertoire of inherent/acquired encodings or information not in current use).

As revealed by chaos theory, the study of dynamic systems, (see Gleick 1987, and Briggs & Peat 1989, for a great introduction) the apparent rationale behind the ubiquitous phenomena of individuality amongst all aspects of nature is to ensure diversity, so that everything won't amass into a ginormous cosmic "clump" by dint of its identical resonance.

The signature frequency of vortical up-or-down spin is what prevents this catastrophe. So you see, this is one powerful indication of a purposeful universal intelligence involved in the creation of the universe.

Signature Based on Golden Ratio Numbers

Signature temporal encodings based on golden-ratio (Pletzer 2010) consist of the most "irrational" set of numbers, ensuring that memories won't jumble together, so that when we go to the kitchen to make a cup of coffee we don't end up with a bowl of Hungarian goulash.

Not only does faithful memory storage and retrieval support the practical stuff of daily life but also the nature of our individuality. In a rather complex way, we are like radio frequencies criss-crossing in social space without conflating our identities. These signature radio frequencies are akin to the FCCP for the human mental system.

The signature existing all around us generates OOPEs at noncritical, or novel levels of arousal. Though we're usually not aware of this subtle evocative contrast of "background noise," this ensures that energy flowing up from the brainstem trickles in sufficient amounts to keep the neocortex in a general mode of awareness (typically in the alpha-beta range).

In this mode of consciousness (hovering in the alpha-low-beta range), we move about tethered to an array of conditioned responses that guide us like intelligent robots through our routines.

Considering the energy-conservation-and-expenditure needs of the organism there'd be no need to do otherwise. For the most part, the organism is somewhat stingy with its energy. Such frugality is saving its capital for a "rainy day," should critical OOPEs unexpectedly emerge.

In the meantime, the organism has little need for intelligence or creativity should OOPEs cease to emerge. If it wants a bit of excitement it might turn on the TV, go to a movie, or call a friend to chat. These instances, based on negative feedback, are apt to occur when the energy bank is overflowing a bit and needs some venting.

Then there are others, with greater needs for novelty, who find pleasure in solving "virtual" or "real" problems. These individuals, for example, solve crossword puzzles, play chess, and delve in the sciences or arts. Nevertheless, when all expressive outlets are lacking the system tends towards boredom: the irking lullaby of life that tends towards sleep.

When the mind-body system transitions from wakefulness towards deep sleep the frequencies decrease, accompanied by a decoupling from the senses (leaving hearing open to some degree). The oscillating neuroglia become more spatially coherent (i.e., "uniformly disordered"), characterized by slow waves of high amplitude, and the wherewithal for resolute consciousness in these low frequencies begins to dip. We need to pause here in order to parse this scenario a tad more closely.

Recent studies have shown that during wakeful states there is a general decoherence consisting of more specified yet comparatively smaller coherent neuroglial clusters, characterized by faster waves of low amplitude. Hence, the system transitions from general coherence to a multitude of specific coherent neuroglial clusters disconnected from one another, which can be progressively entrained by faster frequencies into the larger assemblies of conscious events.

So synchrony has not necessarily left the picture. The chorus has been divided into so many individuals singing to themselves in different keys. Of great importance is the "grand" synchrony consisting of smaller synchronous clusters.

The psychophysiological mechanisms that prompt the system to act in certain ways beyond general arousal are what teases the curiosity at this juncture. Regardless of what they may be, it behooves the individual to be true to their Essence, vibrating in quantum space at a unique eigenfrequency.

THE GENERATION OF SUBJECTIVE MENTATION

The brain-mind (object-subject) is designed to respond to out-of-pattern events (OOPEs). When OOPEs appear that threaten the existence of the PSQ, it triggers the subjective dimension in search of solutions. Therefore, subjectivity is very important to survival.

OOPES can be real or self generated (intentionally or imaginatively); in either case, they produce arousal that increases (doubles) the frequency of certain memory constructs to generate fractals, the basis of subjective permutations.

From a medical perspective it is found in the reference ranges of biochemicals and rates of operation of the different organ systems based on the biological clock situated in the hypothalamus.

In a more generalized way it is found in the vibratory signature of its natural resting state, known as resonance, which I construe to be the vibratory rate of one's Essence (a.k.a., the "child within"), the signature rate of the Self's phenomenal existence.

Negative feedback seeks to maintain a system's identity, so that each metabolic moment of tissue breakdown (i.e., catabolism) and tissue reconstruction (i.e., anabolism) is faithfully repeated. It is responsible for maintaining the PSQ, mediated by the limbic system and/or to a strong memory construct (i.e., belief) that I refer to as an intentionally created master memory construct (MMC).

Information (i.e., configured energy) that does not resonate with the PSQ and does not pose a critical threat is vented psychosomatically through the musculature and psychological defense mechanisms, like the expression and projection of negative emotions.

Knowledge of interference tells one how the mental system interfaces with the inner as well as the outer environments through feedback loops. It tells what the system likes, dislikes, or is indifferent to, whether occurring consciously (using larger assemblies) or non-consciously (using smaller assemblies in fractal time). It also goes towards explaining how the perceptual field is formulated by standing waves (see Lehar 2000).

RATIO OF THE TWO KINDS OF FEEDBACK

In general, negative and positive feedback are needed to co-create a cognitive event and, more specifically, to form a synchronous assembly.

Now: A Complementary Ratio of Opposites

Positive feedback entrains a resonant context while negative feedback "holds" it in place, forming the wherewithal of a standing wave we experience as NOW.

The attentional process is sustained by a complementary balance between negative and positive feedback. Hence, attention is created by a dynamic process of complementary opposites. We experience the enduring equivalence as recurring moments of "stillness." The extent to which we can sustain the balance constitutes our attention span.

We can postulate that positive feedback produces the energy while negative feedback contains it. The rate at which a system oscillates is determined by a combination of negative and positive feedback, forming a balanced ratio of two Fibonacci numbers, producing nature's pattern of growth, which in this case is the neural assembly.

Two Basic Kinds of Equilibrium

Organic systems take advantage of the thermodynamic flow of energy (i.e., biochemical, electromagnetic, etc.), moving through them to power metabolic processes (Ho 2007), which according to the Nobel Laureate Ilya Prigogine take place in "far from equilibrium" states (1984).

Far from equilibrium refers to the lack of entropy, a state of energy in which the constituent forces are so equilibrated that they (supposedly) cannot produce work.

Therefore, for the psychophysiology of an organism to sustain its metabolic complexity, it strives to avoid this hyper-equilibrated state by maintaining non-equilibrium states as energy moves through and entropically dissipates out of the system as heat. In this regard Ho states:

> "The healthy organism excels in maintaining its organisation and keeping away from thermodynamic equilibrium [entropy] – death by another name – and in reproducing and providing for future generations. In those respects, it is the ideal sustainable system" (2007, 1. Bracket added).

Negentropy

By maintaining metabolic "disequilibrium", the organism is producing an anti-entropic process called "negentropy"; this term was coined by the Austrian theoretical physicist Erwin Schrödinger (1887–1961), who founded the study of quantum wave mechanics, and in 1933 was awarded the Nobel Prize for Physics. In 1944 he wrote:

"It is by avoiding the rapid decay into the inert state of 'equilibrium' that an organism appears so enigmatic... What an organism feeds upon is negative entropy. Or, to put it less paradoxically, the essential thing in metabolism is that the organism succeeds in freeing itself from all the entropy it cannot help producing while alive."

Far From Equilibrium: a Strange Kind of Balance

Equating disequilibrium with negentropy doesn't mean that it lacks order. Interestingly, its order is dynamic in nature, and it suggests a process in which the forces that operate to sustain it are not "statically" equilibrated, but in a mathematically expressed ratio that propels growth and development in a proportionate manner. The ideal mathematical construct that comes to mind is the golden ratio (a.k.a., the golden mean, golden section, divine proportion).

The following depiction compares the negentropic-golden-ratio of organized life to the "static" equilibrium of entropy, forming two kinds of equilibrium. Number 1 is based on the *dynamic equilibrium* of the golden ratio while number 2 is based on a "static" *entropic equilibrium.*

$$1). \ a---------------^---------b$$
$$2). \ a-----------^-----------b$$

The golden section (ratio, mean, etc.) is described as a line that is divided unequally, so that the ratio of the larger segment "a" to the whole, 1 (a + b), is exactly the same as that of the smaller segment "b" to the larger one.

Hence, it appears to correlate with the "far from equilibrium" that typifies the dynamics of growth and development quite ubiquitously seen throughout nature as the golden ratio.

For example, the ratio appears: in the chemical bonds in molecules, the DNA helix, snowflakes, in the measurement of the human body, in many plants, branching trees, spiral galaxies, at the quantum level, in our brain waves, and in the branching structure and function of the lungs and cardiovascular system. It is also found in the arts, in music, and in architecture.

Dynamic Equilibrium

This ratio appears as the ideal geometric and mathematical framework to represent the dynamism of "far from equilibrium" states, which paradoxically, have the attributes of a "dynamic equilibrium" of "a" and "b," given that, when divided into one another, they hover around a rather steady constant of 1.618.../.618.... The constant acts like an axis around which the process "equilibrates" and progresses.

The growth sequence tends to follow the additive progression of Fibonacci numbers, such that the first number is added to the next to produce the third one: 1+ 1 = 2, 1 + 2 = 3, 2 + 3 = 5, 3 + 5 = 8, and so on. Here are some examples of Phi:

$$1/1 = 1$$
$$2/1 = 2$$
$$3/2 = 1.5$$
$$5/3 = 1.66666\ldots$$
$$8/5 = 1.6$$
$$13/8 = 1.625$$
$$21/13 = 1.61538461538$$
$$34/21 = 1.61904761905$$
$$55/34 = 1.61764705882$$
$$89/55 = 1.61818181818 \text{ (Bingo!)}\ldots$$

Here are some examples of phi:

$$1/1 = 1$$
$$1/2 = .5$$
$$2/3 = .66666\ldots$$
$$3/5 = .6$$
$$5/8 = .625$$
$$8/13 = .6153846$$
$$21/34 = .617647$$
$$34/55 = .61818181818 \text{ (Bingo!)}\ldots$$

Dinergic

In his book *The Power of Limits: Proportional Harmonics in Nature, Art and Architecture* (1981) the architect György Doczi defined the complementary relationship of these two opposing parameters as being "dinergic", where phi is the "minor" portion and Phi the "major" one.

Positive feedback tends towards accelerative activity while negative feedback counteracts it. The density or viscosity of the medium in which the waves/vortices oscillate (e.g., intracellular and extracellular fluid) is also a contributing factor of negative feedback, since it may "slow" down accelerative effects.

For example, it is easier to stir a cup of tea than a cup of molasses. The uniformity of the medium is also a contributing factor. There are many variables with this dynamic.

When Positive Feedback Reigns

As a general rule, when the ratio of negative feedback to positive feedback is low, the system may lack enough counter force (e.g., grounding) to prevent it from spiraling out of control, like when someone explodes with anger.

When Negative Feedback Reigns

Conversely, when the ratio of negative feedback exceeds that of positive feedback one may feel constrained and compelled to, for example, not speak one's mind.

It is important to not confuse positive feedback with liberalism and negative feedback with conformism, since they can be applied to any value system.

However, the two are needed in some ratio of relative balance to oscillate in sync and form a conscious assembly.

This is biochemically mediated by inhibitory and excitatory neurotransmitters. The attentional process requires a blend of positive to negative feedback to sustain itself. This dualistic complementarity is found throughout nature.

IS THERE CHAOS IN THE BRAIN?

For the mental system, which I also call the "mind," to create cognitive events it needs a state rich in information, like chaos, from which to "choose".

This is its field of cognitive and creative potentials (FCCP). In subliminal millisecond moments the system oscillates between order, where the mind is "made up," and chaos, where the mind is not.

Hence, cognitive moments holistically emerge from the chaotic state and reductively return to it. Therefore, holism and reductionism are but two aspects of the dynamic creative process.

In the orderly state, the mind is committed to specific outcomes. These then feed back into the process, which causes frequency doubling, and then enter a state of chaos. Chaos is the ultimate subjective state of infinite potentials, while order is the limited state of finite potentials that can be altered subjectively.

Although we don't identify these oppositional states in these terms, we do have a notion that the mental system is in free fall when it "freely associates" in night-or-day dreams, or "brainstorms," and when it enters into the orderly state of a thought, image, or perception of the ambient surround. We are rarely aware of the transitions.

However, since the rise of chaos theory (a.k.a. the study of dynamic systems) it has become a concept rich in psychological implications. In the 1980s Freeman used it to understand the activity going on in the olfactory bulb of rabbits processing an odor.

> "When rabbits inhale an odorant, their EEGs display oscillations in the high-frequency range of 20–80 Hz that Bressler and Freeman (11) named 'gamma' in analogy to the high end of the X-ray spectrum! Odor information was then shown to exist as a pattern of neural activity that could be discriminated whenever there was a change in the odor environment or after training. Furthermore the 'carrier wave' of this information was aperiodic [i.e., irregular]. Further dissection of the experimental data led to the conclusion that the activity of the olfactory bulb is chaotic and may switch to any (desired) perceptual state (or attractor) at any time" (Korn, et al, 2003. Italic and line emphasis mine).

THE RELEVANCE OF CHAOS TO THINKING

Though imperceptible to us, this recursive activity reaches a point of intensity where the linearity feeding the loop becomes nonlinear, goes into chaos, bifurcates, and "implodes" (i.e., exceeds its containing negative feedback) into fragments that share likeness to the larger process. This is what happens at the apex of the cognitive vortex: the cognitive crunch just before the vortex reverses its gyration. The fragmentation of the information is referred to in chaos theory as fractals.

Since this conversion effect is ubiquitous throughout nature, we can equate the process by which it emerges to water flowing down a stream. On one occasion the stream has a rather smooth laminar flow. It flows quite linearly. Suddenly, a rainstorm develops upstream. Now the stream begins to progressively flow more quickly. At some point the laminar flow breaks up into white water, which consists of self-similar fragments (i.e., fractals) of the stream.

During this phase the nature of the stream becomes sensitive to its initial conditions and takes on novel changes. Because of the force of the flow, the banks to some degree erode, and rocks shift throughout the bed. It is now not the same stream. Its linear-objective aspect has subjectively morphed into another pattern that could not have been predicted.

The novelties the stream encounters on the way to chaos interact with the "habits" the stream had acquired during its laminar days. Now a new stream holistically emerges that could not be determined by the sum of the effects.

As the storm subsides the energy boost diminishes and the stream returns to a new laminar flow. In this case, linear order has turned into nonlinear chaos, and back again. With each energy-boost the process repeats the marriage of linear (male, yang, etc.) and nonlinear (female, yin, etc.) flows of information.

According to dynamic systems theory (a.k.a. chaology), the acceleration of the stream is due to the cascading effects of positive feedback on the molecules of water (analogous to the memory constructs in the FCCP). To accommodate the extra energy the stream bifurcates, that is, branches into two, which then proceeds to progressively branch into continuous fractals, producing foam, not any different in principle to how twigs and leaves develop, at a different time scale, on a tree.

With each rainstorm, a novel event comes into play, since no two storms are identical; the habituated attributes of the stream are altered, and it proceeds to exist in a new kind of order.

Positive feedback is what accelerates the mental system's information processing rate (as measured by the EEG) so that it may adjust to novel events entering the system. This would be the relevance of chaos to the thinking process. Negative feedback interfaces with positive feedback to give the outcome a "form" or a "border" to help define the outcome in a more linear way.

The process is induced when novel events contextually clash with the PSQ, forming out-of-pattern events (OOPEs). Since novel events are (however nuanced) always to some degree entering the system, they contribute to maintaining a level of conditioned wakefulness: the result of thousands of tiny shocks.

PSYCHOSOMATICS

When either system cannot assimilate an impression they may seek to vent the energy through the musculature, and it often is expressed as subtle discomforts, like itches, tics, or aimless movement.

If the impression critically challenges the PSQ, its energy (remember, all information is vortically configured energy) can be psychosomatically channelled via any of the Freudian defense mechanisms. It may or may not contribute to disease. Often it is expressed as negative emotions, verbally and/or through body language. Typically, these incommensurable constructs find closure in a "complex," like the Z-complex.

EXPECTATIONS/HABITS

Expectations and habits are reinforced memories that operate on autopilot with conventional responses. If memory is to serve the organism in its quest for survival it must take on an enduring structure (essentially this is what memory is all about), and the vortical soliton wave, at the quantum level, is an ideal candidate, both as a mediator of energy and information in the DNA molecule (Gariaev 1994; Meyl 2012a) and as, I propose, an integral force in negentropic (i.e., the creation of orderly form and function) processes.

There's Always A Goal

All behavior is subliminally and/or consciously goal directed by the editing function of the limbic complex. Whether we know it or not, a goal is always being consciously/non-consciously enacted by the mental system. Mainly, it involves preserving the identity of the Self in its predominant state of being. In doing so it draws from various levels of the FCCP that extends from local interactions to nonlocal cosmic ones.

Depending on the degree to which a memory construct is repeated, in thought and/or behavior, it will become a habit. The habit pattern integrates many points of information or fractals that form a history of inputs when the learning occurred (i.e., state-dependent learning) superposed over attractor patterns of the "past" and expectations of the "future" generated in the limbic system, where the PSQ is generated.

The psycho-organism prefers habits because they conserve energy and information, contributing to the regulatory economics of the PSQ. It is more economical from the perspective of the organism to spend most of one's life on "autopilot." Haven't we noticed!

There are instinctive, emotional, and intellectual habits (i.e., reinforced memory constructs). Instinctive habits are based on biochemical homeostasis, which we experience as a generalized vibratory sensation. Emotional habits are based on aggregate vibrations emanating from instinctive needs and intellectual attitudes or beliefs. Intellectual habits are based on concepts one believes in. All are signature vibratory configurations conserved in the FCCP that contribute to the cognitive vortex of NOW at the center of the feedback loop.

They are more localized assemblies, situated in the modular complexity of neuroendocrine plexuses or on the neocortex (e.g., Broca's area concerned with the production of speech). Because of their localization they remain in the non-conscious domain, but they are triggers of specific behavior that complex into larger assemblies.

CYCLING SPEED, & THE TEMPORALITY OF MIND

The rate at which the F-F loop cycles determines its frequency and temporal frame of reference. It also determines how "quickly" the Self will learn and make necessary adjustments.

Now we know that the ability to focus the mind correlates with the ability to handle more information (i.e., memory items or chunks) and accelerate processing time based on the EEG (Weiss, et al, 2003). Therefore, attention span correlates with the size and scope of the assembly and its "intelligence."

If the psycho-organism is to interact intelligently with itself and the environment, it must have the means to perceive what's going on and to react to it in a way that preserves its evolving signature identity.

For it to do so it must have the means by which it can accelerate its psychophysiological metabolism. In other words it must be able to employ a F-F loop that exceeds in speed the things it perceives and the processes it controls.

For example, in playing a game of tennis the mind needs to keep track of the ball at a speed that exceeds the speed of the ball in order to affect the musculature to respond properly. This, of course, requires a degree of focus, since attention span is the accelerating agent of mental metabolism.

Therefore, the notion of "mind" takes on an interesting quantitative aspect: for an organism to function properly in its environment, it must generate a means to perceive and interact with its environment at speeds that must of necessity exceed its cellular structure (i.e., the cell-to-cell communication).

I therefore conjecture that this F-F process would occur in a hierarchical step-wise manner, such that the molecular substrate becomes the cognitive medium of mind for cells, while the electromagnetic field becomes the cognitive medium for molecules (see MacFadden's 2002 paper "Synchronous Firing and Its Influence on the Brain's Electromagnetic Field: Evidence for an Electromagnetic Field Theory of Consciousness").

According to conventional physics, the limit of what we can perceive would be set by the speed of light. However, the PSQ of the brain, situated in the limbic system, acts as a step-down transformer and filter that isolates particular signals from the noise of the ultra-fast electromagnetic interactivity going on around us all the time.

Any entheogenic substance (e.g., the hallucinogenic alkaloid psilocybin or LSD) that blocks or bypasses the PSQ will open the mind to a rapid confluence of unconditioned interactions of habits and incoming OOPEs (see Schartner 2017). In other words, it facilitates "tripping" based on a confluence of interference patterns of subliminal data (see Merrick 2011).

Therefore, mind cannot be an epiphenomenal side effect of cells, but an intrinsic decision-making aspect of anything that needs to "adapt" or "creatively modify" in order to "survive."

This cannot happen by the encumbering solidity of cells trying to move fast enough to avoid an oncoming truck, let alone the metabolic speed of bacteria and viruses that alter their DNA at mind-boggling speeds in their duel with big Pharma.

The FCCP seems to have this hierarchical rationale in the way it organizes learned information into vibratory frameworks based on the progressive frequencies of the EEG and its fractal harmonics.

RATE AT WHICH THE NON-CONSCIOUS PROCESSES ENERGY AND INFORMATION

While the non-conscious aspect (whatever is operating quicker than the conditioned mind) of the mental system operates with smaller assemblies (the size of which may be just as large in the scale invariant realm of fractal time), it carries out its functions at an astronomical speed, which I interpret to occur in the realm of fractal time.

Scientists estimate that the brain consists of 80-100 billion nerve cells, which is about as many stars as there are in the Milky Way. While this astronomical number gives us an idea of the magnitude of the objective combinatorial potentials of the FCCP, it does not include the enormous glial population and its holistic connections, constituting all the memory constructs, including DNA.

Based on information theory for processing data, the clinical and theoretical psychologist Gary Bruno Schmid has made some interesting calculations about the rates at which the conscious and non-conscious minds process information (2016).

1. "The total data processed by the brain is an astronomical 320 Gigabytes (Gb) per second! [Note: each Gb = 1 billion bytes.]
2. Only about 0.01% of all the brain's activity is experienced consciously.
3. In other words, it is as if roughly 10,000 cinema films are actually going on in the brain all at once, while we are only consciously aware of one of them ... through the sensory channels (sight, smell, sound, taste, touch and balance)!" (Bracket and underlines added).

Regardless of the preciseness of these estimates, it is a well-known fact that there's an enormous amount of information processing going on in the brain, the majority of which we are not conscious.

This is in keeping with a research article written by Marianne Szegedy-Maszak entitled "Mysteries of the Mind: Your unconscious mind is making your everyday decisions," in the February 28th, 2005, issue of the U. S. News & World Report. It revealed that the conscious domain occupies a mere 5% of our decision-making activities. This is a bit more than Schmid's estimate, but it still reflects the paltry scope of ordinary consciousness.

> "According to cognitive neuroscientists, we are conscious of only about 5 percent of our cognitive activity, so most of our decisions, actions, emotions, and behavior depend on the 95 percent of brain activity that goes beyond our conscious awareness. From the beating of our hearts to pushing the grocery cart and not smashing into the kitty litter, we rely on something that is called the adaptive unconscious, which is all the ways that our brains understand the world that the mind and the body must negotiate. The adaptive unconscious makes it possible for us to, say, turn a corner in our car without having to go through elaborate calculations to determine the precise angle of the turn, the velocity of the automobile, the steering radius of the car" (Szegedy-Maszak 2005, 57).

This non-conscious control is what would be occurring to a sleepwalker correctly reacting to the environment on autopilot while heading to the refrigerator for a snack. When asked "Who ate my piece of cake?" there'd be an honest and perplexing look on the face with the chocolate icing lingering on the lips.

This has been demonstrated with people suffering from blindsight: the condition of being cortically blind in one eye yet able to correctly respond to stimuli to it.

This pattern has also been observed with subjects who are given deep hypnotic suggestions to perform particular acts, and instructed to not remember them upon coming out of trance. When asked why they behaved in the suggested manner they come up with a false, ego-protective, rationale (see LeCron 1968).

Such cases demonstrate that invisible stimuli, operating in the subtleties of fractal negative space-time, can activate the same neuroglial-and-motor pathways as do visible stimuli. This finding also implicates the idea that there are no singular or direct lines of causation, which has implications for so-called "double blind experiments."

The Non-Conscious is Always Watching

An event I had experienced that has relevance to these findings occurred while in graduate school. I had gotten to my campus apartment late one night and I was quite tired and ready to go to bed. After I fell asleep I had a disturbing dream: an ominous figure was at the door ready to enter the apartment. Startled, I got up, turned on the lights, and went to check the door. I looked through the peephole and there was no one there. When I guardedly opened the door to check further I noticed that had I left my keys in the lock. The non-conscious was watching.

Fractal Time of the Non-Conscious Realm

From this hidden holistic F-F dynamic, which is functional at all scales of the EEG power spectrum, we can instantiate that the integral effects of the psyche are, in keeping with Melloni's postulates, determined by the fractal harmonics of extremely subtle operations, which presumably are not picked up by what the conventional methods of the EEG can detect and amplify.

These subtle operations take place using the harmonics of the EEG parameters, and they may very well, within their fractal temporal frame of reference, extend quite globally throughout the organism via the magnetic field of the heart wave into the meridian network mediated by the fascia (see Bai, et al, 2011). Therefore, conscious and non-conscious operations follow similar functional operations but on different scales of functionality.

These scales may vary in size (from negative to positive numbers), such that, for example in cosmic terms, the solar system could be viewed as a fractal of the Milky Way galaxy, and so on. Fractal progenitors, though larger in size, are made up of multiple self-similar fractals. Therefore, there's an inherent relativity of upscale to downscale fractal relationships that produce different qualitative phenomena.

According to this rationale, the inner most minds—instinctual, movement, intellectual, emotional—would be fractal temporal frameworks that are able to form "large assemblies" in their own right. These "larger" assemblies would be how the psyche (soul, spirit) performs the subtlest levels of mentation that precede the grossest levels mediated by the editing operations of the limbic system to conform with the PSQ.

What we more palpably experience is the end product of the limbic output in the integral construct of a globally distributed assembly that has filtered through the PSQ. Therefore, the interface between the ego and the conscious threshold produces a kind of cognitive compromise that also includes the images from our surround.

In this constantly changing scenario phenomenal consciousness is being formed and reformed in millisecond+ time frames in which the output of the cycle is constantly being enfolded as input. In this sense, the non-conscious domain and the ego are also being revised. A similar hypothesis was developed by the Jungian analyst John R. Van Eenwyk:

> "All this takes place on the interface between the ego and the unconscious, in that region of the mind where consciousness is formed. This is Jung's central hypothesis: that *consciousness is constantly in the process of being created*, that such a process is the result of an intent residing within the psyche—more specifically, in that dimension of the psyche called the 'unconscious'—and that essential to this process are images from everyday life.
>
> Consequently, in the diachronic [Concerned with the way in which something, especially language, has developed and evolved through time. Often contrasted with synchronic] dynamics of individuation ego and consciousness develop together. Through its openness, the ego comes into contact with experiences that increase the categories by which consciousness can assimilate new data, which expands the ability of consciousness to integrate new data with old. In effect, these feedbacks between ego and consciousness, new and old, psyche and experience, individual and surround— all of which occur against a backdrop of unconscious influences—establish *fractal cascades*. Through symmetry building these cascades build consciousness and contribute thereby to the constant refinement of the adaptive capacities of the organism. Folktales are part of this cascade, for they

are both created by psyche and influence its development" (in Mac Cormac 1996, 336-337. Italics and bracket added).

According to the techno-metaphorical collage being developed here, what makes certain "folktales" so auspicious for the development of psychic wholeness is that they have symbolic breadth of many lines of meaning that incorporate large assemblies that facilitate memory formation and recall.

MIND FUNCTIONS LIKE A RADIO

Researchers at the Kavli Institute for Systems Neuroscience and Centre for the Biology of Memory at the Norwegian University of Science and Technology (NTNU) have discovered the mechanism the brain uses to filter out distracting thoughts to focus on specific information. The first author of the study, Laura Colgin, thinks of the mechanism to be like how we "tune a radio to a specific station".

The quote is quite interesting in terms of the thematic being developed in this book. Two phrases are relevant in regards to this observation: "in between stations," and "tuning in."

The Static Between Stations

Though not mentioned in the article, when a radio is in between stations it produces the sound of random "white noise" that we hear as "static." This noise is the chaotic sound of all the radio transmissions coexisting in the electromagnetic field which, according to the model developing here, are the combined sounds of all memory fractals in the FCCP.

Hence, we may consider this noisy field as uniformly disordered potentials from which we may subjectively select an objective event.

White noise is also an attribute of "white light," in which the entire electromagnetic spectrum coexists in a random manner. Using the principle of resonance, organic life selects-deselects from this repertoire of randomly expressed frequencies the ones that are in tune with each particular vibratory frame of reference.

For example, the various shades of green we see on plant leaves indicate that they have absorbed all the frequencies of the spectrum within random light and have rejected the green wavelength. Therefore, what we see on the surface is an excretion, so to speak.

Tuning Into

The phrase "tuning into" has affinity (quite literally so) to the attentional selection-deselection process, which draws specific cognitive assemblies (equivalent to the specific frequency of a radio station) via the principle of resonance.

Hence, all systems in nature oscillate in signature frequencies between order (i.e., the integrated phenomenon) and chaos (i.e., the "field" of potentials from which the phenomenon is formed) in order to sustain their existence.

According to the above study the brain wave frequencies that are responsible for carrying the specific information consist of contextual variations occurring in the gamma range (30-100 Hz).

> "Colgin and her colleagues measured brain waves in rats, in three different parts of the hippocampus, which is a key memory center in the brain. While listening in on the rat brain wave transmissions, the researchers started to realize that there might be something more to a specific sub-set of brain waves, called gamma waves. Researchers have thought these waves are linked to the formation of consciousness, but no one really knew why their frequency differed so much from one region to another and from one moment to the next.

> Information is carried on top of gamma waves, just like songs are carried by radio waves. These 'carrier waves' transmit information from one brain region to another. "We found that there are slow gamma waves and fast gamma waves coming from different brain areas, just like radio stations transmit on different frequencies". (Colgin et al, 2009).

Separating the "Signal" From the "Noise"

According to "information theory" created by the American engineer Claude Elwood Shannon (1916–2001), there is interplay between asynchronous "noise" and a meaningful "signal" synchronously extracted from it.

Shannon was a pioneer of mathematical communication theory. He also investigated digital circuits, and was the first to use the term "bit" to denote a unit of information and the "signal-to-noise" ratio. The term now stands out as a "meaningful" signal above the "noise" (created by the non-selected-uncertain activity of memory constructs reverberating in the FCCP).

Hence, the FCCP as a whole is "noisy" and the selection produces a recognizable signal that stands out from the noise (3a). However, the uncertainty in the noisy field (3b) is rich in possibilities or information while the signal, by its conventional meaning, is not.

Information Capacity of Vortices

By turning attention inward to observe and control an aspect of the Self, the feedback loop accelerates (i.e., steps-up towards gamma+). By extending the dwelling time of attention (i.e., attention span) on a percept, it progressively aggregates and "crunches" the data/information towards the apex and shifts towards the "blue shift" (and beyond?).

If it can hang in there for a period of time it will be able to resonantly connect with the spiritual aspect of the FCCP mediated by its physiological substrate, the pineal gland, which oscillates in the ultraviolet frequency range (see Bosman 2000).

A group of scientists appear to have found the reason nature uses ultra-fast vortical transmission to manage its activities.

> "American and Israeli scientists have combined twisted vortex beams to transmit data at a mind boggling 2.5 terabits (1 trillion) per second [1 trillion Hz]. So far, this constitutes the fastest wireless network ever created — and is likely to be applied in the next few years to vastly improve the throughput of both wireless and fiber-optic systems" (Wang, et al, 2012. Bracket added.).

According to a report by Sebastian Anthony of Extreme Tech:

> "These twisted signals use orbital angular momentum (OAM) to cram much more data into a single stream. In current state-of-the-art transmission protocols (WiFi, and the like), we only modulate the spin angular momentum (SAM) of radio waves, not the OAM. If you picture the Earth, SAM is our planet spinning on its axis, while OAM is our movement around the Sun. Basically, the breakthrough here is that researchers have created a wireless network protocol that uses both OAM and SAM" (2012).

This information goes towards explaining the tremendous speed the spiritual domain uses to process energy and information, many orders of magnitude faster than that estimated by Schmid at 320 Gb/sec. This allows for parallel processing of a greater range of memory constructs in the cosmic zones of the FCCP.

Twisted Vortex Beams Equate to Psiloton Spin

If you picture the SAM of an MC, it would be like our planet self-referentially spinning on its axis, while its OAM is its movement around a MC with greater psychic mass (e.g., sun of the solar system, "Son of man" in Christology).

Hence, the MC uses both OAM and SAM. If we consider the dynamic patterns of the Psiloton Self we'd "find" a rich complexity of objective and subjective potentials flowing in and out of each other at these cosmic speeds. Each configuration would morph to the controlling mechanisms of established habits, intention, and OOPEs.

Psiloton Self-referential Spin Equates to Intelligence

The rate at which the Psiloton cycles energy and information depends on the ratio of flow to resistance. When these two variables are in some degree of parity the system could be said to be in a steady state, and while in the balanced state it would have a relatively high level of psiological intelligence, which could be measured to yield a psiological quotient (PQ), a more holistic interpretation of intelligence and creativity.

I FRACTAL, THEREFORE
I THINK

"Another parable put he forth unto them, saying, The kingdom of heaven is like to a grain of mustard seed, which a man took, and sowed in his field: Which indeed is the least of all seeds: but when it is grown, it is the greatest among herbs, and becometh a tree, so that the birds of the air come and lodge in the branches thereof" (KJHB, 13:31-32).

Fractals are the mathematical equivalent of seeds that are planted (input) in a field (e.g., the FCCP) to grow, given the right nourishment, into something comparatively large, like a tree with a large number of branches and an even greater number of leaves. This fractal development occurs as a result of bifurcations (a division into branches and leaves), none of which are exactly the same. At a certain time in the year the tree develops multiple seeds, and the process repeats in Psilotonic fashion. Note that no one seed will generate a tree identical to its progenitor, but a variation of its "genetic" theme.

Also note that the leaf formations are compressions of the tree as a whole into a predominantly 2D format. In the seed stage the tree reverses the process and replicates itself back into multiple 1D points (relative to the size of the tree), which it scatters back into the field. This dimensional enfolding and unfurling cycle is essentially Psilotonic, and it mirrors the cognitive process.

In ancient Norse mythology it was expressed with a tree-like model of the universe known as the Yggdrasil, the "World Ash."

Hence, the tree has all the attributes of a cognitive system that affect and adapt to the changing circumstances of the field in which it lives and has its being. Other trees in the field (forest/universe) are also doing the same thing, and, as has been discovered, trees have a complex communication system signaled by their roots under the forest floor and their leaves.

"Plants may lack brains, but they have a nervous system, of sorts. And now, plant biologists have discovered that when a leaf gets eaten, it warns other leaves by using some of the same signals as animals. The new work is starting to unravel a long-standing mystery about how different parts of a plant communicate with one another" (Pennisi 2018).

What's so interesting about trees and plants in general is their adaptive and creative intelligence, which has served them for millions of years. They generate micro-ecologies that rely for their existence on the cooperation of many other creatures. They produce ingenious ways to protect and transport their seeds, many of which are embedded in fruits that are tasty to roaming animals or are given aerodynamic "wings" that can be dislodged and carried far into the field by the wind.

In the Psiloton model we can see the two vortices as interfaced tree-like-root-and-leaf bifurcations manifesting as the EEG spectrum.

In Figure 17 we see how the tree-like fractal processes are involved in period doubling and period halving developments of brainwaves in the cognitive cycle.

The fractal structure has particular similarity to the branching of the neural population, the cardiovascular, and respiratory systems. Hence, structure and function share intimate relationships in processing substances into smaller fractal dimensions.

The outer physical mode of the Self is fractal in appearance: we have a trunk and appendages that end in fingers and toes. The physical body would be fashioned from the grossest outer-most spheres of the Psiloton Self, not in appearance but in dynamic functionality. These relationships have been described in terms of the golden ratio parameters, of which more will be given as we go along.

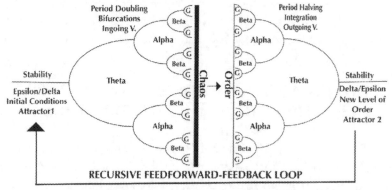

Figure 17: Fractal developments in the thinking process based on the parameters of the EEG (G = gamma).

THE SUBTLEST AUSPICIOUS ZONES

Given that fractal-time in the most central regions of the Psiloton is the most fast (yet similar in relative terms), we can attribute these fractal zones to be where one becomes privy to the subtlest aspect of impressions entering the sensorium at any one moment. If we develop sensitivity to this level we'd have knowledge of what's to come for us personally before it manifests at 6. Because of its ultra-subtlety we'd think of it as intuition or channeling information amplified from a zone of very low volume (e.g., the "small voice within," the "Divine within").

AFFECTS THE READINESS POTENTIAL

This would also account for the "readiness potential" (RP) mentioned earlier, which provides the sense of anticipation of a percept before it blooms into awareness at the 6th stage. Even though fractal time transpires so quickly, it lacks the power of amplitude that would immediately bring it into the borders of the conscious threshold.

It therefore follows that we are already thinking, imagining, sensing at these subtle fractal levels before we are privy to the "results," given that fractals do what their progenitor systems do but in smaller spatial time frames. They operate in negative space-time, which needs to be amplified by the outgoing vortex into positive time-space for all to "see."

As is depicted in the illustration, the ingoing development period doubles towards negative infinity while the outgoing aspect period halves towards positive infinity.

This process is also applicable on larger scales for the enactment of prophecy. The seer is able to not only see into the depth of the soul but also into the depths of the collective soul at the center of the collective Psiloton.

VIBRATORY AFFINITIES OF THE GOLDEN RATIO & THE RESTING STATE AND SYNCHRONY

The Psiloton, memory constructs, Fractals, the Fibonacci numbers, and the golden ratio have inseparable relationships. The seminal discovery was made by the Italian mathematician Leonardo Fibonacci (c. 1170– c. 1250), also known as Fibonacci of Pisa. He made many original contributions in complex calculations, algebra, and geometry; he pioneered number theory and indeterminate analysis, discovering the Fibonacci series.

To have some workable understanding of their nature we need to have an understanding of their quantitative as well as their qualitative affinities.

In mathematics, the Fibonacci numbers are the numbers in an integer sequence, called the Fibonacci sequence, such that the first number is added to the next to produce the third one: $1+1 = 2, 1 + 2 = 3, 2 + 3 = 5, 3 + 5 = 8$, and so on. It is characterized by the fact that every number after the first two is the sum of the two preceding ones.

The recursive looping of the numbers points to the F-F cycle of the cognitive system (7). It is in this sense that when the output is repeatedly fed back to the input that we have the condition for developing an amplifying-positive-feedback process (see Chapter THE MIND IS A CONTROL SYSTEM for details) that contributes to the formation of fractals.

Since fractals have the attribute of self-similarity they quite literally raise the notion of the emergence of "memory" and their combinatorial potentials as memory constructs (MCs).

In their steady state the MCs have "objectivity," given that they repeat an "identity" of sorts with some high degree of "fidelity." Because each MC has signature, that is, not being "identical," they complex in novel ways that support the subjective thinking and imaginary functions of cognition.

The Relevance of Irrationality

It has been proposed that these vibratory affinities are based on the irrational numbers of the golden ratio (Pletzer, et al, 2010). In a mathematical as well as in a physiological sense, the calculations show that the irrationality prevents them from becoming synchronized; yet, they organize in a uniform manner in the "resting state."

> "The classical frequency bands of the EEG can be described as a geometric series with a ratio (between neighbouring frequencies) of 1.618, which is the golden mean. Here we show that a synchronization of the excitatory phases of two oscillations with frequencies f1 and f2 is impossible (in a mathematical sense) when their ratio equals the golden mean, because their excitatory phases never meet. Thus, in a mathematical sense, the golden mean provides a totally

uncoupled ('desynchronized') processing state which most likely reflects a 'resting' brain, which is not involved in selective information processing" (Pletzer, et al, 2010).

In terms of wave interference, irrationality means that the two wave systems (f1 and f2) identified by their numerical frequencies are not "similar enough" to enter into phase, which is the basis of synchrony. In the paper it is shown:

"[T]hat the pattern of excitatory phase meetings provided by the golden mean as the 'most irrational' number is least frequent and most irregular. Thus, in a physiological [as well as in a mathemaical] sense, the golden mean provides (i) the highest physiologically possible desynchronized state in the resting brain, (ii) the possibility for spontaneous and most irregular (!) coupling and uncoupling between rhythms and (iii) the opportunity for a transition from resting state to activity" (Ibid. Bracket added).

The Entraining Effects of Faster Frequencies

What ignites the resting state into activity is the entraining effect of faster frequencies, that is, if they share higher degrees of "similarity" in terms of frequency and structure.

"[E]xcitatory phases of the f1- and f2-oscillations occasionally come close enough to coincide in a physiological sense. These coincidences are more frequent, the higher the frequencies f1 and f2" (Ibid).

Now we have a substantive rationale of how the highly-dense nonlinear point of the ingoing vortex is able to make a selection (3a) by entrainment.

NUANCES OF DIFFERENCE

Note again, that the ingoing vortex is transitioning towards the blue shift while progressively diminishing the scale, yet not significantly altering the information. In chaos theory this is known as scale invariance, which is the attribute of fractals, a curve or geometric figure, each part of which has some degree of similarity to the whole (see Gleick 1987). According to a Wikipedia entry on fractals, fractals may have:

Exact self-similarity: identical at all scales.

Quasi self-similarity: approximates the same pattern at different scales; may contain small copies of the entire fractal that are approximations of the entire set, but not exact copies.

Statistical self-similarity: repeats a pattern across scales, such as randomly generated fractals, like the irregular pattern of a coastline, for which one would not expect to find a scaled segment to be identical to another one.

Qualitative self-similarity: as in a time series, as, for example, in the rise and fall of capital market cycles.

Multifractal scaling: characterized by systems in which a single exponent (the fractal dimension) is not enough to describe its dynamics; instead, a continuous spectrum of exponents is needed.

Multifractal systems are the most common in nature. They include, for example, all kinds of turbulence, meteorology, geophysics, the Sun's magnetic field time series, heartbeat dynamics, brainwave dynamics, human gait, and natural luminosity time series, and much more. Because of this, multifractal modeling constitutes the gold standard for representing the non-Euclidian aspects of nature.

Creative Emergent Arisings

As a consequence of multiscaling attributes, fractal populations may have emergent creative effects, such that the sum of the parts does not equal their novel holistic outcomes. This in and of its self goes towards explaining the creative potential in the cognitive process, especially when the ingoing vortex interfaces with the FCCP.

So it is apparent that fractals have holographic-like-memory properties that can be magnified to reflect a system in a new level of order, with similarities to the previous one, but with significant variation. Because of this:

"Fractals are useful in modeling structures (such as irregular coastlines or snowflakes) in which similar patterns recur at progressively smaller scales, and in describing partly random or chaotic phenomena such as crystal growth, fluid turbulence, and galaxy formation" (NOAD).

Because fractals are not identical to one another, even though they share similarity, they have nuances of difference (see Briggs/Peat 1989). Therefore, the vortices have relative uniformity with degrees of signature that when combined (i.e., entrained) into assemblies provide the nonlinear wherewithal to form creative thinking and imagination.

There's always a nuance of difference that is subject to modifications that can be amplified through iteration. These signature units can, however, be entrained by vortical spin into the synchronous events we experience as objectifications, however complex.

This synchrony provides the "objectivity" we experience all around us with the senses. Nevertheless, this objective state is made up of a multitude of "subjective" recurring patterns from the FCCP.

In this regard, the phenomenal world we experience is quite like a pointillist painting whereas the FCCP in the resting state is more like the drippy artistic renderings of a Jackson Pollock painting that science refers to as fractal attractor basins.

PINK NOISE AT THE CUSP OF ORDER & DISORDER

This symmetric disorder is found to emit, for example, pink 1/f noise where "vortices" are suspended in an orderly animate manner of Brownian motion, described as a "drunken walk" of two steps forward and one back.

The movement of the particles is actually their continuous shifting between KE and PE. This would constitute a rather strange yet basic level of order of the Pluripotent Unified Field (PUF).

Pink noise is also the sound of self-referential movement emanating from the population of MCs vibrating in place in the FCCP. In transformative teachings this background sound is referred to by various terms, such as the "Word," the "life stream," the "aether," and so on. It has a very pleasant effect on the vibratory essence of the Self, quite likely due to its cornucopian potentials to nurture all life forms.

SUPERSYMMETRIC

In physics, for example, this fundamental field would be supersymmetric or at least isotropic, such that when "measured" it would be everywhere the same (see Wolff 2008). In this sense it is linear with the attributes of a scalar volume.

FRACTAL HARMONICS

Similar objects like particles, atoms, molecules, and cells would also qualify as modifiable fields of sorts. By their enormous magnitude these constructs produce other fields by dint of their fractal similarities. Hence, there are fields within fields, like: the electromagnetic field, bio-molecular field, cellular fields, mind fields (i.e., FCCP), etc.

UNIFORMITY = RE-WRITABILITY

Such linear isotropism is what endows a piece of blank paper, a magnetic tape, a digital disc, a holographic film, or the mind of a child (not an emptiness according to the "tabula rasa" doctrine) to be "written on" (i.e., modified) to generate constructs of form, function, and meaning.

THE PSYCHOPHYSIOLOGICAL CYMATICS OF HIGHER FREQUENCIES

I use the term "cymatics," because it shows how vibrations create harmonious patterns in liquids and on surfaces on which powders of different substances have been sprinkled.

> "Cymatics, from Ancient Greek, meaning "wave", is a subset of modal vibratory phenomena. The term was coined by Hans Jenny (1904-1972), a Swiss follower of the philosophical school known as anthroposophy. Typically the surface of a plate, diaphragm or membrane is vibrated, and regions of maximum and minimum displacement are made visible in a thin coating of particles, paste or liquid. Different patterns emerge in the excitatory medium depending on the geometry of the plate

and the driving frequency. The apparatus employed can be simple, such as the Chinese spouting bowl, in which copper handles are rubbed and cause the copper bottom elements to vibrate. Other examples include the Chladni Plate and the so-called cymascope" (Wikipedia: Cymatics).

Ernst Chladni invented a variation of the technique, originated by Robert Hook in 1680, to show how various patterns of a fine sand (or other substance), lightly sprinkled on a metal plate, could be generated by vibrating it with a violin bow. The plate was bowed until it reached resonance, causing the sand to move and concentrate along the nodal lines where the surface is still, outlining the nodal patterns.

The nodal patterns are the result of waves that reflect off the edges of the plate (are also nodal areas) and interfere with one another and form a wide variety of standing waves, now called "Chladni figures."

The bowing adds energy to the plate, producing a positive feedback effect, while the sand, the barrier and density of the plate provide the negative feedback that contains the activity and gives shape to the waves.

In Figure 18 we see the same principle, using vibrations of electromagnetic sine waves to create a standing-wave pattern in cornstarch.

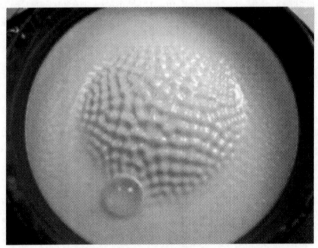

Figure 18: Cornstarch and water solution under the influence of sine wave vibration.

An interesting phenomenon occurs when the frequency of vibratory input changes its rate. Lower frequencies tend to produce simple fundamental patterns, but when they are increased the complexity of the pattern also increases. They seem to be priming the FCCP to explore the permutational potentials of the slower wave forms in the faster patterns of waves.

PERMUTATION OF COMBINATORIAL POTENTIALS

This contextual exploration during moments of high arousal would be in keeping not only with the need to increase the rate at which the mind needs to process information but also with the need to quickly explore alternatives based on the learning the system has already in place.

Hence, as the system accelerates along the EEG spectrum it is enhancing its creative and subjective potentials by an ultra-fast process of "permutating" combinatorial possibilities that support the act of thinking.

WHERE DOES THINKING BEGIN?

Because "smaller" is "faster" but not "stronger" (in the sense of amplitude), thinking begins in the subtlest fractal zones of the Psiloton.

If we keep the meaning of these zones in terms of the rate at which they process energy and information, we come to realize that cognition is happening in the subtleties of "fractal time" in negative space-time way before it is amplified by the outgoing vortex into positive time-space.

If we include the conventional terms for these zones we'd need to address them in terms of how cognitive moments relatively develop from spirit to mind and then to body. In this quite rational scenario stimuli are processed by spirit before mind and body become privy to what has transpired. On the illustration of the Psiloton (Fig. 8) the cognitive progression would follow a FZ0==>FZ7==>FZ0 recursive trajectory, following the in-out flow.

This fractal fast-time process would go towards explaining the experimentally observed "readiness potential" (RP), where the non-conscious primes the neural system to behave with anticipation, expectation, and the like. It also goes towards providing a rationale for intuition (a.k.a. satori: sudden enlightenment in Zen).

Because the dynamic nature of the Self is motivated to speed up its cognitive processing rate in proportion to its encounter with OOPEs, particularly its interpretation of "great danger," creative types often place themselves at the "edge of criticality," in physical as well as in mental scenarios, in order to access the subtlest dimensions of fractal time, ever brimming with novelty and insights.

This kind of going against the conditioned responses of the PSQ creates a subtle tension, a kind of "eustress" the physiologist Hans Selye discovered in his research on the "stress adaptive disorder." However, the term "edge of criticality" is born from chaos theory to indicate that this is where opposites, paradox, and states of physical and mental chaos precede a new kind of order.

While creative types feed off this kind of confusion, there are those who pursue them mainly through physical involvement for the sake of experiencing a "thrill," in which adrenalin and dopamine play an important role.

Thrills may lead to insights, but in and of themselves they are limited. Nevertheless, a number of thrill seekers may encounter the mind-expansive implications of "transcendental states," of which the OBE has great significance for the realization that there's continuity of life in the fractal dimensions of the Self.

As an undergrad student at the Inter American University I had to take mandatory courses in religion, and although it was the first time I ever read the Bible and was instructed as to its "meaning" I was never so inspired by its mention of spirits and angels as I was affected by the OBEs as the actual experience of being spiritual. It was from that point forward, especially after the VW-vs-Mack Truck incident, that I took a deeper interest in the Bible and other associative readings.

On a number of occasions I was told by those with religious and even secular training in these matters that my experiences were illusions brought on by evil spirits. At one time I may have believed them, but after all of what I experienced I knew they were in their institutionalized PSQ minds, judging others in order to protect their doctrinal and academic investments.

It seemed quite plausible to me that memory is the fractal equivalent of life existing on different scales where the intrinsic information of the Selves is preserved. However, since each MC is a Psilotonic system in its own right it has all the attributes of life. Yet, to this day the concept of memory and where it may be found is still a great mystery.

WATER, WATER, EVERYWHERE

Back in the 1920s the Harvard psychologist Karl Lashley tried to resolve the mystery. He trained rats to learn a maze, and then gradually excised different parts of their brains to see how it affected their behavior. Surprisingly, he discovered that memory wasn't impaired no matter where brain tissue was taken out. The key factor was how much of the total brain was excised. This led him to conclude that memory was not localized but globally distributed throughout the brain, as in a field.

While there's evidence that more primitive reflexes (e.g., knee jerk, eye blink, pupillary dilation/contraction) are localized in neurological clusters, more complex processes, such as the findings of neural synchrony support, are more broadly distributed throughout the brain.

Nevertheless, memory has also been found to be localized in individual cell clusters in the brain. According to Freeman this occurs at the micro level of individual neurons, but must expand in scope to enrich the complexity of the outgoing assembly at 5.

However, to hold that memory is globally distributed doesn't reveal the nature of the medium in which it is conserved.

ELECTRO-MAGNETIC TRANSFER OF GENETIC SIGNATURE TO WATER

The clue comes from research done by Luc Montagnier, the French virologist who was awarded, together with Francoise Barré-Sinoussi and Harald zur Hausen, the 2008 Nobel prize in Physiology/Medicine for discovery of the AIDS virus (HIV).

About eight years prior to this discovery Montagnier noticed an interesting aberration when he went through an aqueous-filtrate procedure to take out a small HIV-like bacterium (Mycoplasma pirum), about 300 nm (nanometers) in size, from viral particles whose size is about 120 nm.

According to this standard laboratory procedure, there should be no trace of the unwanted bacterium in the filtrates. However, when the filtrate was incubated with human lymphocytes (previously controlled for not being infected with the mycoplasma), the mycoplasma with all its characteristics was repeatedly recovered.

Astonished at what had transpired, he and his colleagues wondered: how did the information in the test tube containing the DNA fragment of the mycoplasma make its way to the sterile aqueous filtrate?

This query sparked an extensive investigation on the physical properties of DNA and its effects in water. It brought back memories for Montagnier of the work done by the French immunologist Jacques Benveniste (1935-2004), who was shunned and scorned by the scientific community for performing verifiable homeopathic-like experiments with viruses, in which dilutions of an active substance or organism can reach to over 1 part per quintillion ($1/10^{18}$).

He decided to explore Benveniste's findings by duplicating the dilutions of the mycoplasma in pure water to the calculated level of 1/100,000,000,000,000,000,000 ($1/10^{20}$).

According to scientific standards, there should be no molecular or electromagnetic trace of the mycoplasma at all at these levels of dilution. To find out, he proceeded to probe the samples to see if they emitted the electromagnetic signal (EMS) he detected in the sample tube prior to it being watered down.

It was found that the mycoplasma at these subtlest of levels, as well as other bacterial and viral DNAs, emit low-frequency EMSs in some water dilutions of the filtrate while in an ambient electromagnetic field.

To determine if the signature DNA information was mediated as an electromagnetic signal (EMS) from the ambient field further experiments were performed. The results were published in the *Journal of Physics* (2000). Figure 19 is an abridged schematic of the experiment.

According to the protocol described in the article, the highly-diluted fragment of DNA is held in one container (sealed test tube), which is placed inside a solenoid coil that is shielded from the ambient electromagnetic field by a mu-metal cage (a nickel-iron soft magnetic alloy that shields against static or low-frequency magnetic fields) to keep it from muddying the results. Another tube(s) filled with pure water, that is not emitting an EMS, is placed in the coil with it. Then a weak electromagnetic current produced by an external generator, pulsed at 7 Hz, is sent through the coil. The produced magnetic field is maintained for 18 hours at room temperature. The current is then amplified from each tube, and finally analyzed in a laptop computer using specific software (see Figure 19).

It was shown that the tube (he used several as controls) containing pure water emits signature EMS, at the dilutions corresponding to those positive for EMS in the original DNA tube. This result shows that, upon a 7 Hz excitation, the transmission into pure water of the oscillation of nanostructures initially originated from DNA was achieved. To rule out other factors, control experiments were done. They revealed that the EMS transmission in the water tube(s) were suppressed when:

1. The time of exposure of the Two tubes was less than 16-18 hrs.
2. There was no coil.
3. The generator of the magnetic field was turned off.
4. The frequency of the excitation was less than 7 Hz.
5. There was absence of DNA in the "specimen" tube.

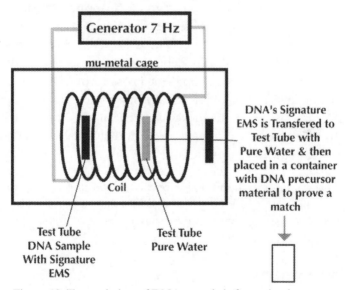

Figure 19: Transmission of DNA genetic information into water through electromagnetic waves. Though more test tubes were used, here the effect is illustrated simply to convey the concept.

According to Montagnier and his team, this strongly supported the notion that DNA emits its own electromagnetic signals that imprint its structure on other molecular substances (e.g., water), a process akin to quantum teleportation of genetic information (originally proposed by the

French virologist Jacques Benveniste—1935-2004—in his controversial experiments with homeopathy).

In other words, DNA can project and copy itself from one cell to the next in the intracellular medium of water, which by the way, constitutes about 70% of the human organism (and about 80% in the brain). How then is the genetic information sustained in the water molecules?

Obviously, the answer to this question has enormous implications not only for the specific transfer of genetic information but also for issues concerning memory in general, since EMSs, pulsed with the 7-8 Hz range (the theta-low-alpha EEG frequency) and water do coexist in the brain, and the holistic neuroglial system.

To Montagnier and associates it all pointed to a coherent effect of the water molecules occurring at the quantum level. However, at that level water is not a very stable medium, since liquid water molecules cannot be assumed to be bound by purely static interactions.

According to their interpretation of quantum theory (for the technical details see the article), coherence among molecules is counteracted by thermal collisions. The noisy-wet-hot brain would not seem to allow it. The competition between electrodynamic attraction and thermal noise produces a permanent crossover of molecules between a coherent regime (order) and a non-coherent one (chaos). Hence, the water molecules are constantly flickering between these two states, which is not conducive to the conservation of information. Nevertheless, this property holds only for bulk water. However:

> "Near a surface, the attraction between water molecules and the surface could protect the coherent structure from the thermal noise, giving rise to a stabilization of [a] coherent structure" (2010, 6 pdf).

Coherence Domain

In this condition the ensemble of molecules interacting with the EMS (i.e., radiative EM field) would acquire a new minimum energy state, in an extended region called a *Coherence Domain* (CD) where all the molecules enclosed within it oscillate in unison in tune with EMSs trapped within it.

In this scenario the molecules are brought into phase with the EMS through entrainment (i.e., through the iterative effect of the coil). Once again we encounter the phenomenon of entrainment, synchrony, and, to my surprise, the dynamics of vortices.

> "CDs store externally supplied energy in form of coherent vortices. These vortices are long lasting because of coherence, so that a permanent inflow of energy produces a pile up of vortices; they sum up to give rise to a unique vortex whose energy is the sum of the partial energies of the excitations, which have been summed up. In this way water CDs can store a sizable amount of energy in a unique coherent excitation able to activate molecular electron degrees of freedom; this high-grade energy is the sum of many small contributions, whose initial entropy was high" (Ibid., 6, pdf).

The paper does not mention entrainment; however, I suspect that it is a contributing factor. As has been discussed, entrainment is facilitated by iteration of a signal (e.g., in the 1-to-7 cycle), which in this case is the information of the DNA fragment that is transmitted by the signature EMS.

Note that when the tube containing the DNA sample is put into the copper coil housed in the mu-shielded enclosure, it is isolated from the slow waves of ambient electromagnetic field. Therefore, the ambient field is taken out of the equation. Now it's up to the dynamics of the coil to do what the external field would not provide: iterating the genetic information by its repeating vortical structure.

This has affinity to how a copper coil wrapped around an iron rod can magnetize it by entraining the molecules into coherent domains. In this experimental condition the iteration needs to be sustained for 16-18 hours. As I see it, the coiling effect is what traps and conserves the information as self-referential vortices in the quantum state, and since solitons have been detected in the coherent domains of magnets, I deduce that the conserved information occurs in vortical soliton units.

Although Montagnier and his associates arrive at this conclusion through a more sophisticated interpretation of quantum field theory, the idea of entrainment is a simple way to express how molecular chaos is transformed into the order of biological systems, including the mental system.

DNA replicating itself through "ghost-like imprints" mediated by electromagnetic waves, rather than the usual cellular processes, is a radical notion.

TRANSFER OF GENETIC
INFORMATION VIA EMAIL

As would be expected, Montganier's work has been received with great skepticism. Recently, he and the physicist Guiseppi Vitiello, and others, have performed experiments that not only replicate the results but also show that the electromagnetic signature can be transmitted as a file from one computer to a distant computer through email and used to regenerate the molecular sample (Vitiello 2009). This is one of the most astonishing effects of the experiment.

Theta: the Memory Frequency

What makes Montagnier's research so compelling is that it correlates with studies that show how the formation of memory as a cognitive potential is facilitated by the theta frequency range of about 7-8 Hz (Michelson, et al, 2016), which has affinity to Montagnier's use of an ambient-shielded weak electromagnetic field pulsing at ~7Hz.

Active⇌Passive Oscillations of the EEG

Freeman's observations show that the cortex oscillates abruptly from a receiving state (i.e., ingoing vortex) to an active transmitting state (i.e., outgoing vortex). According to the physicist Guiseppe Vitiello:

> "Spatial AM patterns with carrier frequencies in the *beta and gamma ranges* form during the active state and dissolve as the cortex return to its receiving state after transmission. These state transitions in cortex form frames of AM patterns in few ms, hold them for 80-120 ms, and *repeat them at rates in alpha and theta ranges of EEG*" (Vitiello 2009. Italics added).

Hence, it seems that the higher frequencies attained through cognitive crunching by the ingoing vortex are modulated into lower frequencies when it inverts to the outgoing vortex.

Technically, 99.9% of the Organism Consists of Water!

Also, electromagnetic emissions not only appear to facilitate DNA information transfer into biological water, but also are endemic of the neuroglial activity; this would allow for transmission of information to the water substrate of the whole organism via the broad pulsing electromagnetic heart wave and the fascia enveloping every organ system in the body. There are so many compelling implications.

Conventional knowledge has held on to the notion that our bodies are about 70% water. However, if we assess the trillions of cells from its molecular composition we'd find that the body is actually composed of 99.9% water, and these units cooperate to create the human organism.

IS THE SEA PROGRAMED BY HIGHER MIND?

It has been postulated that organic life on Earth emerges from the sea. Preceding this watery emergence, lightning bombards the very salty and conductive sea. Is information being sent by some HI into the water? The amount of water of the Earth is about 75% of its "solid" matter, approximately the same ratio of water to solid matter in the human organism. We are gestated in a womb filled with amniotic fluid. Very interesting!

HOLY WATER?

From these conjectures the whole issue of homeopathy comes to mind. Maybe prayer-infused "holy water" is not so superstitious, and may be based on an ancient knowledge that has been ignored or lost to the modern world.

BORN OF WATER

In the King James Holy Bible we find a rather compelling affinity to Montagnier's discovery.

> "Except a man be born again, he cannot see the kingdom of God. ... Except a man be born of water and [of] the Spirit, he cannot enter into the kingdom of God. That which is born of the flesh is flesh; and that which is born of the Spirit is spirit" (John 3:4-6.).

The notion of being "born of water" implicates that the information (i.e., memory) of the entire organism mediated by DNA can be conserved by electromagnetic transmission, pulsed at around 7 Hz, in water. Note that brain cells are comprised of about 80% water. Hence, the quantum 1D or 2D imprint of the whole mind-body system can be amplified by attention to form a physical vessel as well as a spiritual one. Since both aspects are endowed with circular feedback, they display consciousness and creative intelligence in their operations but at differing rates of operation.

CELLULAR MEMORY THEORY

Support for the capacity of structured water to conserve information also comes from a number of compelling cases involving heart transplants.

For example, it has been reported that the cells in a transplanted heart remember information of the donor, and influence the mind and behavior of the recipient. Consider the following study done at the School of Nursing at the University of Hawaii in Honolulu:

> "Researchers sought to evaluate whether changes experienced by organ transplant recipients were parallel to the history of the donor. Researchers focused on 10 patients who received a heart transplant and found two to five parallels per patient post-surgery in relation to their donor's history. The parallels that were observed in the study were changes in food, music, art, sexual, recreational, and career preferences, in addition to name associations and sensory experiences. In the study, a patient received a heart transplant from a man who was killed by gunshot to the face, and the organ recipient then reported to have dreams of seeing hot flashes of light directly on his face" (Borreli 2013).

Perhaps one of the most dramatic and better-known cases implicating cellular memory is that of the heart recipient *Claire Sylvia* who received the organ from an 18-year-old male who died in a motorcycle accident. After the transplant she began experiencing unusual cravings for beer and chicken nuggets.

> "[She] also began to have reoccurring dreams about a man named 'Tim L.' Upon searching the obituaries, Sylvia found

out her donor's name was Tim and that he loved all of the food that she craved, according to her book *A Change of Heart*" (Borreli 2013).

Many critics rejected the unorthodox explanations of the phenomenon. They were inclined to see it as sheer coincidence, or as a side effect of the immune suppressant drugs she was under at the time.

However, the journalist John Brown interviewed (1997) two University of Arizona psychologists—Gary Schwartz and Linda Russek—who had a different, though controversial, explanation of what may have transpired in Sylvia's case.

It was based on a hypothesis they formalized in a paper, entitled *The Origin of Holism and Memory in Nature: The Systemic Memory Hypothesis*, subsequently published in Temple University's scientific journal Frontier Perspectives (1998).

According to the Systemic Memory Hypothesis (SMH), memory occurs when two similar systems enter into a state of resonance and exchange energy and information via feedback.

A simple example is the interaction of two tuning forks of the same frequency. When vibrationally coupled, they form a system held together by a recurring feedback loop, which forms the basic principle of how more complex systems conserve and transmit and receive energy and information: memory.

Self-sustaining Feedback Loop

According to the hypothesis, Tim L's love for chicken nuggets and beer was conserved in his heart cells as a self-sustaining feedback loop. When the heart was transplanted the memory went with it, affecting Sylvia's behavior once her organism retrieved the implicit information.

Let's look at another equally compelling case of the functionality of memory in patients lacking a substantial part of the brain.

NO BRAINER: THE RESEARCH OF DR. JOHN LORBER WITH HYDROCEPHALIC PATIENTS

Montagnier's revolutionary postulate requires additional research to substantiate. However, there are other cases in the clinical literature that lend some credible support.

While the brain consists of 85% water, within and outside of the cells, we know that the distribution of electromagnetic charges (see Pollack 2001) in it determines whether cell polarizations take place, but otherwise there is no other cognitive role given to it.

One report that is beginning to gnaw at this view comes from the British neurologist John Lorber, whose specialty involves patients with hydrocephaly, a birth defect causing abnormal fluid buildup within the interior of the brain. One case involved a 26 year-old man who was referred to him for a brain scan.

After the scan Lorber noticed that this man's cerebrum had been squashed by fluid pressure to a thickness of less than one millimeter (0.039 inch). On average, the walls of the cerebrum should be around 4-5 millimeters (0.157-0.196 inch) thick.

Brain scans of hydrocephalic brains showed fluid replacing large amounts of the brain tissue. On seeing these scans one would think that the fluid filled brains would have little or no intellectual capacity.

Quite surprisingly, Lorber's patient, as well as many others like him, functioned intellectually and socially quite well. He had an IQ of 126, and had earned top honors in mathematics. While most of the cortical neurological architecture was missing, it did not affect his intelligence. The case raised eyebrows about the neuron doctrine. One reporter postulated that the brain may actually function like a hologram.

> "If the way the brain functions is similar to the way a hologram functions, that one-millimeter sliver might suffice. Certain holograms can be smashed to bits, and each remaining piece can reproduce the whole message. A tiny fragment of this page, in contrast, tells little about the whole story" (Lewin, 1980, 1232).

While electromagnetic transmission beyond the brain is not likely to move objects and imprint information in water, the subtle currents inside of it could be the means by which cognitive information is conserved in the brain's profuse liquid environment.

The water inside every cell, which on average is about 10 micrometers (one millionth of a meter) in size, may actually be a cytoplasmic linear medium on which cognitive holograms are stored in 2D formats, while the quantum EMS would be in 1D formats.

Upon being stimulated by micro-electromagnetic currents they are synchronously conjoined and projected as 3D standing waves in the volume of the cortex (see Lehar 2004).

That tiny amount of water (i.e., gel-like cytoplasm, excluding the nucleus) in each cell is a uniform crystalline lattice consisting of nanoclusters that could be modified with "biophotons" in the infrared band carrying nonlinear information, as Fritz-Albert Popp has proposed (1998).

After all, we write and draw on uniform surfaces, and quite similarly conserve information on the linearized domains of magnetic tape, discs, and optical devices; why not in any kind of uniform linear substance?

THE WHOLLY GHOST

Montagnier's experiments strongly implicate intracellular water as the medium in which information in the form of signature electromagnetic signals (EMS) can be preserved in quantum coherent domains (as Psiloton units, MCs), using ~7 Hz electromagnetic pulsations as the encoding parameter, and from which they can be resonantly selected into synchronous assemblies by frequencies of the outgoing vortex. This suggests that intracellular-structured-water is the molecular substrate of configured quantum electromagnetic formats of the FCCP.

OBEs BORN OF WATER

If this is indeed the case, it is not inconceivable to deduce that a greater portion of the FCCP could be evoked with a much higher frequency, say in the gamma 200+ Hz range, that will entrain the watery fascia substrate and project its signature information (i.e., DNA and acquired memory constructs) via the heart wave out of the body into the quantum space medium as OBEs.

Of course, the scope of what is being pulsed would quite likely need to exceed the population of nerve cells in the brain, by using the heart wave to select and energize the meridian substrate of the fascia to include the ~50 trillion non neuronal cells of the organism.

The notion that the entire formative-and-functional information of the Self's psychophysiology can be transferred into another state of being and consciousness in a "subtler" medium has some support from experiments done with DNA referred to as "DNA phantom effect."

DNA PHANTOM EFFECT

The DNA phantom effect was first observed in Moscow at the Russian Academy of Sciences as a surprise while performing experiments to measure the vibratory modes of DNA in solution by Peter Gariaev and Vladimir Poponin (1992), using a sophisticated MALVERN (technology developed in England by Malvern Instruments in 1971) laser photon correlation spectrometer (LPCS).

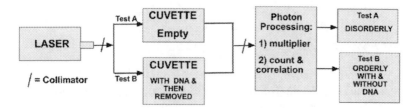

Figure 20: Illustrated summary of the process performed by the Laser Photon Correlation Spectrometer.

Two separate comparative measurement procedures are performed with the LPCS: Test A and Test B. The first measures the coherence of the photons in a laser-affected empty cuvette (a straight-sided, optically clear container for holding liquid samples in a spectrophotometer). The second measures the coherence of photons in the laser-affected cuvette in which a DNA sample has been inserted. In both cases the mild laser beam is modified by a collimator (a device for producing a parallel beam of rays or radiation) that is placed before the cuvette and after it. As the illustration shows, after interacting with the cuvette the photon beam undergoes two processes: 1) by a photomultiplier tube, and 2) by a photon count and correlation process before the results are displayed on a computer screen.

The first test of the empty cuvette showed a background random noise count by the photomultiplier, proving that the photons are not organized. The DNA sample is then placed in the cuvette and subjected to the same procedure, and then taken out.

A QUANTUM IMPRINT

Much to their surprise, when the DNA sample was taken out, leaving the cuvette undisturbed, the DNA left a kind of quantum "imprint," a certain order, lingering in the field of the cuvette. Many more experiments were performed with similar results. The experimental data suggests

> "[T]hat localized excitations of DNA phantom fields are long living and can exist in non-moving and slowly propagating states. This type of behavior is distinctly different from

the behavior demonstrated by other well known nonlinear localized excitations such as solitons which are currently considered to be the best explanation of how vibratory energy propagates through the DNA" (Poponin 1992, 3).

Moreover:

"According to our current hypothesis, the DNA phantom effect may be interpreted as a manifestation of a new physical vacuum substructure [negative space-time?] which has been previously overlooked. It appears that this substructure can be excited from the physical vacuum in a range of energies close to zero energy provided certain specific conditions are fulfilled which are specified above" (Ibid, 4. Inquiring bracket added).

Furthermore:

"This suggests that the electromagnetic phantom effect is a more fundamental phenomenon which can be used to explain other observed phantom effects including the phantom leaf effect [i.e., of Kirlian photography: a technique for recording photographic images of corona discharges and hence, supposedly, the auras of living creatures] and the phantom limb" (Ibid, 4. Bracket mine).

Meyl had independently proposed that these coronal discharges are actually eddy currents mediated by counter vortices in the ambient surround (2012b), and that they would be more pronounced if the medium from which they manifest consists of a poorly conducting dialectic (e.g., like water, air). Since biological systems are embedded in a watery matrix, it is proposed that vortical memory constructs are conserved in quantum "coherent domains" (indicated by Montagnier 2010).

Moreover, since attention is a synchronizing agent by dint of how it repeats a percept in the 1-to-7 cycle, it is not inconceivable that it would act to produce coherence like the laser in this experiment to stimulate aspects or the whole of the Self to project in the "phantom" (i.e., mental or spiritual) state.

While this typically involves the Self turning its attention on specific thoughts, feelings, and sensations emerging in the recursive Psiloton flow, it would also activate greater swaths of information when it is turned inwardly. Depending on **how much** of the Self is included in the self-observation, this may very well be the dynamic that "quickens" the psychophysiology to release and project the "spirit" to produce an OBE.

HAROLD SAXTON BURR'S LIFE FIELD

Harold Saxton Burr (1889-1973) was E. K. Hunt Professor of Anatomy at Yale University School of Medicine and a researcher specializing in bioelectronics. From his research he discovered that living organisms are structured according to electromagnetic fields that serve as templates conveying organizational principles. He was able to detect these fields with a standard voltmeter. With it he measured the growth patterns of salamanders, eggs, seeds, and other biological forms.

Burr discovered that all organic life—from singled-celled creatures, plants, to mice, and human beings—are formed and controlled by electro-dynamic fields that could be measured and mapped with standard voltmeters. He referred to them as "fields of life," or L-Fields, as being the basic blueprints of life. From his research he found that varying L-Field voltages could reveal specific physical and mental conditions, as well as the overall health of the organism; he proposed that these measurements could be used for improved diagnoses and treatments of disease conditions.

He felt that these ubiquitous fields are ever present, which in this book are held to exist in negative space-time, and to form the immortal blueprints of life. According to Burr the L-Fields are similar in principle to the field of a magnet. In his book *Blueprint for Immortality* (1973, NIB) he writes:

> "Electro-dynamic fields are invisible and intangible; and it is hard to visualize them. But a crude analogy may help to show what the fields of life - L-fields for short - do and why they are so important. Most people who have taken high school science will remember that if iron filings are scattered on a card held over a magnet, they will arrange themselves in the pattern of the 'lines of force' of the magnet's field. And if the filings are thrown away and fresh ones scattered on the card, the new filings will assume the same pattern as the old. Something like this happens in the human body. Its molecules and cells are constantly being torn apart and rebuilt

with fresh material from the food we eat. But, thanks to the controlling L-fields, the new molecules and cells are rebuilt as before and arrange themselves in the same pattern as the old ones" (1973).

Since the time of Burr, the mysterious field of magnets has been subject to experimentation that reveals that they display the attributes of biological metabolism (see Davis 1974), adding another organic dimension to his discoveries.

IMAGERY & THE WHOLLY GHOST

Research has revealed that imagination affects the same neuromuscular, and other, pathways involved in behavior by default (Decety 1996).

High performance athletes typically use this to improve their performance. Research has shown that by going through imaginary practice they are able to improve their skills while postponing the use of physical behavior (see Highlen, et al, 1979).

The phenomenon has also been found in the dream state. Martin Dresler and associates have discovered that dreamed movement also elicits activation in the sensorimotor cortex (2011). Imagery, whether intentionally formed in the waking or dream state has the same effect on neuromuscular and other pathways.

IMAGERY REQUIRES LARGE AMOUNTS OF COMPOSITE MEMORY

The psiological rationale behind this is that imagery requires a large number of MCs in its make up, and that by engaging them, an associative magnitude (as "psychic mass") is formed that will affect subsequent behavior in the imagined manner. In this regard not only quantity but also "coherent quantity" has transformative power.

Therefore, envisioning one's Self in all of its modes of being integrates the associative MCs into a powerful integral unit. This can initially be attained by being fully relaxed and not expressing the energy through movement while doing so.

Evoking the Whole Equalizes its Polarities

If we are able to hold on to this image for an "extended" period of time, something interesting happens to the musculature. It enters into a paradoxical state of equanimity, which is experienced as a kind of tension (pseudo-paralysis) in which a subtle internal energy embodiment appears to be "solid," producing a state akin to the Wholly Presence, which is another term for the Wholly Ghost *in situ*.

It is assumed that all the intrinsic polar information contained in the DNA constructs of the physiology merge by the "tension" of sympathetic/parasympathetic pressure to create and release/attract a more subtle mode of being. Hence, as has been argued all along, this is the basis for generating and having out-of-body projections, or any partial projections of the Self that might be involved in lucid dreams or remote viewing.

The MC with enough psychic mass to prevent such OBEs is known in conventional psychology as the default mode network (DMN), presented earlier. We need to take a closer look at this network in regard to its resistance to the Wholistic Response.

DEFAULT MODE NETWORK (DMN) AND CONDITIONED STATES OF CONSCIOUSNESS

The network typically activates "by default" when a person is not involved in a task. For example:

> "Studies have shown that when people watch a movie, listen to a story, or read a story, their DMNs are highly correlated with each other" (Wikipedia: DMN).

It is interesting to note that the DMN has been shown to be negatively correlated with other networks in the brain, such as attention networks, or if there is a lack of order to what is being perceived. Consequently:

> "DMNs are not correlated if the stories are scrambled or are in a language the person does not understand, suggesting that the network is highly involved in the comprehension and the subsequent memory formation of that story" (Ibid).

225

This strongly suggests that the DMN instantiates widespread processes that provide an individual with a sense of conventional "order" that spontaneously arises when intentionality is lacking (akin to absorption). According to the psiological view, this conventional sense of order could be viewed as a broad yet limited repertoire of MCs in the FCCP.

When the conventional sense of order is by some means disturbed, the PSQ is apt to be bypassed, thereby opening an opportunity for greater access to a much larger interactivity with MCs in the FCCP.

This is obviously related to the hypnotic technique the psychiatrist Milton Erickson used with his patients to induce them into trance and to access the non-conscious domain where there is a greater range of MC connectivity to generate more powerful therapeutic assemblies.

Comparatively speaking, the DMN functions with a conditioned set field of connections that transformative systems of psychology seek to bypass, which has been experimentally accomplished with the use of psychedelic drugs (Carhart-Harris 2012).

Low Connectivity

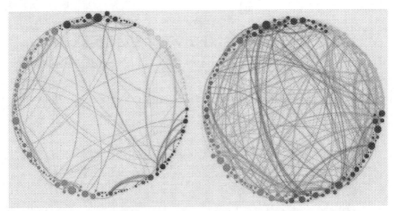

Figure 21: Communication between brain networks in people given psilocybin (right) or a non-psychedelic compound (left).

Figure 21 shows the "low connectivity" of the DMN (left) compared to when it is bypassed by the use of psychedelic drugs (in this case psilocybin), and other transformative activities, including meditation.

It is important to note that the same effects could be correlated with the practice of meditation (as well as other similar iterative rituals), but the reaction time of the substance is considerably quicker and more conducive to experimentation.

WHAT HAPPENS AT THE TIME OF "DEATH"?

At death, the organism undergoes entropic stasis in which the biological system can no longer do work. It reduces to entropic equilibrium ($\div$), which produces a node that draws in the life force (a.k.a. "entropic force" see Waser 2004) to invigorate the memory stores in the watery matrix to generate a higher being body (spirit). This historic information is compressed by an ingoing vortex into a MC (∞) in the negative space-time of the FCCP, in what constitutes the aforementioned "PSILOTON" of one's being and consciousness.

Since the organism can do no further work (i.e., it reductively decays and does not oscillate at the receptive rate in "far from equilibrium" golden ratio vibes) it is not conducive to forming the outgoing vortex to reestablish a life-giving control cycle for the organism.

Hence, it completes or stalls the 1-to-7 cognitive cycle at 3, where the intrinsic information is "crunched" to become a spiritual construct being poised to respond to attracting oscillations from "above" or "below," depending on its level of development.

The ingoing vortex transitions towards the blue shift with an accompanying faster gyration. Therefore, the mind accelerates at the time of physical or simulated death to ensure the Self's continuity in negative space-time.

MIND SPEEDS UP AT TIME OF "DEATH"

We go through life in temporal frames of reference which, according to Ayurvedic medicine (see Lad 1993, 58), begins for the baby in the womb with a pulse of 160 beats per minute (BPM). After birth, it reduces to around 140 BPM, and at one year old it dips to 130 BPM. Over the years its normal resting pulse gradually slows down to the adult range of about 72 BPM. Interestingly, at the time of death the pulse rises to the rate it had in the womb of 160 BPM. This should be evident just before the heart stops beating. However, something interesting happens upon cardiac arrest.

Cardiac arrest is the conventional indicator of death. When it occurs, the brain is assumed to be hypoactive (i.e., straight lines on the EEG). However, this assumption has never been systematically investigated.

To verify the hypoactive claim, Jimo Borjigin and associates have created a study to test this assumption with laboratory rats. They performed continuous electroencephalography on rats undergoing experimental cardiac arrest and analyzed changes in power density, coherence, directed connectivity, and cross-frequency coupling.

Quite unexpectedly, they identified a transient surge of synchronous gamma oscillations that occurred within the first 30 seconds after cardiac arrest. These oscillations were more coherent than those found in the waking state, suggesting a heightened conscious process at near-death.

> "Gamma oscillations during cardiac arrest were global and highly coherent; moreover, this frequency band exhibited a striking increase in anterior–posterior-directed connectivity and tight phase-coupling to both theta and alpha waves. *High-frequency neurophysiological activity in the near-death state exceeded levels found during the conscious waking state.* These data demonstrate that the mammalian brain can, albeit paradoxically, generate neural correlates of heightened conscious processing at near-death" (Borjigin, et al, 2013, 14432. Italics added).

THE HISTORICAL SEED OF ONE'S FUTURE LIFE

Because state-dependent memory is recorded and coded according to the physiological temporal frame of reference when the learning took place, we can deduce that at the time of death it is vortically transformed and compressed to form a memory construct conserved in negative space-time constituting the historical seed of a "future" life.

This seed reincarnates when resonantly prompted by a compatible fetal/embryonic receiver (regardless of sex orientation) in positive time-space. It may also regenerate towards existence in attracting spiritual milieus.

In positive time-space temporality is necessary "to keep," as Einstein stated, "everything from happening at once." According to this rationale, our time frames, expressed via the pulse, are like a classification system of our entire history contained in the Psiloton.

APOLLONIUS OF TYANA

Apollonius of Tyana (c. 15 - c. 100 AD) was a Greek Neopythagorean philosopher from the town of Tyana in the Roman province of Cappadocia in Anatolia, making him, in time and place, a contemporary of Jesus.

Biblical scholar Bart D. Eherman relates that in the introduction to his textbook on the New Testament, he describes an important figure from the first century without first revealing that he is writing about the stories attached to Apollonius:

> "Even before he was born, it was known that he would be someone special. A supernatural being informed his mother the child she was to conceive would not be a mere mortal but would be divine. He was born miraculously, and he became an unusually precocious young man. As an adult he left home and went on an itinerant preaching ministry, urging his listeners to live, not for the material things of this world, but for what is spiritual. He gathered a number of disciples around him, who became convinced that his teachings were divinely inspired, in no small part because he himself was divine. He proved it to them by doing many miracles, healing the sick, casting out demons, and raising the dead. But at the end of his life he roused opposition, and his enemies delivered him over to the Roman authorities for judgment. Still, after he left this world, he returned to meet his followers in order to convince them that he was not really dead but lived on in the heavenly realm. Later some of his followers wrote books about him" (in Wikipedia: Apollonius of Tyana).

Although almost everything about Apollonius grabs at one's attention, what's so fascinating about his writings is the way he described the immortality of the Self in rather scientific terms, quite prescient for the times. For example:

> "There is no death of anyone, but only in appearance, even as there is no birth of any, save only in seeming. The change from being to becoming seems to be birth, and the change from becoming to being seems to be death, but in reality no one is ever born, nor does one ever die. It is simply a being visible and then invisible; the former through the

density of matter, and the latter because of the subtlety of being—being which is ever the same, its only change being motion and rest. For being has this necessary peculiarity, that its change is brought about by nothing external to itself; but whole becomes parts and parts become whole in the oneness of the all. And if it be asked: What is this which sometimes is seen and sometimes not seen, now in the same, now in the different?—it might be answered: It is the way of everything here in the world below that when it is filled out with matter it is visible, owing to the resistance of its density, but is invisible, owing to its subtlety, when it is rid of matter, though matter still surround it and flow through it in that immensity of space which hems it in but knows no birth or death" (Apollonius of Tyana in Meade 1901, 149-150).

PRINCIPLES & PRACTICE OF TRANSFORMATIVE PSIOLOGY

"They know not, neither will they understand; they walk on in darkness: all the foundations of the earth are out of course. I have said, Ye [are] gods; and all of you [are] children of the most High" - Psalms 82:5-6

"I said, you are Gods" - Gospel of John 10:43

"Now it is permitted to enter intellectually into the mysteries of faith." - Emanuel Swedenborg

"All matter originates and exists only by virtue of a force. We must assume behind this force is the existence of consciousness and intelligent Mind. The Mind is the matrix of matter." - Max Planck

Ancient forms of spiritual development were never without a kind of "science" from which they formulated their models of the universe, their principles, and their developmental methodologies to open the door to the hidden aspects of our cognitive reality.

Essentially, science is the study and measure of fundamental vibrations, since science can only know what it measures. In physics a vibration is a measured movement that oscillates as a wave. It is observed in a medium (e.g., quantum, plasmic, gaseous, fluid, solid) whose equilibrium of potential energy (PE) has been by some means disturbed to produce kinetic energy (KE) moving as waves in a certain direction in a regular frequency or rhythm.

Modern science has revealed that it is the nature of how things vibrate that determines their behavior: how they will interact with other systems, how they exchange energy and information, and how they will bond into various configurations, creating phenomenal contexts. All this, as has been presented, is encompassed by the principle of interference.

By employing the knowledge of wave interactions, modern science has achieved its most technical advances. However, in this chapter the term psychology will be replaced with that of psiology, which is the integral study of cognitive processes that incorporate objective as well as subjective potentials, albeit in ratios of preponderance.

These ratios of preponderance instantiate a midpoint in which a "healthy" state of mind is sustained, that neither veers towards the extremes of "objective compulsion" or "subjective compulsion."

So strongly affected by materialistic science, modern psychology seeks objective outcomes from its experiments, which are by default suppressive of any subjectivity. Yet, there's actually no cut off border of the scientist's mind and the experiment being performed. In other words, it is illusive to think that subjectivity can be wiped from the picture with methodological rigor.

The same can be said for those in the grips of subjective compulsion. This mind set points to the miraculous and denies materialistic science altogether.

While modern psychology is benefiting from this vibratory technology to understand the nature of the kinetic aspects of the mental system, these advances have not revealed the secrets of transformative esotericism, which Gurdjieff stated was sitting under everyone's nose, and is presented in this book as psiology.

Psiology is a term revealed to the author that aims at returning psychology to its initial ethos rooted in ancient systems of spiritual development that have been distorted with the rise of beliefs characterized by incommensurable dualities, such as: mind-body, life-death, good-evil, religion-secularism, normal-paranormal, the mind as "epiphenomena," that have continued with pernicious effects into the post-modern era.

These notions, so entrenched in modern psychology, are fundamentally detrimental to creating the Wholistic Response and all that pertains to this "holy" state, a divine potential inherent in every human being.

The Fourth Way was Gurdjieff's attempt to revive the essence of this "science." While he spoke of the teaching as being imbued by an "objective" set of principles, there is little doubt that accompanying the copious amount of knowledge he presented was a hidden subjective dimension; this intimated a special revelation or "gnosis" that eluded all forms of objectification and could only be "objectified" in the form of esoteric symbols, especially to those that are produced by the higher mental center.

While at certain points in history the secret principles and methodologies would come to the surface, and transformative forms of psiology would rise under different facades, they would eventually fall into institutionalized oblivion, as Gurdjieff averred that at some point everything eventually becomes "mechanical."

During these revelatory periods the powers of the human psyche would be reinvigorated by those who employed the principles, only to once again succumb to the collective habituation of the times, and become, according to Ouspensky, a "forgotten science."

What obliterates the powers of psiology are the vehicles used to express it. This typically occurs when the mental system sacrifices its subjective dimension while in search of objectivity. Modern psiology, as well as modern science in general, has because of this skewed form of inquiry developed an entrenched kind of "objective compulsive disorder" (OCD).

Conversely, however, when spiritual teachings avoided and denied the curiosities and experimentation of the scientific method, they too entered into a rather irrational "subjective compulsive disorder" (SCD). Both kinds of compulsion radically strayed from the "golden mean" that ensures stability as well as freedom.

ABANDON THE SYSTEM

It appears that it is for this reason that Ouspensky in his last days advised his pupils to "abandon the system" and begin anew. The night before his passing he said:

> "You must start again. You must make a new beginning. You must reconstruct everything for yourselves—from the very beginning" (Collins 1971, 14).

Rodney Collins realized that if the system were to be institutionalized that it would endure as a mechanical doctrine, and would lose the seminal novelty and mystery that opens the mind to new energies and understanding. Collins states:

"This then was the true meaning of 'abandoning the System'. Every system of truth must be abandoned, in order that it may grow again. He had freed them from one expression of truth which might have become dogma, but which instead may blossom into a hundred living forms, affecting every side of life.

Most important of all, 'reconstructing everything for oneself' evidently meant reconstructing everything in oneself, that is, actually creating in oneself the understanding which the system had made possible and achieving the aim of which it spoke -actually and permanently overcoming the old personality and acquiring a quite new level of consciousness" (Ibid, 14).

Though not quite the mission I had in mind when I began this book, the scientific principles used to explain psiospiritual states and information processing provide a "fundamental" aspect to the system on which a wide variety of subjective "dressings" could be added. Yet, underlying the outward appearances, beyond the masks, we find the unhindered language of waves whispering realities to the universe. This notion erases all notions of a bored Creator who shuns "its" own creations.

IN THE LANGUAGE OF WAVE DYNAMICS

Hence, the nature of the "old personality" and the acquiring of a new MMC as the "true personality," with which a new level of being and consciousness could be entrained, would be more scientifically discerned as occurring at the level of the psychophysical language of waves. While this suggests objective criteria, we need not forget that there are the attributes of subjective nonlinear interference (O & S) in every psiological equation.

Wave interference reveals some of the key principles of the "forgotten science" in a new light. Gurdjieff and Ouspensky brought this old science into the ethos of modern science. Collins attempted to translate the system he learned from Ouspensky, using notions of resonance available at the time. It formed a step in the right direction. So let's have a brief view of what the "forgotten science" was all about.

RECONSTITUTING THE FORGOTTEN SCIENCE

Back in the 1970s when I was reading Ouspensky's writings, I took particular interest in the book entitled *The Psychology of Man's Possible Evolution* (1974), which consisted of lectures he had given on the basic principles of The Fourth Way (a.k.a. the System, the Work) that he had learned from his own research and from Gurdjieff.

This form of psiology was immersed in esotericism, such as: Hermeticism, Neoplatonism, mystery schools, esoteric Christianity, Cabala, Free Masonry, Theosophy, Rosicrucianism, and Anthroposophy, perhaps extending back to ancient Greece, particularly to the legendary Pythagoras, and the revelation of secret knowledge (logos) by the Gnostics. Ouspensky referred to it as "esoteric psychology."

According to Ouspensky, this form of psychology existed for many hundreds of years as a science before it was relegated to a level of insignificance by a variety of social forces. Because it seeks to enact transformations of being and consciousness in individuals, I thought of it as transformative psiology, and the systems that practice it as the transformative tradition.

Traces of its existence can be found in almost every aspect of the pre-modern era. Yet, Ouspensky writes,

> "By one reason or another psychology always was suspect of wrong or subversive tendencies, either religious or political or moral, and had to use different disguises" (1974, 4).

Moreover:

> "For thousands of years psychology existed under the name of philosophy. In India all forms of Yoga, which are essentially psychology, are described as one of the six systems of philosophy. Sufi teachings, which again are chiefly psychology, are regarded as partly religious and partly metaphysical. In Europe, even quite recently, in the last decades of the nineteenth century, many works on psychology were referred to as philosophy. And in spite of the fact that almost all subdivisions of philosophy such as logic, the theory of cognition, ethics, aesthetics, referred to the work of the human mind or senses, psychology was regarded as inferior to philosophy and as relating only to the lower or more trivial sides of human nature" (Ibid, 4).

While still not identified by its proper role, psychology had an even longer relationship with religion.

> "This does not mean that religion and psychology ever were one and the same thing, or that the fact of the connection between religion and psychology was recognized. But there is no doubt that almost every known religion...developed one or another kind of psychology teaching connected often with a certain practice, so that the study of religion very often included in itself the study of psychology" (Ibid, 4-5).

A form of this psychology can be found to have existed in early Christianity, and its tenets can be found in a compendium of commentaries under the name of the *Philokalia* (Kadloubovsky 1977).

Modern psychology is based on the probabilistic study of the collectively conditioned mind. Within this statistical context, Ouspensky claims that psychologists:

> "...study man as they find him, or such as they suppose or imagine him to be" (Ibid, 6).

On the other hand, psychology systems which focus on the human evolutionary potential:

> "...study man not from the point of view of what he is, or what he seems to be, but from the point of view of his possible evolution" (Ibid, 6).

Having abandoned its ancient origins, modern psychology has been viewed as a new science. Yet, Ouspensky emphasizes:

> "This is quite wrong, psychology is, perhaps, the oldest science, and unfortunately, in its most essential features a 'forgotten science'" (Ibid, 3).

To have an initial understanding of what all this might mean, Gurdjieff's resurrection of the "forgotten science" must be viewed in the broader social context in which it took form.

We can only glean a meagre slice of what this environment was like. There's no doubt that it was written for the Western mindset of the early twentieth century affected so powerfully by the rise of industrialism, democracy, science, and humanism, which challenged the hold conventional religion had on people's minds, especially in Europe where the rituals of conventional religion were becoming insignificant.

This would especially effect intellectuals, artists, and other independent thinkers to search for more meaningful forms of spiritual involvement. It was a period caught between the elan vital of the romanticists, lingering from the likes of Goethe and Blake, and the consolidation of scientific discoveries led by the Newtonian and Einsteinian paradigms.

An influential esoteric organization at the time was the Theosophical Society, founded by the Russian philosopher and occultist Helena Blavatsky. A number of these ideas were adapted into Gurdjieff's system.

In science, physics was shifting from the classical Newtonian three-dimensional paradigm separate from time to Einstein's radical theories of relativity, which centered around the synthesis of space and time (the fourth dimension) and the curvature of space as the cause of gravity. Quantum physics was in its infancy.

While it seems that Gurdjieff was informed about these scientific developments, Ouspensky appeared to have a keener interest in regard to their possible relationship to consciousness (*The New Model of the Universe* was about this relationship), particularly the "fourth dimension" introduced by Einstein with his theory of "relativity." Nevertheless, the Gurdjieffian model of the universe, and his system of development based on it, is replete with mythical overtones that reflect this scientific ethos.

One could say that it paradoxically forms a kind of panpsychism with a blend of Deism (i.e., the Absolute creates the laws of the universe that are expressed without further intervention), described with technical metaphors relating to: Matter, Vibrations, Octaves, Hydrogens, Forces, Evolution, Dimensions (worlds, cosmos), Magnetic Center, Density, Energy, Chemistry, Speed, and relative space-time frames of existence.

Everything in the universe is alive, manifesting on varying scales of cosmic existence, to some degree conscious, and thinking intelligently, all organized according to a variety of vibrations Gurdjieff classified into a hierarchy of "hydrogens" (H) obeying the laws of triads (involving positive, negative, and neutralizing forces) and octaves (based on the solfeggio musical scale of frequency doublings). The whole system is couched in quasi-scientific terms which I found to be conceptually exciting.

The idea that captured my imagination and initially drew me to these works is the notion that everything in the universe, including our inner states of being and consciousness, can be explained to some objective degree by the nature of vibrations. The aim to bring the mind-body system into a state of harmony later brought to mind its correlation to neuronal synchrony.

It was obvious that Gurdjieff built his system intuitively and from his knowledge of inner vibration existing in other transformative teachings he encountered in his journeys in search of spiritual knowledge. Let's touch on some examples of systems that expressed use of this vibratory ethos in their principles and practice.

THE YOGA OF VIBRATION

We find an expression of this notion with the "Yoga of vibration and divine pulsation" that forms the basis of Kashmir Shaivism, which holds that

> "Cit - consciousness - is the one reality. Matter is not separated from consciousness, but rather identical to it. There is no gap between God and the world. *The world is not an illusion* (as in Advaita Vedanta), rather the perception of duality is the illusion" (Wikipedia: Kashmir Shaivism. Italics added).

The essence of this system is captured in the following comment by the scholar and practitioner of this system Paul E. Muller-Ortega on the nature of the Shaivite vibratory principle of spanda.

> "In the modern period, we have grown accustomed to the notion that sound and light are vibratory phenomena, and that indeed all of physical reality is composed of the solidification of vibratory energies. Long before the discoveries of modem physics, the Shaivite concept of spanda intimates a view of reality as composed of a vibratory web of infinite complexity. *Tantric Shaivism would have us understand that the vibratory energies that compose our physical reality are themselves condensations of ultimate consciousness.* Moreover, the Shaivite tradition suggests to us a unifying continuity between our physical reality, the activities of sense perception, and all forms of interior awareness. All of these are seen as

phenomenal manifestations of the ultimate consciousness, enmeshed in a complex vibratory matrix" (in Singh, 1992, xvii. Italic emphasis added).

If we embrace this broader inclusiveness we can see that the physical organism, the mental system, and the spiritual state of being and consciousness are aspects of consciousness in its various modes of integral vibration.

PYTHAGORUS

One of the earliest formal studies of vibrations was done by the legendary Ionian Greek philosopher, mathematician, mystic, and scientist Pythagorus (c. 570 - c. 495 BC). Most of what we know about Pythagorus is attributed to the school he founded, known as the Pythagoreans, and the writings of those who studied there or had contact with Pythagorus himself or his teachings. Pythagorus' religious and scientific views were inseparably connected to the principle notion that "all is number." The following quote from the scholar of esotericism Manly Hall summarizes the essence of this philosophy.

> "The Pythagoreans declared arithmetic to be the mother of the mathematical sciences. This is proved by the fact that geometry, music, and astronomy are dependent upon it but it is not dependent upon them. Thus, geometry may be removed but arithmetic will remain; but if arithmetic be removed, geometry is eliminated. In the same manner music depends upon arithmetic, but the elimination of music affects arithmetic only by limiting one of its expressions. The Pythagoreans also demonstrated arithmetic to be prior to astronomy, for the latter is dependent upon both geometry and music. The size, form, and motion of the celestial bodies is determined by the use of geometry; their harmony and rhythm by the use of music. If astronomy be removed, neither geometry nor music is injured; but if geometry and music be eliminated, astronomy is destroyed. The priority of both geometry and music to astronomy is therefore established. Arithmetic, however, is prior to all; it is primary and fundamental" (2003, 209-210).

Everything in the Pythagorean paradigm is related to the vibratory scheme he discovered while experimenting with the monochord, an instrument with one string made taut by placing it over a moveable bridge. By moving the bridge he was able to determine that certain ratios of the string produced consonant or dissonant sounds.

This is known as "Pythagorean tuning" in which the frequency relationships of all intervals are based on a 3:2 ratio. This interval is chosen because it is one of the most consonant. We could deduce that this is an example of the vibratory nature of the "pleasure principle."

It was also averred that he was the father of music therapy by dint of how he applied his knowledge of musical ratios to human behavior. The discovery of golden ratio is attributed to a member of his school, and because it was viewed to be divinely auspicious it was shrouded with secrecy.

> "Concerning the secret significance of numbers there has been much speculation. Though many interesting discoveries have been made, it may be safely said that with the death of Pythagoras the great key to this science was lost. For nearly 2,500 years philosophers of all nations have attempted to unravel the Pythagorean skein, but apparently none has been successful. Notwithstanding attempts made to obliterate all records of the teachings of Pythagoras, fragments have survived which give clues to some of the simpler parts of his philosophy. The major secrets were never committed to writing, but were communicated orally to a few chosen disciples. These apparently dated did not divulge their secrets to the profane, the result being that when death sealed their lips the arcana died with them" (Wikipedia: Pythagorus).

THE I CHING

In China the vibratory concept is implicit in The I Ching (Book of Changes), the origins of which are attributed to the legendary Fu Xi around 2,800-2,737 BC, and serves as a means of divination.

While the I Ching does not directly address its model in terms of vibrations, it is indisputable that it is formulated on the notion of energetic movements and how they are resisted, advanced, or brought into equilibrium by the relationship of two principles represented by two sets of trigrams: yang (☰) plus yin (☷) trigrams, yielding permutations of 64 different hexagrams.

The book describes change in terms of motions governed by the permutations of yin and yang. Yang is conceived to be the more energetic aspect of the two. We can understand their complementarity by how they can become one another, not any different in principle to how portions of the atmosphere condense (yang) into clouds, which eventually dissolve (yin) back into it. In this regard yin is thought of as "space" and yang as its condensation. A key idea behind the work is that variety and complexity can be derived from a few essentials.

The possibilities have psiological implications that allow for permutations of thought and imagination. The use of chance derived from the casting of sticks and other objects, as the means by which oracles were derived, could be seen as a method to focus the mind by the out-of-pattern effect of each novel configuration.

With this, two contrary conditions meet: the linear orderly movement (objective aspect) of the hexagrams and their contextual suggestions (subjective aspect), and the unique, chaotic, patterns of the sticks. Hence, the future emerges at the cusp where order (predictability) and chaos (chance) meet.

I was surprised to see how this use of the I Ching closely resembled complexity theory, the contemporary study of how novelty and prediction emerge at the cusp of order and chaos (see Loye 2000). It appears that the I Ching was ahead of complexity theory by over two thousand years.

VIBRATIONS DORMANT IN THE HEART

We experience waves as vibrations that exceed a threshold in some aspect of the Self. In Sufism the concept is central to the spiritual development of its students. The following quotation describes the role of cultivating certain vibrations dormant in the physical heart of the Sufi student.

> "The direction in which the Sufi student is going is called 'The Path of Substantial Evolution'. The given exercises and lessons, individual or general, in which a serious discipline is demanded, help the Sufi student to develop his substance, which is hidden in the corner of the physical heart. One can compare this evolution of the substance with the development of a little bird in an egg. In the beginning there is only a little seed, captured in a closed environment. The

241

vibrations coming from outside, make this seed grow and become a little bird. The closed environment, which is the shell of the egg, is its protection. However, when the bird is strong enough, it can break through the shell and it can learn to fly. Already with this comparison we are approaching the fundament on which the Sufi school is built: 'the mystic rhythms and their inner vibrations'. The relationship between warmth and growth, illustrates the relation between mystic rhythms and substance. To develop the hidden light of the real self, good and positive vibration is necessary. The inner music of the mystical poetries carries these positive vibrations."

Knowledge of how these vibrations affect the human organism was incorporated in poetry and music.

"According to the great Sufi masters, the use of rhythms has a very important role in the teachings, spiritual evolution and self-awareness. The inner vibration inside the music of the mystical poetry is extremely important because it possesses a secret language that affects without knowing the words or vocabulary and it directly influences the heart. World famous mystic poets like Rumi, Hafez and Shah Nematollah Vali describe this path in their own unique way. The method of using mystic rhythms for substantial evolution was first developed by Shah Nematollah Vali and contains a series of 720 rhythms, divided in 24 families. Each rhythm corresponds with the spiritual state of the Sufi student" (Nematollahi Sultan Ali Shahi Gonabadi Sufi Order: www. sufism.ws/sufischool.php)

Having been taught by Sufi masters, Gurdjieff used this knowledge to develop, in conjunction with the composer and pianist Thomas de Hartman, an "objective" form of music, more attuned for Westerners; he used it in the choreography of sets of "Movements" that were enacted over the enneagram drawn on the stage. The purpose of these exercises is to inculcate the knowledge of the system into the "unused parts" of the movement and emotional centers.

While Gurdjieff lent more value to the notion of "objective" we cannot exclude the fact that these pieces involved the creative use of "subjective" potentials by this mutual collaboration.

In the School, the music of Johann Sebastian Bach and Antonio Vivaldi were deemed by the Teacher to have this objective design, intended to stimulate the Wholistic Response (this term was not used in the School).

VIBRATIONS AS METABOLISM OF "FOODS"

According to Gurdjieff, these energetic terms define how the body, mind, and spiritual states of being process energy and information, analogous to ingesting and metabolizing food, which he deemed involved regular foods and liquids, air, and, impressions—the smallest units of thoughts, feelings, and sensations. Therefore, the path of human evolution incorporates progressive rates of psychophysiological metabolism.

As mentioned earlier, Ouspensky defined these rates based on the speed of the intellectual center, which he designated as X. The movement-instinctive center functions at 30,000 faster than X. The optimal speed of the emotional center, the sex center, and the higher emotional center is 30,000 times that of the movement-instinctive center while the higher mental center (the symbolic "brain" of the logos) exceeds them by the same amount.

This accelerative effect had much affinity to the vibratory effects I had experienced in my various altered states of mind, especially with formation of OBEs, which Gurdjieff referred to as "higher being bodies." It also correlated with Yogananda's assertion of "stepping-up of the mind's vibratory rate."

A PSIOLOGICAL KIND OF INQUIRY

While the notion of vibrations is a step towards adding a more scientific dimension for understanding and enriching the principles and practices of the transformative tradition, it has limitations if it is not brought into the more measured understanding of wave dynamics, of which some key examples have been thus far presented. Otherwise one would drown in the often-confusing symbolism that comprises the vast history of transformative psiology (TPsi), though this confabulation of meaning is not without a "hidden" purpose.

To help overcome this conceptual divisiveness I found it quite useful to categorize what I had learned of its philosophy and methodologies by asking the following four questions of a transformative system:

What is its goal or purpose?
What are the obstacles to attaining the goal?
What methodologies are used to overcome the obstacles?
How does one know if the goal is being met?

To help answer these questions I will use what has been presented thus far as our guide. We have looked at some personal experiences, a variety of transformative concepts, and a Symbol-guided selection of scientific and other kinds of references. All have been merged to form a universal dynamic-cognitive model based on the cognitive (Subject-Object) process of what we call the "mind."

It has been proposed that this 1-to-7 cognitive cycle is linked by a F-F loop that makes decisions using the principle of wave interference (its gradients of being in-or-out of phase to constitute its "yes" and "no" and anything in between).

The rate at which it is able to process a certain amount of information drawn from the FCCP, a vast memory involving local and nonlocal potentials, determines its level of phenomenal consciousness and intelligence. As the speed of the 1-to-7 cycle increases, the amount of energy and information the system can garner and process also increases exponentially.

It has also been proposed that what is commonly referred to as "body, mind, and spirit" are not separate entities but different emergent gradients of this psychophysiological processing. In this regard, the mind emerges from the body when its metabolism is accelerated beyond a certain threshold, and the spirit emerges from the mind in a likewise manner. Each one has all three aspects in degrees of subtlety. They co-exist in different ratios of development as gradients of body-mind-spirit potentials in their own right. When fully matured they can transcend their progenitor vessel and merge with the temporal frame of reference of another psiological milieu in the Psilotonic cosmos.

Yet all are immersed in an isotropic primal field, referred to as the PUF, from which they originate and to which they return. Hence, they develop in the field as self-similar fractal energy configurations, and the PUF, in all of its fractal extensions, becomes conscious of these modifications.

When the speed of processing exceeds a certain limit, these signature fractal states may merge back into the PUF and lose their signature identity. In that grand merger the fractal state of consciousness is transformed into cosmic or universal consciousness mediated by the isotropic nature of the PUF.

Accelerating a system in this progressive manner requires a certain amount of energy, some of which needs to be transferred from its normal consumption by superfluous thought, activity, and behaviors.

While this is a step in the right direction, the Self needs to draw from a "higher" source of energy in order to complete the transformation from one level of being and consciousness to one occurring in a faster and more comprehensive frame of reference. As will be discussed in this chapter, this requires learning to bring the whole system into an auspicious equilibrated state. Let's begin the inquiry with what constitutes an evolutionary goal/purpose.

GOAL/PURPOSE

Much of the emphasis in transformative systems is on the divisive effects of "duality," attributed to the PSQ (a.k.a. the ego, sin, Satan, ahamkara). The PSQ tends to think in terms of either "yes" or "no," "good" versus "bad," or dithers in between.

However, the issue can more explicitly be described in terms of the pernicious effects of division, and its disempowering effects on the Self in terms of degree of vibrations.

The goal/purpose can therefore be defined in quantitative terms that facilitate the unification of larger assemblies, and consequent subjective creative transformations, based on the principle of emergence.

According to this psiological principle, the potential of the physical mode of the Self begins at the most spiritual level that condenses into mind, which in turn condenses into the physical mode of being over aeons of creation.

Hence, the mind and spirit are complexed as potentials that the practitioner can evoke and evolve through goal-based iterative practices, in which the Self enfolds back on its essential nature, using accelerative positive F-F enactments to increase its rate of processing energy and information.

Not Conducive to Creating the Wholly State

The key divisive effect of the PSQ is related to how it isolates and dwells on specific impressions at the expense of the "whole." Therefore, while useful for practical issues, it is not conducive to creating a unifying assembly sufficiently large to elicit the Wholistic Response. It is a quantitative as well as a qualitative issue.

Wholistic Versus Holistic

The term "wholistic" instead of "holistic" is used in the psiological context to differentiate between all inclusiveness, with the former, as opposed to referring to something intimately interconnected and explicable only by reference to the whole, as in the latter.

While both interface as intrinsic to the system and form important aspects of the science of the Self, the difference is dramatized by how much of the system (including the non-conscious) is involved when a moment of perception is produced in the 1-to-7 cycle. Holism doesn't clearly define that variable amplified moment.

THE TRANSFORMATIVE MEANING OF "BALANCE"

It is important to note that the EEG appears to manifest serially in the early neuropsiological studies, and subsequently it was discovered that it actually functions in a conjunctive, "cross coupling" manner in which the hypo and hyper temporal aspects of the spectrum enter into some kind of auspicious ratios based on the golden mean (see Pletzer 2010).

We conventionally notice the transitions occurring sequentially as appears with the sleep-wake cycle, but we are rarely aware of the auspicious-cross-coupling conjunctions revealed by neuroscience, which the seers of TPsi long ago introspectively discovered to be the gateways through which various gradations of the life force could be tapped, and other dimensions could be revealed (i.e., amplified into awareness).

From my experiences of lucid dreaming I intuited, later substantiated by research (see Hobson, et al, 2010), that when the frequencies of gamma (30-100+ Hz) merge with the dream-zone of theta (4-8 Hz) we have lucid dreams, a more phenomenally conscious state.

However, when lesser frequencies, like low beta (14-20 Hz) merge with theta, the dreams are not so lucid (i.e., the assemblies formed are relatively "smaller") and are more likely to be forgotten. It seemed obvious that there is an ideal ratio between the slower and faster frequencies to enact the lucid states.

It subsequently made sense that when the high frequencies of the gamma range connect with those in the lower pre-alpha frequencies, larger assemblies are formed, and hence the production of more lucid states of phenomenal consciousness while dreaming or awake.

This cross brainwave coupling effect was disclosed by research on lucid dreaming:

> "The authors review electrophysiological findings in the night sleep of subjects reporting a peaceful inner awareness-witnessing/transcendental consciousness during dreaming and deep sleep and the implications for lucid dreaming research. Findings included EEG tracings of theta alpha (7-9 Hz) simultaneously with delta during deep sleep stages 3 and 4, decreased chin EMG, and highly significant increased theta2 and alpha1 relative power during stage 3 and 4 sleep as compared to controls. The authors discuss alpha synchrony during witnessing deep sleep and gamma during lucid dreaming" (Hobson, et al, 2010, 28. Italic and underlines added).

It revealed early on that it is a conjunctive state that correlates with rapid eye movement (REM) involving the theta and beta-gamma range of frequencies, called a "paradoxical state."

Subsequent research revealed that the faster gamma frequencies were also involved. Frequencies in the gamma-theta state, as I have proposed, would contribute to lucid dreaming.

It is lucid because the Self as observer maintains its personal identity in the midst of the dream scenario. This strongly suggests that the lucid state is facilitated by the attentional frequency operating at a rate that is able to bind and entrain contexts modulated in the frequency of theta into states of phenomenal consciousness.

It is proposed that non-lucid dreams occur with the conjunction of beta in the theta rhythm, while lucid ones would take place in conjunctions with the higher frequencies of gamma.

It is also conceivable to conclude that the gamma frequency is not just coexisting with theta, but actually entrains a portion of its information into an amplified and more resolute state. These lucid states have a "real life" resoluteness and brightness not seen in non-lucid dreams.

The inclusion of the gamma frequencies, being the "intentional" end of the spectrum, appears to contribute to the experience of a "separate" or "dissociated" identity in the context of the dream scenario. This sense of separateness is not due to the rejecting effects of destructive interference but is quite likely due to the sharpness of the resolution where borders are more clearly defined.

If the Self is the amplifying agent it will maintain its identity, whereas if the scenario is the amplifying force the Self will be absorbed into its context (a.k.a. attachment, identification).

Furthermore, in following the progressive dynamics of this rationale, it is also not inconceivable that these gamma modulations on the theta, and quite possibly in the delta range, are what facilitate nonlocal communication (telepathy, distant healing, etc.) and out-of-body states (OBEs, "higher being bodies") of being and consciousness.

According to this rationale, the slower frequency is more field-like and constitutes a modifiable field of being-and-cognitive potentials that are connected with and amplified into awareness by the faster frequencies, especially in the gamma range.

By altering the theta frequency to its rate it forms, by entraining "force," the condition of phase in which the conjoined information pulses in synchrony as the cognitive vortex gyrates outwardly.

While modern neuroscience implicates this passive-active balance in their discovery of frequency "cross coupling", the notion that the couplings could occur in ratios that are conducive to higher states of consciousness is not mentioned in the observations of these scientists. However, their work is quite valuable and highly suggestive.

On the other hand, the ancient seers knew that there was a conjunctive auspiciousness to these higher states, and in the transformative tradition it was often referred to symbolically as a "marriage," which psiologically implicates the conjunction of opposites.

In almost every case the seers held that they needed to enact a state of sustained balance in which the phenomenal Self is equipoised between an active phase and an opposing receptive phase.

At the cellular level the oscillation is periodic between the two values (positive and negative). Hence, there's a continuous process of charging and discharging, and between these two extremes, there is a brief period in which the cell recharges, and when it reaches a balance of the charges, it is referred to as capacitance. The postulate has been presented that during the capacitive-balance phase the cell becomes still for a brief period, and thereby attracts the life force, which is sent forth by the discharge to perform metabolic work.

Gurdjieff referred to this dynamic as the "law of three," being mediated by a "neutralizing force."

While Gurdjieff (and the transformative literature in general) allude to attracting this subtle energy from external as well as internal sources (i.e., prana, chi, hydrogens, od, orgone etc.), precisely why and how this occurs in the equilibrated state has until now remained a mystery, to which the phrase "Be still, and know I am God" points but cannot of itself resolve.

Why, we might ask, is the equilibrated state so auspicious for drawing in the life force, and, moreover, how does this occur?

Serial Versus Multi-faceted Psiloton Processes

When we look at the Psiloton graphic we are apt to notice that the process is not only a series of linear steps, but also a nonlinear integral enfoldment of the entire sphere that repeats in millisecond+ time frames. It is, therefore, spatial and temporal.

As it flows in and out of itself, the self-referentiality is creating emergent effects from the MCs of its inherent and acquired nature. It is not by any means "static" or haphazardly erratic. It quite rationally instantiates chaotic and orderly interactions that go towards supporting the objectivity and subjectivity of the system to facilitate cognition with objective and subjective parameters.

TAPPING OF "FREE ENERGY"
AS THE LIFE FORCE

Clues on the science of how this paradoxical oscillatory activity occurs have been emerging since the days of Nikola Tesla (1856-1943) to provide "free energy." A compelling clue of how the psycho-organism taps into this subtle energy stream is quite surprisingly found in the technical literature of the "free energy" movement, about which the late physician Hans Nieper wrote extensively (1983).

He wrote of this energy in terms of neutrinos (subatomic particles with a mass close to zero). However, as the work of Meyl (1996) and others attest to the singular nature of waves and vortices, neutrinos can be understood to be ultra-fractal equivalents of vortical progenitors prevalent throughout the cosmos.

Runs Above Unity

From a technical perspective, what constitutes "free energy" is when the energy input to run a machine is less than the output. Since the output exceeds the input it is proposed that the excess can be fed back to the generator in a kind of "perpetual motion machine." It is said to run "above unity."

The only thing that would stop it is the wear and tear of the parts by mechanical friction. Nevertheless, many experimenters have continuously run generators for months and years with the sole use of an antenna. You can see why the power-producing moguls would be against this due to loss of profits, despite the dire effects on the planet.

The Same Energy in Fractal Dimensions

We can rely on this source because the energetics of the universe do not appear to differ from one system to another, except in terms of scale or gradients, whether "animate" or "inanimate." The same energy that activates a motor also works with biological systems like an amoeba, neurons, or "automata" in computer algorithms devised by Stephen Wolfram.

> "Wolfram Mathematica, conceived by Stephen Wolfram and developed by Wolfram Research of Champaign, Illinois, is a modern technical computing system spanning most areas of technical computing—including neural networks, machine learning, image processing, geometry, data science, visualizations, and others. The system is used in many technical, scientific, engineering, mathematical, and computing fields" (Wikipedia: Wolfram Mathematica).

What characterizes the free-energy movement is the invention of generators that are designed to resonantly tap the alleged enormous amount of scalar energy (a.k.a., zero point energy consisting of longitudinal waves that sport a volume with no direction) from the spatial surround and convert it into electricity (a movement with a directional force). The scientist Meyl has, in rekindling Nikola Tesla's genius with actual working models, described this life-giving-scalar energy in vortical terms (1996).

Some of the key principles employed to tap this energy come from the discoveries of Tesla, particularly from the design of his "self-cancelling coil" technology.

Tesla's Bifilar Self-Cancelling Coils

The notion of "self-cancelling" can be explained by the principle of interference, particularly the destructive effects produced by the merging of similar electromagnetic waves moving towards one another in opposite directions due to the antithetical crisscrossing of two wires wound around an iron rod.

The net result is zero ($>0<$), a low pressure point called a node, where the electromagnetic field as Hertzian waves has been cancelled (somewhat like the effects of a Faraday cage), allowing zero-point energy (the life force?) to enter that space and be resonantly drawn in (since nature abhors a "vacuum") and converted to produce electricity.

ACCESSING QUANTUM FIELDS WITH SELF-CANCELLING COILS FOR HEALING

The FCCP extends to anywhere in the negative-space-time universe where information is conserved as elemental and/or electrically-encoded memory constructs in the ambiance of strange attractor basins embedded in subtle fractal complexities.

Within this broad framework, physicists have become aware of the existence of anomalous subtle "energy fields" with attributes that cannot be explained by the classical equations developed by James Clerk Maxwell (1831-1879) used by electrodynamics, or by quantum equations developed by Erwin Schrödinger (1887-1961).

Free Energy?

These unorthodox fields were researched and associated with "free energy" as an attribute of quantum fields involving superconducting properties where there is zero resistance between the elemental "objects" to interact and communicate. This is a property I also attribute to negative space-time akin to David Bohm's "implicate order" (1980), although I view it more as a superfluid, similar to liquid helium-4.

> "A superfluid is a state of matter in which the matter behaves like a fluid with zero viscosity. The substance, which looks like a normal liquid, flows without friction past any surface, which allows it to continue to circulate over obstructions and through pores in containers which hold it, subject only to its own inertia" (Wikipedia: Superfluid helium-4).

Spin Forever

An object, such as a vortical soliton wave, spinning in a frictionless superfluid could theoretically spin "forever." Vortices were found to exist in other superfluids.

For example:

> "In the 1950s, Hall and Vinen performed experiments establishing the existence of quantized vortex lines in superfluid helium. In the 1960s, Rayfield and Reif established the existence of quantized vortex rings. Packard has observed the intersection of vortex lines with the free surface of the fluid, and Avenel and Varoquaux have studied the Josephson effect in superfluid helium-4. In 2006 a group at the University of Maryland visualized quantized vortices by using small tracer particles of solid hydrogen" (Ibid).

Using Tesla's paradoxical windings that cancel the electromagnetic field and produce volumetric scalar waves appears to be the essence of his discoveries.

Superfluid, a Boson-Like Ectoplasm

Ectoplasm is, in spiritist circles, a supernatural viscous white substance that is said to exude from the body of a medium (though this is not always the case) during a spiritualistic trance and form the material for the manifestation of spirits (see Greber 1932/2007; Kardec 1857/1996). This appears to be similar in principle to the activity of certain white blood cells.

If you've ever used a dark field microscope to watch the activity of phagocytes (a type of immune cell within the body capable of shape-shifting to engulf bacteria and other small cells and particles) projecting, rather more like growing, and wrapping its "pods" around a foreign substance, you might think you were watching the ectoplasmic plasticity of a spirit manifesting before your eyes.

The formation of the superfluid has affinity to the formation of a massless Bose-Einstein condensate. This is made obvious by the fact that superfluidity occurs in liquid helium-4 at far higher temperatures than it does in helium-3. Each atom of helium-4 is a boson particle, by virtue of its zero spin, which I interpret to exist in negative space-time. While it lacks spin-based animation it is nevertheless, according to the postulates developed in this book, self-referentially animated, and may be the key aspect in the molecular self-referential thinking process behind phagocytosis (how a killer white cell knows how to target a foreign substance).

Helium-3, however, is a fermion particle, which can form bosons only by pairing with itself at much lower temperatures, in a process similar to the electron pairing in superconductivity (quantum entanglement?).

A QUANTUM ENERGY HEALING MODEL

This property has been also found to exist in biological systems. Glen Rein, the director of the Quantum Biology Research Lab, in Miller Place, NY, writes:

> "Recent findings in biology indicate that certain bio-molecules act as superconductors and biological systems in general exhibit *nonlocal*, global properties, which are consistent with their ability to function at the quantum level" (1998, 16. Italics added).

This implicates that superconducting quantum fields exists endogenously within the framework of organic life.

Rein incorporates the findings of a bio-energy field into his "Quantum Energy Healing Model." He proposes that it consists of three levels nested within one another involving at least three kinds of energy contained in:

> Classical electromagnetic (EM) force fields.
> Potential fields identified as the bridge between the classical EM force fields and the quantum "subtle energy fields.
> Quantum fields, identified as "subtle energy" fields (presumably containing the "life force").

These three fields are nested (i.e., fractal embedded) within one another, such that they:

> "Can be defined using Bohm's model (1980) of the implicate order, which is *embedded within the explicate order.* In this case classical EM fields exist at the level of the explicate order, which has embedded within it the potential field" (Ibid, 17. Italics added).

According to the Bohmian hypothesis the implicate order could also be characterized as nested attractor basins of progressive fractal subtlety:

> "Composed of a series of levels each embedded with the next, where each level is increasingly more subtle and fundamental" (Ibid, 17).

The quantum level is, therefore, the most fundamental, which is considered by Rein with the common term "subtle energy," that eludes the measuring instruments of scientists.

> "Even Einstein himself used the term subtle to refer to energy that could not be measured" (Ibid, 17).

Quantum as Spiritual

Rein proposes that because experiments reveal that the healing potential is more evident when the quantum subtle energy field is tapped, he then conceives of this "most fundamental level" in the implicate order as being "spiritual," which:

"Cascades into the outer layers of increasing energy density eventually reaching the electromagnetic domain. Thus *quantum fields act as a bridge between the higher dimensional energies of the spirit and classical EM field*" (Ibid, 17. Italics added).

Therefore, healing with spiritual energy occurs by infusion of energy from this external source by dint that it:

"Resonates [note the intimation of the principle of "interference"] with the level in the bio-field according to how subtle it is" (Ibid, 17. Bracket added).

Moreover, healing (having an etymological relationship to becoming "whole") can also occur:

"Through internal sources of energy generated from the individual *in a meditative state of consciousness*" (Ibid, 17. Italics added).

Tapping the "Spiritual" Field With Self-Cancelling Coils

Healing, according to Rein's model, has compelling correlation with the theta band of frequencies discovered by Beck (1978) in a wide range of "healers", while healing other people and also with the hypothesis being developed in this book: that conjunctive states of consciousness at all levels of the EEG have the same self-cancelling properties to access higher levels of energy and information in the FCCP across the entire spectrum. According to Rein:

"Relatively little attention, however, has been given to the role of endogenous (internal) EM fields or their role in the body's remarkable self-healing capabilities" (1998,16).

However, EM fields, regardless of their important roles in biological processes, may not be the most fundamental negentropic control systems.

"The concept that these local EM fields might be part of (a subset of) a more fundamental, global bio-energy field inside the body has not been incorporated into mainstream bio-electro-magnetic thinking. Nonetheless, this idea has become a basic tenet in the new emerging field of energy medicine, where it is proposed that such endogenous fields regulate biochemical processes, where changes in the bioenergy field precede the physical/chemical changes that manifest as disease, and where early diagnosis and treatment of disease can be achieved by field" (Ibid, 16).

One of the most outstanding pieces of Rein's research is how he was able to tap these subtle bioenergy fields using Tesla's self-cancelling-coil technology, which ties us back to Meyl's EM waves converting into vortices that account for scalar waves.

"It has been previously proposed that quantum fields can be generated (in combination with potential fields) from [Tesla] self-cancelling coils with unique winding geometries. These coils were used to demonstrate that quantum fields inhibit neurotransmitter uptake into nerve cells, stimulated the growth of human lymphocytes (white blood cells), and alter the absorption of UV light by water treated with quantum fields. In all cases the quantum fields produced larger effects than classical electromagnetic force fields" (Ibid, 16. Bracket added).

This new evidence is used by Rein as the basis for his Quantum Energy Healing Model, which introduces the concept of a bio-field composed of layers comprising force fields, potential fields, embedded (as fractals?) within one another.

"The model further proposes that healing information [quite likely coming from the FCCP?] originates from a higher dimensional source [preserved as "memory" in negative space time?] and is transformed [by outgoing vortex?] into biologically usable electromagnetic energy as it propagates through the layers of the bio-field" (Ibid, 16. Inquiring brackets added).

REICH'S ORGONE BOX

Although by a different process, the "orgone box" developed by the psychiatrist Wilhelm Reich (1897-1957) appears to also block the electromagnetic field via a "Faraday cage" effect, consisting of a wrapping of metallic mesh (e.g., steel wool) as one of the outer layers, and, thereby, blocking EM waves and allowing the subtler more fundamental life force he referred to as "orgone" to filter through. It appears that orgone is the EM wave morphed into vortical-soliton scalar waves at the near field of the box and inside the unit.

Reich supposed that sexual energy consisted of the life force distributed throughout the universe, and that it can be collected and stored in the "orgone box" (e.g., like a capacitor or battery) for therapeutic use.

Here, the capacity to trap and conserve orgone (a.k.a. SuperLight, chi, prana, od, etc.) is attributed to the infinite looping of the vortical soliton scalar wave, which is drawn into the biological system via the equilibrated capacitance of cells.

CAPACITANCE: BIOLOGICAL BALANCE

The only biological equivalent that intimates this state of equilibration is when the nerve cell in particular, and cells in general, separate the positive and negative charges (alleged to be ions of potassium and sodium, but challenged by the "fourth state of water" hypothesis of Pollack 2001) in a state of capacitance, like a charged battery poised to discharge (as proposed by Yogananda and Gurdjieff).

However, this resting state of the cell only indicates to the researcher that it has become polarized and poised to fire again (depolarize). According to the standard biological paradigm, there's nothing auspicious or advantageous about this state, other than a recurring, though important, mechanism of "action potentials."

The idea that a broader state of cellular capacitance can be intentionally more broadly evoked and sustained as a potential force throughout the psycho-organism is missing from the medical as well as the psychological literature.

ORGONE ARK OF THE COVENANT?

The Ark of the Covenant is the wooden chest that contained the tablets of the laws of the ancient Israelites. Carried by the Israelites on their wanderings in the wilderness, it was later placed by Solomon in the Temple at Jerusalem. No one actually knows, though various claims have been made, of its current whereabouts.

Speculation has suggested that the design of the Ark of the Covenant, with metal overlays on acacia wood as a nonconducting insulator, would have all the components of a "capacitor" which would gather and store "spiritual" energy.

A capacitor is a device used to store an electric charge (like a battery), consisting of one or more pairs of conductors separated by an insulator. The positive and negative charges are separated on either side of the insulating material and are discharged upon making contact with conducting material.

However, if we closely examine the "common" design of the Ark drawn in many many illustrations we notice two golden cherubim with their wings pointing towards each other, tip-to-tip, very suggestive of the antithetical (>0<) condition for self-cancelling the electromagnetic field picked up by the overlay of highly-conductive gold around the entire apparatus. This accumulation of the EM field on the surface could deliver a fatal shock to anyone not "prepared" to touch it.

Like Reich's orgone box, this would allow the valuable life force (a.k.a. "vital fluid") to filter through and be contained by the insulating wooden walls inside in the box, not any different in principle to bio-cellular membranes that would act as mini-Faraday cages (see Klee 2014), allowing the life force to enter and accumulate, leaving behind an electromagnetic field as intense as a bolt of lightning (Sato 2007). However, it is not known if this field has a magnetic component greater than the electrical one.

By orgone having direct contact with the tablets, their context would be imbued and enlivened as a spiritual force that would resonate with those practicing the commandments. Of course, this connection would occur at a very subtle fractal level, say, in the Z4 region of the Psiloton.

CONJUNCTIVE STATES OF BEING & CONSCIOUSNESS

We can extend this insight to propose that all states of consciousness share this conjunctive process to some degree, such that, for example, the Maharishi's "alert hypo-metabolic" paradox can be applied to the entire EEG spectrum, as an indicator of degrees of contextual neutrality.

We learn from this that attention can be a selective psycho-accelerant of memory constructs in all levels of the FCCP in a dualistic manner (either this or that) and it can also select both to generate a broader involvement or a cancelling effect.

This cancelling generates a node through which a finer level of energy, a gradient of the life force, can be drawn in to invigorate another level of being and consciousness to process energy and information at a faster rate, thereby facilitating a transformative entrainment effect.

Hypo-metabolic resting states progressively lead towards the proposed negative space-time (a more space-like dimension somewhat "free" of temporal constraints, where instantaneity is progressively the "norm") of the non-conscious field, while the hyper-metabolic ones incrementally occur as a mirror image in the proposed positive time-space (a more time-like dimension manifesting objectively with a signature frequency and span of time that, as Einstein averred, prevents everything from happening at once) in the opposite direction.

When the alert band conjoins with any of the pre-beta part of the spectrum, more intentional states of being and consciousness can be enacted while in these conjunctive states.

This is supported by the psychological literature as "cross-frequency coupling" referred to previously. With this in mind, the EEG spectrum can be modeled as a progression of conjunctive bands of being and consciousness, in Fig. 22.

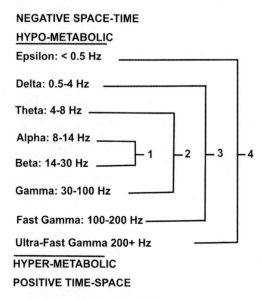

Figure 22: Conjunctive states of Being & Consciousness based on the parameters of the EEG.

It's important to note that all cells of the organism use this conjunctive auspiciousness on entering the equilibrated state of capacitance to draw in the life force. However, this occurs in an aperiodic synergistic manner that prevents an all out synchrony conducive to the Wholistic Response.

Considering that all brainwave frequencies are to some extent present in every EEG measurement as a vortical spectrum, we need to address each state as a moment in which the frequencies cross-couple in a balanced manner to progressively produce larger and more encompassing assemblies via the 1-to-7 cognitive cycle.

Accordingly, each conjunction represents a predominant state of phenomenal consciousness that incorporates all the conjunctive levels that precede it. These would be the "levels" in which the Self turns attention inward to create a F-F cycle with the non-conscious domain, the FCCP that progressively expands as the EEG conjunctively transitions from gamma+ towards epsilon, from locality of the outer world merged with the non-locality of the inner world.

Symbolism Of Conjunctive States

From the foregoing we can quite readily tell that the two symbols are portraying polar equilibration and self-cancelling effects based on the principle of destructive interference. See Figure 23.

Merging the Inner with the Outer

Earlier it was explained how the brain is divided into two broad networks: the extrinsic network and the intrinsic or default mode network (DMN). The extrinsic portion of the brain is activated when individuals are focused on external tasks, like playing ping pong or replacing a light bulb. On the other hand, the DMN turns on when people reflect on matters that involve themselves (autobiographically) and their personal emotions.

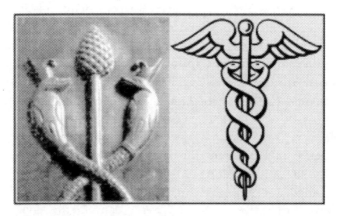

Figure 23: Staff of Osiris (Left). This pine cone staff in the Egyptian Museum, Turino, Italy is a symbol of the solar god Osiris. Modern depiction of Caduceus (right) as a symbol of commerce and medicine.

Interestingly, the two networks typically operate in an either/or fashion such that when one is on, the other is to some degree suppressed. They rarely work together at the same time. Like a seesaw, when one is up the other is down. This allows individuals to concentrate more easily on one task at any given time, without being diverted by distractions like daydreaming.

However, research beginning in 2008 with longstanding Buddhist meditators (by Dr Zoran Josipovic, an adjunct professor of New York University and a Buddhist monk himself), found something different. Using a functional magnetic resonance imaging (fMRI) machine to study metabolic blood flow to different regions of the brain he discovered:

> "...that some Buddhist monks and other experienced meditators have the ability to keep *both neural networks active at the same time during meditation* - that is to say, they have found a way to lift both sides of the seesaw simultaneously" (Danzico 2011. Italics added).

Dr Josipovic believes that this ability to simultaneously evoke the internal and external networks may be responsible for the harmonious feeling of oneness these monks and long-term meditators experience with themselves and their environment.

The neural and other pathways that mediate this paradoxical effect appear to be those responsible for "being in the zone" as well as the effects produced when "dividing attention." It is known that turning attention inwardly shifts the mental metabolism towards the parasympathetic mode, while including attention of the outer environment shifts the mental metabolism towards the sympathetic mode, a paradoxical state I have equated to the Maharishi's "hypo-metabolic alert" state and to Jung's "conjunction of opposites."

In that conjoined state the wave nature of the mind appears to undergo a self-cancelling effect due to destructive interference (the clashing of out-of-phase incommensurables), and it has been hypothesized that this to some degree cancels the contextualized configuration of the energy, leaving behind the experience of the "pure potential" of the un-configured-ground state of being (i.e., the PUF). This is the state in which the mind is "not made up." In Zen it is said one goes from "zero to hero."

While Dr Josipovic discovered the pattern using fMRI detection of metabolic blood flow to both networks instead of one or the other, the conjunction has also been detected using the EEG.

Conjunction of the EEG

Antoine Lutz and associates compared the EEGs of long-term Buddhist practitioners and non-practitioners of meditation. Both used the meditative technique of unconditional loving-kindness and compassion, described as an "unrestricted readiness and availability to help living beings".

> "This practice *does not require concentration on particular objects, memories, or images,* although in other meditations that are also part of their long-term training, practitioners focus on particular persons or groups of beings. Because "benevolence and compassion pervades the mind as a way of being," this state is called "pure compassion" or "*nonreferential* compassion" (Lutz, et al, 2004. Italics added).

Note that "pure compassion" has a more generalized aspect that may include particular individuals, yet provides all the attributes for the formation of large assemblies when thinking of, say, "humankind" as a "whole."

The results showed that the long-term practitioners displayed a conjunctive ratio pattern between the high oscillatory gamma activity and the slow high-delta-theta activity.

> "In addition, the ratio of gamma-band activity (25–42 Hz) to slow oscillatory activity (4–13 Hz) is initially higher in the resting baseline before meditation for the practitioners than the controls over medial frontoparietal electrodes. This difference increases sharply during meditation over most of the scalp electrodes and remains higher than the initial baseline in the post meditation baseline. These data suggest that mental training involves temporal integrative mechanisms and may induce short-term and long-term neural changes" (Ibid).

Conjunction of the Autonomic Nervous System

Back in the early 1970s a model of the physiological processes involved in meditation focused, almost exclusively, on the two branches of the autonomic nervous system: the sympathetic and parasympathetic (Gellhorn & Kiely, 1972. NIB). The model suggested that intense stimulation of either branch could lead to spiritual states.

A number of subsequent studies showed a predominance of resting physiology of the parasympathetic system. This made much sense since meditation is often associated with decreased heart rate and blood pressure, decreased respiratory rate, and decreased oxygen metabolism (Travis, 2001. NIB). However:

> "A recent study of two separate meditative techniques suggested a mutual activation of *parasympathetic* and *sympathetic* systems by demonstrating an increase in the variability of heart rate during meditation (Peng et al., 1999. NIB). *The increased variation in heart rate was hypothesized to reflect activation of both arms of the ANS. This notion also fits the characteristic description of meditative states as involving a sense of overwhelming calmness as well as significant alertness.* Also, the notion of mutual activation of both arms of the ANS is consistent with recent developments in the study of autonomic interactions (Hugdahl, 1996. NIB)" (Newberg, et al, in Paloutzian 2005, 208. Italics added).

These gross neural effects appear to manifest as a cascade of subtler EM and biochemical processes that elicit a strange kind of equilibrium. (See: "Correlation between Pineal Activation and Religious Meditation Observed by Functional Magnetic Resonance Imaging - Chien-Hui Liou.)

LOCALITY & NON-LOCALITY OF THE FCCP

With each transition towards epsilon-ultra-fast gamma, the extent of the FCCP and access to it expands exponentially to include the entire cosmos (the notion of nonlocality and quantum superposition in physics). This greater access to the FCCP poses a transformative principle, which can be thought of as magnitude. See Figure 24.

Conversely, as the transition shifts toward the alpha-beta conjunction, the FCCP becomes more limited and local. In the more localized state the psycho-organism is practical: it doesn't need cosmic consciousness to peel carrots (though this is not necessarily an either/or issue).

Interval Between Cause and Effect Shrinks

As the faster frequencies of the EEG progressively conjoin with the slower ones, the interval between "cause" and "effect" is reduced to the point where "desire" and "manifestation" appear "instantaneous." In these ultra-fast transactions, giving the impression that no "time" transpires, transmission and reception become projective and prophetic.

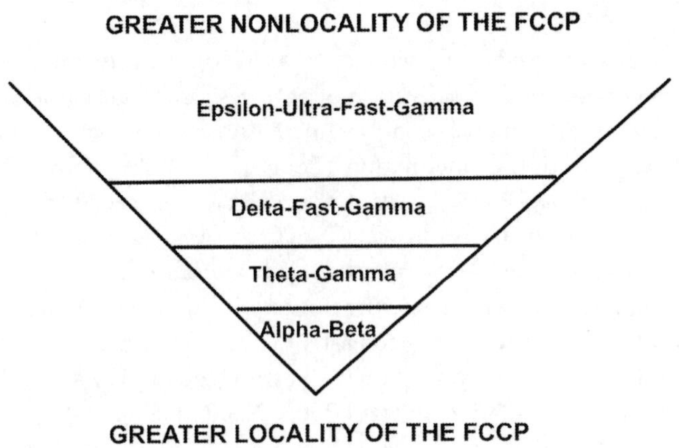

GREATER NONLOCALITY OF THE FCCP

Epsilon-Ultra-Fast-Gamma

Delta-Fast-Gamma

Theta-Gamma

Alpha-Beta

GREATER LOCALITY OF THE FCCP

Figure 24: EEG Correlates to Locality/Nonlocality of the FCCP.

Degree of Relaxation

It is assumed from the research of Montagnier that a broader memory field is distributed by the electromagnetic pulsation of the heart wave throughout the organism into the fascia. Hence, the more relaxed the practitioner is during the enactment of spiritual exercises, the greater will be the reach into the physical substrate, in this case water, to be activated into subtler levels of being and information processing.

The Presence of the Past

However, it is important to note that we are always immersed in the negative space-time of the FCCP where the "presence of the past" always exists (see Sheldrake 1989). We can say that it consists of the "stuff of space." Hence, the nervous system is designed to sculpt moments of presence from this enormous informational vastness.

MANIFESTATION OF A "THIRD THING"

We now have a model—conjunctive states of being and consciousness (CSBC)—to understand how transformative systems of psychology aim to make "the unconscious conscious," as Freud, Jung, Yogananda, the Maharishi, and Gurdjieff, amongst many others, have averred.

By conjoining the faster frequencies with the slower ones, we intimate Jung's "union of opposites" and the wherewithal of "creative imagination" as the means to subjectively modify one's objective reality in relationship to the world.

For Jung the concept of the psyche includes a transcendence of opposites. In this regard he writes:

> "The confrontation of the two positions generates a tension charged with energy and creates a living, third thing—not a logical stillbirth in accordance with the principle *tertium non datur* but a movement out of the suspension between opposites, a living birth that leads to a new level of being, a new situation. *The transcendent function manifests itself as a quality of conjoined opposites**" (in Buser/Cruz 2015, 97. *Italics added).

Furthermore:

> "But if a union is to take place between opposites like spirit and matter, conscious and unconscious, bright and dark, and so on, it will happen in a third thing, which represents not a compromise but something new, just as for the alchemists the cosmic strife of the elements was composed by the stone that is not stone, by a transcendental entity *that could be described only in paradox*" (Ibid, 97-98. Italics added).

The CSBC model involves the interaction of the entire EEG spectrum in which brain waves modulate one another according to the principles of linear-and-nonlinear wave interference, and in keeping with Freeman's "mass action."

Such interactions involve the coupling and vortically integrating the different frequency bands, which create our diverse moments of contextual and non-contextual states of being and phenomenal consciousness.

The equivalent of Jung's conjunction of opposites is intimated by two behaviors: the intentional iteration of a thought, image, or activity, and maintaining it as the organism transitions towards sleep-zone in delta; hence, the development of the 4th state of consciousness, and the forming of a master habit via the "Holy" 1-to-7 cycle.

Let's now take a closer look at each level of these conjunctive states.

THE FIRST STATE: ALPHA-BETA

The alpha-beta conjunction is the most predominant state of functioning where people are somewhat alert and relaxed. This is the state meditators find themselves in when they begin to practice while avoiding distractions picked up by beta.

The beta frequency is what we use to scan the environment for OOPEs. On finding them it accelerates just enough to extract a response from memory constructs with the alpha-signature encoded in the FCCP.

When unable to find OOPEs the center becomes bored, and it will generate imaginary scenarios or prompt one to perform acts (e.g., like working on a cross-word puzzle) that will keep it "awake."

If this does not occur, the mental system will go into permutation mode to scan the FCCP for something to dwell on. If that fails it will go to sleep.

However, if one relaxes the musculature into a tonic state and keeps the eyes open to observe the surround, the first state of transformation can be achieved. If sustained for a period of time, say for at least five minutes, the system will naturally transition to the theta-gamma level, and imagery will appear as the mind scans the FCCP for a response. By not responding to the imagery the system equilibrates and settles in the theta-gamma level.

Interval Between Cause and Effect is Largest

In this state there are many competing forces between a stimulus and a response. As a consequence, the gap between desire and manifestation is the largest of all the states of being and consciousness.

THE SECOND STATE: THETA-GAMMA

Typically, the second state is evoked when the mental system in the first state encounters a critical/novel event that it cannot process with the PSQ. In this case, the mental system cancels and connects to the influx of the theta-gamma conjunction in search of a solution.

In transformative systems this forms the basis of methodologies that use critical OOPEs, like shocks and confusion, akin to the methods the psychiatrist Milton Erickson used to induce his patients, or, hyper-auto-stimulation to befuddle, particularly the vibratory boundaries of the intellectual center and evoke this deeper connection to the non-conscious domain in the FCCP.

In this state the mental system becomes very sensitive to inputs by the observer or from an external source (e.g., hypnotist, subliminal messages from the environment).

People with lots of skepticism and mistrust have difficulty being comfortable in this zone, and tend to hover in low-alpha and high-beta frequencies. These represent the bulk of low-induction subjects.

If one maintains the equilibrated state in this level, it will automatically connect to the next higher zone of influence, which puts one in the delta-fast-gamma non-conscious domain.

Interval Between Cause and Effect Begins to Shrink

In this state there are not as many competing forces between a stimulus and a response. As a result, the gap between desire and manifestation begins to shrink as the practitioner begins to sense some control over personal events.

Mytho-Symbolic Language

The language of this state is mainly mytho-symbolic, simply because of the magnitude of anthropomorphic information it processes (a.k.a., the "astral" domain) from the repertoire of the FCCP in the mode of "spatial logic." This generates the mythic stories of the culture.

Theta: the Healing Zone

Studies done by Dr. Robert Beck (1978) revealed that the theta frequency of 7-8 Hz predominated in the EEGs of a broad sampling of sensitives performing a variety of "paranormal" activities and healing. The sample ranged from: charismatic Christian faith healers, authentic Hawaiian kahuna, practitioners of Wicca, Santeria, radiesthesia and radionics, seers, ESP "readers," and psychics. He quite unexpectedly found that:

> "A striking number of these authentic "sensitives" exhibited a nearly identical EEG signature when reporting that they were in their "working" state of consciousness, a state persisting from one to several seconds intermittently during the "altered state of consciousness" phase of the paranormal activity. The signal appeared to be an almost pure sine wave of up to 25 µV [microvolts] (read monopolar, frontal to occipital, midline) and 7.8–8.0 Hz in frequency [on the "alpha-theta" borderline]" (1978, 48. Bracket added).

Marching to the Same Cosmic Drummer

It was interesting how Beck found a common electromagnetic pattern amongst the diverse disciplines, even though each practitioner denied the authenticity and effectiveness of the other systems. Beck states:

"[A]lthough the subjects were practising opposing disciplines, and came from totally disparate teachings and held opposing viewpoints, and would barely acknowledge the existence or authenticity of practitioners outside their belief systems— they all appeared to be marching to the sound of the same "cosmic drummer" when in an altered state of consciousness" (Ibid, 48).

To confirm his findings, Beck replicated the experiment in a number of different sites thousands of miles apart, and came up with the same results. This caused him to wonder how:

"[T]hese exceptional people, unknown to one another, were falling into the same "cosmic consciousness" mode, what or where was the signal to which their brain waves became entrained? What was the "drummer"? Was there a "clock" or "synchronizing" signal present?" (Ibid, 48.)

Theta-Gateway of the Spirit World

Using a special hypnosis technique to reach the hidden memories of subjects, the psychologist Michael Newton discovered some amazing insights into what happens to us "between lives."

In his ground breaking book *Journey of Souls: Case Studies of Life Between Lives* (1994/2003) he presents the record of 29 people who recalled their experiences between physical deaths. Through these extraordinary stories, they consistently reveal:

- How it feels to die
- What you see and feel right after death
- The truth about "spiritual guides"
- What happens to "disturbed" souls
- Why you are assigned to certain soul groups in the spirit world and what you do there
- How you choose another body to return to Earth
- The different levels of souls: beginning, intermediate, and advanced
- When and where you first learn to recognize soul mates on Earth
- The purpose of life

While these fascinating inner journeys have been with Western culture since the Eastern transformative traditions began to enter the West, and particularly when the deep-trance medium Edgar Cayce entered the scene to invigorate "holistic forms of healing," here I just want to point out that access to these states requires taking the subject into the theta-gamma (2nd state) or event the delta-fast-gamma (3rd state) realms of the FCCP. In his subsequent book *The Destiny of Souls...* Newton writes:

> "I use a systematic approach to reach the soul mind by employing a series of exercises for people in the early stages of hypnotic regression. This procedure is designed to gradually sharpen my subject's memories of their past and prepare them to analyze critically the images they see in Life in the spirit world. After the usual intake interviews, I place the client in hypnosis very quickly. It is the deepening that is my secret. Over long periods of experimentation, I have come to realize that having a client in the normal alpha state of hypnosis is not adequate enough to reach the superconscious state of the soul mind. For this I must take the subject into the deeper *theta* ranges of hypnosis" (Newton 2000, 9. Italics added.)

THE THIRD STATE: DELTA-FAST-GAMMA

The moment the mental system shifts towards the delta-fast-gamma coupling, it interacts via linear-and-nonlinear interference with the memory constructs of the bio-field that contain the intrinsic information of organic life conserved as 2D vortical Psiloton waves, oscillating between .5 and 4 Hz. These waves have been documented to have deep phylogenetic origin in the instinctive center (see Schutter 2012).

It is a well-known fact that deep deltic sleep is the most auspicious state for healing, by dint that there is significant dissociation from the sensory-motor strip in the brain.

Its combinatorial potential is conserved in the DNA molecule, and it forms the basis for the creative evolution of the various organic systems populating the universe.

Healers able to tap this realm in the lower regions of the theta-delta border can attract its enormous healing repertoire, especially when they consciously approach it through theta, as Beck discovered (1978).

Bio-morphic Symbolic Language

The language of this state is mainly symbolic, simply because the magnitude of its intrinsic-bio-information processes from the DNA helix, used in fashioning Life across the organic tree (fractal branching relationships), of the FCCP.

Interval Between Cause and Effect Becomes Evident

In this state the competing forces between a stimulus and a response are at a minimum. Hence, the state has a greater degree of freedom from the two previous ones. As a consequence, the gap between desire and manifestation is close enough to impart a sense of assurance and power to the practitioner.

When Heart & Brain Waves go into Sync

At the 1 Hz delta range, brain waves and heart waves go into resonance (e.g., 60 BPM = 1 Hz) and exchange information.

In the third state one becomes aware of information entering the magnetic field of the heart. Since the heart wave permeates and surrounds the entire body there is also a cosmic connection at the border of epsilon.

THE FOURTH STATE: EPSILON-ULTRA-FAST-GAMMA

The epsilon frequency of less than .5 Hz occurs when the brain is in its most restful and, often, deep-non-conscious state. It is hardly moving. A long slow wave is observed in the EEG, which is held to have affinity to deep mystical states (Thompson 2009). In sleep yoga (see Norbu 1992) it is referred to as the "ground state of being." It also has affinity to the "undground" of the German mystic and theologian Jakob Böhme (1575-1624).

Generally speaking, it is the deepest trance state in which the Self is most consciously "separated" (i.e., vibrationally dissonant) from its physical mode of being, and takes on the attributes of being essentially spiritual while being in a more subtle body with an equally faster mind, with incredibly powerful transformative powers and expanse of consciousness (by its "closeness" to the isotropism of the PUF).

Activation of the Pineal Gland

When the entire system equilibrates it produces the Wholistic Response in the form of the Presence. By its polar tension, the conjunction ignites the pineal gland to output its oppositional chemical constructs: seratonin and melatonin. This leads it to mediate a "download" from a higher level of order (since a lower level cannot create of itself a higher one, but only become susceptive to the influence of a HI and its energy).

The language of this state is mainly symbolic compressions, simply because of the magnitude of information it processes from the entire repertoire of the FCCP. When these symbols are triggered they unfurl into a more condensed form of specific "reality" that the observing system can, given that it is to some degree tethered to the body, handle.

A Fractal State of "Higher Mind"

While still a "fractal" state of Higher Mind it performs what Higher Mind does, but on a smaller and limited scale. Similarly, it does what lower minds do but at a much finer and faster processing rate. Note that to encompass more swaths of memory in the FCCP, the frequency of the mental system must increase while the organism is very relaxed.

The processing rate and the idiosyncratic context of one's belief system combine to form a signature sub-Psiloton world in the planes/spheres of existence (see chart below), that on a grander refined scale is part of the FCCP.

However, to enter a "higher world or sphere", the state of conjunctive equilibrium is required (note that it may occur fortuitously or intentionally) to "unlock" the gates to other planes.

In the fourth state the quickened-spiritualized mind is able to exchange energy and information with the cosmic aspect of the FCCP, in terms of "worlds" or "spheres," and entities in those negative space-time domains.

One's Place In the Universe

The entire system intimates a developmental scale based on the degree to which the vibratory nature of one's being is in harmony with itself and the social, physical, and cosmic context. This implicates a resonant-relationship to "worlds" defined by their media and the signature rate at which they vibrate in Psiloton envelopes, which on a collective level can be seen as "societies" in the cosmic dimension of the FCCP (embedded in the PUF).

These ideas are found through many different transformative systems, although not quite in the vibratory terms of Yogananda or the Fourth Way. In the Christian tradition they are called "kingdoms" in which there are "many mansions." In Hindu lore they are called "lokas." In much of the New Age literature they are often referred to as "planes."

George Gaskell in his *Dictionary of All Scriptures and Myths* (1960) produces a more comprehensive comparison of the world's major religions and spiritual belief systems going back to antiquity. He found that, regardless of their monotheistic or polytheistic structure, they all were organized into five planes of hierarchical existence. With the hypothesis developing in this book, we could describe these levels in terms of degrees of intelligence, creative freedom, inner-most, and integrality. Accordingly, six models as examples can be offered to reflect these attributes as those being the Highest (**H**) and those being the Lowest (**L**) in the micro as well as the macro planes of all systems:

> **PSILOTON**: **H** FZ0, FZ9, FZ8, FZ7, FZ6, FZ5, …FZ1 **L**
> **GREECE**: **H** Zeus, Apollo, Hera, Hermes, Hades, Hestia **L**
> **INDIAN**: **H** Para-atma, Buddhi, Mannas, Kama, Sthula **L**
> **CHRISTIAN**: **H** God, Son of God, Holy Ghost, Mental Faculties, Lower Emotions, Sensations **L**
> **EGYPT**: **H** Ra, Osiris, Isis, Thoth, Set, Nephthys **L**
> **CHINA**: **H** Heaven, Water, Fire, Metal, Wood, Earth **L**

However, to enter a "higher world or sphere" the state of conjunctive equilibrium is required (note that it may occur fortuitously or intentionally) to "unlock" the gates to other planes.

In the fourth state the quickened-spiritualized mind is able to exchange energy and information with the cosmic aspect of the FCCP, in terms of "worlds" or "spheres," and entities in those negative space-time domains.

All Levels in Worlds Only Differ by their Processing Rates

Therefore, each fractal scale-invariant manifestation has its comparable objective time-space and a more subjective negative space-time. They are all self-similar in this scale-invariant respect (though not identical) and functioning in signature frames of reference. To use a current term, we can say it is a kind of "holographic" unit.

The Higher Can See The Lower But Not the Inverse

As a general rule, a system processing information at faster rates can "experience" a slower one, albeit by entrainment, but not the inverse. It requires two systems operating as similar frequencies to exchange energy and information. So, a person receiving information from a higher level needs to "step it down" to make it accessible to one functioning in a lower processing rate, or entrain the lower system into a higher frequency.

Interval Between Cause and Effect is Indistinguishable

In this state the competing forces between a stimulus and a response are almost nil. As a consequence, the gap between desire and manifestation is the smallest of all the states of being and consciousness. Hence, it is an auspicious zone for "magic" and "miracles" to manifest at will. Anyone endowed with such powers can project thoughts, images, feelings, etc., to others anywhere in the universe.

It is important to note that these projections need to be such that they will not affect the wholistic state of the sender and cause damaging psycho-fragmentation.

Post Deep-Trance Amnesia

Depending on the degree to which the "separation" is defined, the events that transpire are often "forgotten" upon coming out of trance. Usually this appears to occur when the epsilon wave overpowers and dampens the ultra-fast-gamma waves or totally absorbs it.

However, there is a more stable conjunctive state in which the fourth state remains even during normal wakefulness. In yogic lore this is known as samadhi. However, there are two general levels of stability.

The Fourth State & the Two Kinds of Samadhi

These deep fourth states are very suggestive of Yogananda's notion of "suspended animation" (known as sabikalpa samadhi) of advanced yogis, in which all physiological processes are slowed down or held in the capacitive state.

> "In the initial states of God-contact (SABIKALPA SAMADHI) the devotee's consciousness merges with the Cosmic Spirit; his life force is withdrawn from the body, which appears 'dead,' or motionless and rigid. The yogi is fully aware of his bodily condition of suspended animation" (Yogananda 1946, 175).

I suspect that in coming out of this state the yogi is subject to post deep-trance amnesia, as is the case with many deep-trance mediums (see LeCon 1968). Nevertheless:

> "As he progresses to higher spiritual states (NIRBIKALPA SAMADHI), however, he communes with God without bodily fixation [catalepsy], and in his ordinary waking consciousness, even in the midst of exacting worldly duties" (Ibid, 175. Bracket added)

He claims that this ultra-dissociative technique was also known by St. Paul.

> "St. Paul knew KRIYA YOGA, or a technique very similar to it, by which he could switch Life currents to and from the senses. He was therefore able to say: 'Verily, I protest by our rejoicing which I have in Christ, I DIE DAILY.' *By daily withdrawing his bodily life force, he united it by yoga union with the rejoicing (eternal bliss) of the Christ consciousness [the supersymmetry of the PUF].* In that felicitous state, he was consciously aware of being dead to the delusive sensory world of MAYA" (Ibid, 175. Italics and bracket added).

Ecstasy

During the 4th state, degrees of ecstasy (a.k.a. bliss, love, euphoria) are often experienced. These are here understood to be due to profound moments of being in phase (based on constructive interference) with an "object" or "subject" of worship. The duration of such moments may be brief or lasting for months and more. From my own experience it has strong resemblance to the Presence in a state of love and compassion in which the whole of one's Self is vibrating in sync (absorbed) with itself and the universe.

It stems from the expansion of ordinary moments of being happy, pleased, etc., in which the input and the output are merged in a positive F-F psicle.

Edgar Cayce

Cayce, known as the "sleeping prophet," performed most of his enormous amount of work in successfully healing many individuals with prescriptions for each signature identity. He also produced some insights on our spiritual history and some prophesied material.

Like many deep-trance mediums (e.g., Jane Roberts for the entity Seth), he'd go into deep trance to verbalize his findings to a transcriber, and upon awakening he had no recollection of what he had witnessed. If it weren't for the transcriber there would be no substantive record of his work.

Living On Prana From Heaven

Since remote antiquity transformative systems have revealed that life, in all of its variations, is sustained by the life force (i.e., Psilotonic units). It can be derived indirectly from fresh food and liquids; yet, it can be drawn directly from the surround, especially when any/or all modes of the Self are in a state of entropic equilibrium, mimicking a death-like experience (DLE).

One contemporary case of solely living by attracting the life force is that of Prahland Jani (a.k.a., "Mataji" by his followers).

Born in 1929, Jani is an Indian fakir and follower of Jainism and the Hindu goddess Amba. He claims to have consumed no liquids or food since he was eight years old (i.e. for 74 years as of 2010). Claims of this kind are known as inedia, the ability to live without food.

To test his claims, Indian military doctors put him, at 83 years old, under round-the-clock observation during a two-week hospital stay.

During that time he didn't ingest any food or water, and never moved his bowels or passed urine, yet remained perfectly healthy (see Rajshree 2010).

They tested his heart and kidneys, examined his brain functions, and performed a DNA analysis and hormone tests. He only came in contact with liquid when he was gargling and bathing. After the end of the two-week round-the-clock monitoring one of the doctors, the neurologist Sudhir Shah said with astonishment "We still do not know how he survived."

Jani himself believes in a miraculous gift from the goddess Amba Mata, and that he is fed a mysterious fluid called "Amrit" nectar through a "hole" in his palate. While no hole was clinically found, I conjecture that it was a nodal point in the palate through which the life force flowed.

From a technical perspective, the opposing wave forms produce a cancelling effect, which is essentially nodal points of stillness where the up-flowing parallel undulating waves criss cross. These nodal points are like "empty spaces" towards which subtler energies will be drawn (from inner as well as from outer resources), since nature abhors a vacuum (see Müller on the scientific explanation of nodes, in Waser 2004).

Hence, the kundalini (life force) is drawn up as it is attracted by the linear arrangement of nodes that are situated along the equilibrated centers or chakras. The equilibration of the centers prevents the life force from laterally activating the centers and being used by their activity. So in this sense negative feedback enters to prevent the deviation.

By equalizing the oppositional forces, the life force enters to quicken the rate at which the mind processes energy and information (which are actually two fundamental forms of nourishment), and in that transition it transforms into spirit, which exists as a inherent potential of the Self.

In this regard, there is no need for Jani to eat food or drink water, as long as the equilibrated condition remains, which means that his positive and negative energy configurations in the FCCP are in a polarized state of unity, and more importantly, that life is sustained by this force.

However, entering into the fourth state of being and consciousness allows one to tap powers that the modern scientist, influenced by the standard model in physics, finds impossible to take seriously, for instance, the psychokinetic power to levitate things by communing with the spirit world in negative space-time.

Levitation of Heavy Objects by Deep-Trance Mediums

In his wonderful book *The Occult*, (1973) Colin Wilson presents a somewhat complete history of mediums and their spiritual powers to support his theory of a hidden creative force he referred to as "faculty X," the underlying basis of a "sixth sense," which he believed would eventually be the modus operandi of a more evolved human race. He claims that the paradox is that we already possess it to a large degree. But:

> "[We] are unconscious of possessing it. It lies at the heart of all so-called occult experience" (Wilson 1973,10).

Wilson highlights the modern medium Daniel Dunglas Home (1833-1886)—(pronounced Hume)—as an exemplar producer of carefully-studied spiritist events. Though there were many skeptics, Wilson felt that these criticisms were totally unfair.

> "Not only was Home never 'exposed' or shown to have used trickery, but scientific observers repeatedly verified that *he could float through the air and make heavy items of furniture move*. This was not done only on one or two occasions but on hundreds of occasions over some forty years. Home also performed these feats in broad daylight, and with none of the medium's usual paraphernalia" (Wilson 1973, 209. Italics added).

Moreover:

> "Home travelled on to Italy, where the English community awaited his coming with intense expectation. In Florence his powers were stimulated by the scenery and adoration; *a grand piano rose into the air while the Countess Orsini was playing on it, and remained floating throughout the piece. Tables danced, chandeliers gyrated, spirit hands serenaded the visitors on concertinas or shook hands with the sitters, who observed that they felt warm and human*". (Ibid, 542. Italics added).

Saint Joseph of Cupertino

Because the 4th state is a high degree of conjunctive coherence, the molecular and electromagnetic fields are in exquisite equilibrium.

Equilibrated thusly, the organism has the wherewithal to some degree escape the gravitational force of the planet and, as a result may levitate.

Overcoming gravity is hypothetically held, according to the Presence, to be due to a cessation of fractal development during steady-state cycling, given that the input and output are equal, thereby self-cancelling the force. According to this notion, the gravitational force is the result of numerous ingoing vortices micro-extending towards multiple implosions in a concomitant manner.

Note that in the 3rd state this phenomenon is not under the control of the subject, but in the 4th state it is (such as in "nirbikalpa samadhi"). Having it under control, the subject may move about in the air with intention. Such has been the case of Saint Joseph of Cupertino (1603-1663), an Italian Franciscan friar honored as a Christian mystic.

Though it was said he was remarkably not clever, he was prone to intense ecstatic visions and the ability to levitate. Wilson, in his book *The Occult*, (1973) writes this of him:

> "He seems to have been a curious but simple case; floating in the air when in a state of delight seems to have been his sole accomplishment. *The ecstasy did not have to be religious*; on one occasion, when shepherds were playing their pipes in church on Christmas Eve, he began to dance for sheer joy, then flew on to the high altar, without knocking over any of the burning candles. Unlike Daniel Dunglas Home, St. Joseph seems to have been able to control his flights" (Wilson 1973, 237. Italics added).

The control he exhibited involved precise movement through the ambient air, and also the ability to pick up and move heavy objects.

> On one occasion, when he had flown past lamps and ornaments that blocked the way to the altar, his superior called him back, and he flew back to the place he had vacated. When a fellow monk remarked on the beauty of the sky, he shrieked and flew to the top of a nearby tree. *He was also able to lift heavy weights; one story tells of how he raised a wooden cross that ten workmen were struggling to place in position, and flew with it to the hole that had been prepared for it.* He was also able to make others float; he cured a demented nobleman by seizing his hair and flying into the air with him, remaining there a quarter of an hour, according to his biographer; on another occasion, he seized a

279

local priest by the hand, and after dancing around with him, they both flew, hand in hand" (Ibid, 237. Italics added).

Here again, we find the 4th state involved in overcoming fundamental forces in physics. In this most wonderful of states, the cognitive system is capable of nonlocal communication, prophecy, and the feats of anti-gravity. It has all the appearance of spirit over mind, and mind over matter. What would materialistic scientists make of such mystical phenomena? They could in all likelihood dismiss the historical records as a pack of lies or as mass hypnosis without a twinge of conscience. They would most likely cringe at the reported "miracles" of Saint Dunstant of Glastonbury who is recorded to have "changed the position of a [massive stone] church by pushing it" (Wilson 1973, 238. Bracket added).

In support of the claims on Saint Joseph, Wilson argues:

> But the evidence cannot be dismissed; it is overwhelming. His feats were witnessed by kings, dukes and philosophers (or at least one philosopher – Leibnitz, *one of the inventors of the infinitesimal calculus*). When his canonisation was suggested, the Church started an investigation into his flights, and hundreds of depositions were taken. He became a saint four years after his death" (Ibid, 238. Annotated italics added).

Mirabelli: A Grand Master Medium of Brazil

Negative space-time, as postulated, is where universal memory endures as a more spatially oriented existence. Life in positive time-space is constantly breaking down (i.e., entropy) and being rebuilt (i.e., negentropy) from its intrinsic blueprint in negative space-time.

In spiritist circles (a psycho-transformative system) the transformation of memory (spirit) becoming physical is expressed, according to psiological reasoning, as "materialization," and conversely, when it returns to negative space-time it is said to "dematerialize" (see Greber 2007; Kardac 1857/1996).

However, accounts of Life in negative space-time (e.g., "heaven") describe a positive time-space-like existence, except that its laws transcend it by many degrees of "freedom" (see Greber 1932/2007). This supports the intuited notion that all vibratory milieus are self-similar but in signature fractal frames of reference, operating in faster-or-slower frequencies like radio receivers/transmitters.

These transitions are mediated by a pluripotent "substance" referred to as the "life force" (a.k.a., odic force, chi, prana, manna, hydrogens, etc.). Here, the life force equates to the PUF, and its fractal-self-similar constructs in the act of Creation.

It shifts, in millisecond+ time frames, between information existing in spatial or physical formats. There's a kind of "morphic resonance" between the two realms that strongly suggests the "presence of the past" as an enduring interaction with the "habits of nature" (see Sheldrake 1989).

The materialization-dematerializations are mediated by high degrees of mental concentration practiced by "deep trance mediums." This is in keeping with my illustrated rendition of Freeman's cognitive-vortex model (see Figure 2), and the ability to extend attention span for prolonged periods of time.

By focusing on a thought or image in this super-concentrated manner, the medium is able to more explicitly select it from the FCCP, particularly in the range where the "collective unconscious," proposed by Jung, resides in cosmic space (which I have described as the "fourth state of conjunctive consciousness").

While we are alluding to mediumistic powers it must be kept in mind that this cyclic dynamic is responsible for all transformative powers (e.g., siddhis) and acts of creation mediated by various methodologies, like prayer, meditation, ritual, and so on. It can even engender "deep silence" and/or a pure "void-like" space using a "contextless context," like focusing on "empty" space (see Fehmi 2009).

While we are more familiar with deep mediums like the American psychic Edgar Cayce (1877-1945), we're not so versed in the versatile Brazilian spiritualist mediums like Carlos Mirabelli (1889-1951). Upon reading about his PSQ-challenging feats, my understanding of such powers was extended. As the word implies, mediums are conduits of energy and information coming from "higher as well as "lower" worlds," who are able to produce the self-cancelling effects that lead to the 4th state of consciousness.

Whenever Mirabelli entered into deep trance he also was unable to remember what had transpired. What makes him such a fascinating subject to study was his broad psychic powers, observed by many qualified witnesses under rigorous experimental conditions, even in broad daylight, and the precision of the recorded events (e.g., photographs, psychic writings, and anecdotal accounts).

Mirabelli's gift endowed him to act as a speaking and writing medium, a physical medium, an apport (a material object produced supposedly by occult means, especially at a seance) medium, and as a materialization medium. Here are some quite astonishing examples:

> "He is a speaking medium. While in a state of trance, he speaks, besides his mother tongue and several of the local dialects: German, French, Dutch, English, four Italian dialects, Czech, Arabic, Japanese, Russian, Spanish, Turkish, Hebrew, Albanian, several African dialects, Latin, Chinese, modern Greek, Polish, Syrio-Egyptian dialects, and ancient Greek. In his normal state he knows only his native language. While in a trance he holds lectures on subjects about which he as a human being knows nothing. These lectures deal with the fields of medicine, jurisprudence, sociology, political economy, politics, theology, psychology, history, the natural sciences, astronomy, philosophy, logic, music, spiritism and occultism, and literature.

> According to the medium's own statement, *nothing that he utters while in trance comes from him, but from spirits that speak through him and whose names he willingly gives. He calls them his spirit guides".* (Greber 1926, 218-219. Italics added).

As a physical medium, Mirabelli once materialized the spirit bodies of a marshal and a bishop, both long deceased, and both were instantly recognizable to many who had assembled for the seance.

In these events a skull elevates, violins play, billiard balls move of themselves:

> "Once when Mirabelli visited a pharmacy, a skull rose from the back of the laboratory and came to rest on the cash register. Before a gathering of doctors, who lent their names to a deposition, Mirabelli caused a violin to be played by spirit hands. To exhibit spirit control, Mirabelli caused billiard balls to roll and stop at his command" (Ibid, 221).

Mirabelli as an apport medium invokes the spirits to generate motion or production of an object without apparent physical agency. For example:

"From the residence of Pinto de Queiro in So Paulo a revolver that had been locked in a trunk was apported into the residence of a Mr. Watson, after the announcement had been made that this would be done. Furthermore, a picture was apported in broad daylight from Mr. Watson's residence over a distance of several kilometers into the office of an insurance company, where it fell to the floor with a crash, causing immense excitement" (Ibid, 222).

Why not levitate an automobile to mystify the critics:

"Levitation seemed almost to be a specialty of the medium, and *witnesses in broad daylight once observed him levitate an automobile to a height of six feet, where it was suspended for a period of three minutes*". (Encyclopedia of the Unusual and Unexplained. Italics added).

Mediumistic Power May Be Lost

Eventually Mirabelli lost his powers. Prior to this misfortune, Greber, who had discovered spiritism as the authentic way to embrace the original meaning of the Bible, had foreseen that this was likely to happen:

"Mirabelli endeavors to attract only the good spirit world and to serve as its instrument, as proven by his prayers for Divine assistance, but the fact that he also lends himself to seances held only to satisfy the scientific interests of the participants, and in many cases merely their sensationalism, is an error on his part that gives the evil spirit beings a great hold on him. Were he to confine his activities to religious gatherings exclusively, thus devoting his mediumistic talents only to the cause of good, the forces of evil would have no power over him and the low and vulgar exhibitions he makes of himself would never occur. Moreover, his mediumistic power would be maintained, while it is to be feared that it will dwindle little by little if he continues to lend himself as a medium for worldly purposes. The weakening of his odic power by the evil spirits will assume such proportions in the long run that he will fail utterly as a medium, and also lose his physical health entirely and possibly be driven to something even worse" (Ibid, 220).

The Secret of Pyramids and Coral Castle

If deep-trance mediums can levitate a heavy cross, pianos, and automobiles, and shift the position of a massive stone church by pushing it, why not the megalithic materials used in the architecture of the ancient world like Stonehenge and the pyramids?

Pyramids of all sizes have been found throughout the world. The most famous is the Pyramid of Giza in Egypt. However, a rather recent discovery of one greater in size than the one at Giza is the Sun Pyramid found in Bosnia, which is virtually the size of a mountain.

One of the mystifying attributes of all these pyramidal, as well as other megalithic structures, is the question they raise of how these massive stones could be quarried, transported, lifted, and shaped with such precision, supposedly lacking the proper tools (which do not exist today) to do the whole job, that could explain how it was done.

Currently, there are many who believe these ancient peoples had access to an unknown technology, perhaps coming from the assistance of extra terrestrials (ETs) on how to use vibratory methods to raise them (see Bruce 2010, "Acoustic Levitation of Stones"). In other instances a race of giants were thought to be involved.

The spiritualist history, of which so much is utterly suppressed by modern science, as well as by religious institutions and the conventional media, including conventional psychology (Russia, however, may be the exception, see Ostrander/Schroeder 1970), has so many compelling studies that leave little doubt of their veracity that the spirit world exists, appearing to process energy and information at rates that boggle the conditioned mind.

The aforementioned strange levitations and apports may have relevance to understanding how the pyramids and other ancient megalithic structures were built. If communication with spirits can lift grand pianos and cars into the air that remain suspended for a period of time (feats no "trick" can explain), why not the quarrying and lifting of multi-tonnage stones and apport them into far distant places? When many indigenous peoples are asked about the nature of such feats they typically attribute them to spiritual beings or the gods.

An interesting example of a single man performing such prodigious architecture is that of Coral Castle.

Coral Castle is a mystifying oolite limestone structure created by the Latvian American Edward Leedskalnin (1887-1951). It is located in Leisure City, Florida, in Miami-Dade Country.

The structure consists of numerous megalithic stones, mostly formed from coral, each weighing many tons. These are claimed to have been sculpted and built single-handedly by Leedskalnin. However, non-spiritual-intervention theories abound.

> "It is currently a privately operated tourist attraction. Coral Castle is noted for legends surrounding its creation that claim it was built using reverse magnetism [the inversion of positive time-space into negative space-time?] or supernatural abilities to move and precisely carve numerous stones weighing many tons" (Wikipedia: Coral Castle. Bracket added).

OBSTACLES TO PARADOXICAL BALANCE

To attend to something and carry out a task with any degree of efficiency is a selective process utilizing a very small part of the mind in particular and the Self as a whole.

Because the ratio between the selected portion and the rest of the Self is not balanced (according to the golden ratio of 1.618…) it does not fully draw on the amount of spiritual energy to fuel the Wholistic Response.

What conventional psiology and TPsi refer to as the "conscious mind" is actually a fragment of the mind operating with "working memory" that utilizes 7, plus-or-minus 2, chunks of memory at any one time (see Miller 1956; Bolles 1988).

Studies have shown that people ordinarily operate with about 3-5 memory chunks while those who use more, up to 9, correlate with higher degrees of awareness and intelligence (Weiss, et al, 2003). A medium with sufficient psychic power would correlate with a greater degree of FCCP involvement.

Fear-Based Psycho-fragmentation

The flow of energy and interactivity of MCs in the FCCP can be scientifically understood in terms of available energy, its potential to flow, and what resists this from happening in a life-fulfilling manner. Careful analysis of such resistance finds that they are quintessentially

MCs with some form of fear as the ruling context. As explained, the fear dynamics appear to have affinity to Ohm's law (see below).

As a general rule, fear is undoubtedly at the root of all obstacles to ones's inherent and/or acquired physical, mental, and spiritual potentials.

The inculcation of fear-based memory has inherent roots as well as those acquired in this "life time." Given that no two individuals (including identical twins) are identical but share similarities, the nature and context of the fears will have some degree of signature. Nevertheless, they all hover around a number of common themes, of which some key roots are as follows:

1. Fear of not being loved.
2. Fear of not being worthy.
3. Fear of death: loss of one's signature identity.

From these three all other kinds of fears emerge. For example, the first one leads to wanting to be accepted by others to the point where one becomes nonexistent to other-direction at the expense of one's Essence (the embodied source of all of one's potentials). The second category compels one to think and do things that generate admiration from others in a kind of group-think meritocracy. The third category is rooted in the inherent need for nourishment and protection, but often turns out to be the governing rationale as one pursues existence in the so-called "real world."

Conventional psychology views memory as a an "unreal" figment of imagination, an epiphenomenon not having life or intelligence, like memory in a computer, but psiology sees it as a MC, a Psiloton writ small, with all the cognitive processing powers of the mental system itself. Because of that, it can in the spatial medium in which it finds itself, complexify with other MCs to form "complexes" that have considerable impact on a person's life.

PSQ Shrinks the Wholistic Potential

Whenever a selection is made from the FCCP, the PSQ intervenes to edit what will be selected (at 3a). Consequently, a huge portion of the memory field is deselected.

This lopsided result prevents the Wholistic Response from being activated, since the selected ensemble of thoughts, images, feelings, and sensations, is the memory precursor to what we experience.

Laws of 7 and 3

We can equate this dynamical learning process to the transformative notion of the "Law of Seven," and the notion that the rate at which the process can be advanced or hampered equates to the "Law of Three." The former tells us what the process is, and the latter tells us how fast and harmoniously it can operate.

Hence, the Law of Seven involves a kind of algorithmic flow that can be activated exclusively at the level of imagination and/or as actual behavior, while the Law of Three addresses issues of the amount and kind of energy the process has at its disposal and to those of resistance. It has compelling affinity to Ohm's law, mainly used in electronics.

Psiological Interpretation of Ohm's Law

Ohm's law was named after the German physicist George Ohm, who in 1827, described measurements of applied voltage and current through simple electrical circuits containing various lengths of wire.

The law states that the current through a conductor between two points is directly proportional to the voltage across the two points. Introducing the constant of proportionality, the resistance, one arrives at the usual mathematical equation that describes this relationship:

> "$I = V/R$: where I is the current through the conductor in units of amperes, V is the voltage measured across the conductor in units of volts, and R is the resistance of the conductor in units of ohms (Ω) More specifically, Ohm's law states that the R in this relation is constant, independent of the current" (Wikipedia: Ohm's Law).

While this is a rather linear equation, it does serve as a compelling techno-metaphor for understanding the acquisition and use of energy in its various formats, used to "fuel" processes in the Self.

Even in quite convoluted nonlinear interactions on all levels of processing, Ohm's law, though difficult to calculate nonlinearly, remains a useful interpreter of energy flows.

Ohm's Law has Wide Psiological Potential

Everything in the Self, and any recurring identity in the universe, is learned, either from an ongoing species-wide evolutionary process that is being "incrementally" updated by personal experience in relation to the cosmic, physical and social environments, or through sudden "quantum evolutionary leaps" that tend to "explode" into the scenario at the edge of criticality.

The modifications of the PUF covered thus far are quintessentially units of intelligence that provide (on numerous scales of existence) the wherewithal to make adjustments as the complexities of the creative process unfolds. Smaller systems need to adjust to changes occurring in the larger systems in which they are embedded. For example, we adjust to weather conditions, not the other way around, even though it is possible though to some degree improbable.

When, for example, a sentient system encounters a "life-threatening" challenge (imagined or "real"), its rate of mental processing accelerates in order to process energy and information at faster rates.

With this automatic maneuver it is increasing its intelligence. This, I deem, is the basis of "instantaneous" evolutionary changes that answer the issue related to the notion of "punctuated equilibrium," the hypothesis that evolutionary development is marked by isolated episodes of rapid speciation between long periods of little or no change.

Ignorance of Management of Energy

We all understand the experience of "energy," and we know when it's available and to what degree. However, it is not commonly known that learning to manage it is a prerequisite to the development of the Self in its various modes of being.

Wasting it through aimless movements, rambling imagination, and the expression of negative emotions about anything is considered by most transformative teachings to be a serious assault on the development of the Self, simply because to increase the rate at which it processes information requires an extra amount of energy.

Transformative systems have brought forth numerous methodologies to deal with this very issue.

Ignorance of the Transformative Powers of Tolerance, Understanding, and Love

However, regardless of the form the methodology takes, it must include three progressively ideal states of mind that will ensure that energy is conserved and not wasted by impulsive thinking and rash behavior.

When confronted with strange and/or offensive behavior, the practitioner of spiritual science must learn to transform the aroused energy into something useful for oneself and others. This often turns out to be an attempt to withhold judgment and seek understanding. This is also what is meant by "non-attachment."

However, the learning curve that is almost always laid before the practitioner begins with tolerance, advances to understanding, and culminates in unconditional love, because from a technical view, this state produces the largest and most auspicious assemblies.

If PSQ is Not Transformed

The reason for this is that the PSQ is an energy-configured complex that needs to be stripped of its defining moral-and-ethical contexts in order to release its large amount of energy to the new Essence-supportive way of thinking. Therefore, this new personality can resonate with Essence to draw in the life force and help bring the entire system into a state of ONENESS.

If the PSQ is merely bypassed or suppressed and remains untransformed it will retain its energy and present a continuous struggle to divide and fragment the Self.

Blockage of the Wholly Presence

An example of the encumbering effect of the PSQ, as ahamkara, is found in the Bhagavad Gita:

> "...Lord Krishna says to Arjun that ahamkara must be removed—in other words, it should be subordinated to the lord. The reason for this is that the Self is not (cannot be) present when one is in a state of ahamkara" (Wikipedia: Ahamkara).

There are two important meanings that we can glean from this statement. The first entails the understanding that ahamkara prevents the total presence of the Self (i.e., the Wholistic Response). The second is that it should be under the control of the "lord," which means that it needs to be transformed, by a new ruling belief system, such as a MMC, which I interpret to be one that resonates with one's Essence, to facilitate its "quickening," bringing one vibrationally closer to the 4th state.

Inability to Sustain Mental Focus at Will

In keeping with our cognitive model of the attentional process (Fig. 4), the ability to keep the mind on target requires that the 1-to-7 cycle be faithfully repeated for a prolonged period of time. That ability defines attention span on a particular goal. This means that the ability to focus one's mind on a particular goal needs to become a key component of the MMC. If not, the mind will keep scanning the environment for OOPEs, and break the needed continuity.

The inability to sustain attention on any or all lower centers for an intentional period of time beyond the needs of ordinary perception does not allow an accelerative (i.e., quickening) process to take place.

Remember, attention is a psycho-accelerant, extending into the high gamma+ range of frequencies. When obstacles are removed, the system becomes "cleansed," "purified," and the centers will begin to operate at their natural speed. This disempowering resistance is called false personality (i.e., the PSQ), conditioned existence, Satan, or the malevolent asuras in the Vedic literature.

In such a state the centers are thwarted from performing right work, which involves the proper transformation of three kinds of food— regular food/water, air, and impressions—into the finer substances, referred to as hydrogens (a.k.a. life force, etc.), needed by these higher bodies to develop. Without the proper food a center cannot operate at its optimal speed or vibration.

It is held that every morning each person has enough refined food to feed the development of a higher body, but that it is wasted on aimless thoughts, movements, emotions, and rambling imagination. The aim of observing the centers is to gain control over this wastage. One of the key energy wasters is the expression of negative emotions, which excretes the finest hydrogen lower centers produce, namely hydrogen 12 (H-12 in Gurdjieff's cosmology).

Not Understanding the True Nature of Balance

As has been fully discussed, the extent to which the psychophysiology is in a state of balance, in which its polarities have become equilibrated, constitutes the secret of the wholly science. A scientific understanding of how this takes place has been revealed here for the first time.

The notion of "surrendering the ego" and its associative meanings makes little sense to many attempting the pursue Essence's urge for WHOLENESS, for the ego is often misunderstood as being selfish, greedy, non-compassionate, and the like, but while this may be the case it is essentially a system for stepping-down the speed for processing information in a dualistic manner to deal with the dimensional constraints in positive time-space.

Because of this attribute it hates contradictions, and as a consequence, is not conducive to learning about and/or generating states of paradoxical balance, let alone the "bizarre" nature of "paranormal" events. Therefore, it takes a person (primed by personal experience due to the emotional down swings of destructive interference) between Essence and the acquired persona.

Inability to Sustain Neural Synchrony

Studies (in Kopell 2009) using EEG and MEG to detect neural synchrony strongly suggest that several clinical conditions, such as schizophrenia and autism are associated with the inability to sustain synchrony and coordination of distributed brain processes.

Non-clinical Multiple Personalities

Gurdjieff contended that most people go about their lives in a robotic state of "sleep," a common fragmented psiological condition he described as a kind of non-clinical multiple-personality disorder (now known as "Dissociative Identity Disorder") in which identity fluctuates based on the salient stimuli or cues of the moment.

In this condition one is interacting with oneself and the world through one of many fluctuating disconnected small selves (personas) "residing" within a center he referred to as "I's", or memory constructs (MCs) as I prefer to call them (I will use both terms interchangeably).

Each 'I' consists of a configuration of energy into a certain context. Because many of these 'I's are incompatible (i.e., lack consonance), when one 'I' takes center stage, a small neuroglial cluster is formed, while others are inhibited. According to this either-or (dualistic) "normal" way of perceiving and behaving, the sum total of one's potential energy invested in the 'I's is rarely available for evolutionary purposes due to internal conflicts and disagreements.

Transformation, Not Dissociation

Allowing contradictory MCs to share the same psychic space facilitates the principle of destructive wave interference to take effect. This, over time, will strip the MCs of their negative-or-positive contexts and release the energy they contain in a neutral manner.

By rejecting certain MCs due to their uncomfortable or painful contexts, the either-or nature of the PSQ is reinforced as well as the process of dissociation. Dissociation not only separates MCs from exchanging information with one another but also their valuable energy. This divisiveness contributes to some degree of lethargy and memory loss.

Identification/attachment is another form of dissociation. It indicates that one has a positive or negative attitude towards experiences one has had and can have in the "future." Memories associated with these experiences will be blocked, contributing to keeping varying swaths of information from being used. Since energy flow hates being pent up it will find psychosomatic pathways of expression.

Hemispheric Imbalance and Pathology

Ultradian rhythms of alternating hemisphere dominance and the nasal cycle are tightly coupled with each other. Following up on these ancient postulates, research on alternate nostril breathing is revealing in a more objective format what the seers of old intuited from their meditations and clinical observations (see Lad 1996). As the tenets of Swara Yoga (see Johari 1989) reveal, imbalance of the breath of the nostrils points to mental and physiological problems.

A study done by Serap Yildirim and associates finds that predominance of the right brain hemisphere over the left one links to schizophrenia. They used an assessment of nostril airflow to determine nostril dominance in 83 schizophrenics to compare with 64 healthy non-schizophrenic controls. They found that:

"The rate of left nostril dominance was significantly higher (65.1 %, n=54, p=0.00) in patients with schizophrenia. In healthy controls, the rate of left nostril dominance was 25.0 % (n=16), while the majority (53.1 %, n=34) had no lateralization in nasal cycle" (2017,1).

They concluded that:

"Schizophrenia might be associated with reversed cerebral lateralization in terms of nasal cycle. Left nostril dominance, which would correspond to greater sympathetic tone and lesser electrical activity in the left hemisphere, may be a functional measure worsening the left hemisphere dys-or hypo-function. Our findings support the lateralization abnormality and left hemisphere dysfunction hypothesis in schizophrenia"(Ibid.).

Disempowerment

Divided thusly a person becomes disempowered (i.e. lacks the contextual coherence of will) to make the necessary changes to be in harmony with themselves and with the physical and social environments. A fragmented psiology has difficulty making and keeping promises. In this regard Ouspensky quotes Gurdjieff:

"Man has no permanent and unchangeable I. Every thought, every mood, every desire, every sensation, says 'I'. And in each case it seems to be taken for granted that this I belongs to the Whole, to the whole man, and that a thought, a desire, or an aversion is expressed by this Whole. In actual fact there is no foundation whatever for this assumption. Man's every thought and desire appears and lives quite separately and independently of the Whole. And the Whole never expresses itself, for the simple reason that it exists, as such, only physically as a thing, and in the abstract as a concept. Man has no individual I. But there are, instead, hundreds and thousands of separate small I's, very often entirely unknown to one another, never coming into contact, or, on the contrary, hostile to each other, mutually exclusive and incompatible" (Ouspensky 1949, 66-67).

Therefore, the interpretation of disempowerment in the transformative tradition is based on a quantity: how much of the human psychophysiological system is not working in sync. How the mind-body system can be brought into alignment forms the methodological basis of the transformative tradition.

The theme of psycho-fragmentation is often addressed in the transformative tradition through allegory and metaphors. For example, in the King James Holy Bible the Self is likened to a kingdom (i.e., the "kingdom within"). In his Gospel St. Matthew quotes Christ:

> "Every kingdom divided against itself is brought to desolation; and every city or house divided against itself shall not stand: And if Satan cast out Satan, he is divided against himself; how shall then his kingdom stand?" (12:25-26).

The Buddhist rendition of psycho-fragmentation is sometimes referred to as the "monkey mind," that jumps from one thought to another.

In the energetic terms of science this condition could be explained as the lack of psychophysiological coherence (since it takes the whole system to generate thought, feeling, and sensation) in which the "I's" as waves (i.e., "undulations" or "spirals" or vrittis in Yoga) are not in phase.

A system predominantly in phase would acquire a state of resonance. However, since the aim is to bring the entire system, or the greatest portion into a state of synchrony, groups of "I's" may display this characteristic in smaller clusters.

With this bit of technology I could see how the physically predominant Self, when exposed to a coherent level of energy, can be entrained into higher being bodies (OBEs) consisting of degrees of subtlety, and higher states of consciousness. By incorporating this kind of "materiality", the quest to resolve incommensurables like subject-object, mind-body, matter-spirit simply dissolve.

The Laws of Attraction & Distraction

In this divided scenario things happen. They are not planned, although subconscious "choices" do take place. On the surface, Life bobbles along like a cork floating on the ocean. Now a strong wave comes, then it's the wind, a boat passes by producing huge waves that shifts its direction, and due to these random unplanned forces one may end up on some strange shore. Bouncing around from one event to another, we may come to realize that the non-conscious rules human Life (i.e., about 95% of the time; See Szegedy-Maszak 2005) through the "law of accident" (LOA), a term coined by Gurdjieff.

The acronym "LOA" is identical to the one representing the "law of attraction" (currently based on a popular book), but there's an important difference. With the law of accident things just happen, and whether it brings "good" or "bad" effects they are the result of chance. Some might call it fate or destiny.

On the other hand, the law of attraction is based on intention, the notion that human Life is endowed with the potential to enact advantageous self-transformations by iterating positive thoughts and beliefs that, via the principle of psychic resonance (the notion that thoughts are made from "pure energy," and that like energy attracts like energy), are able to attract them from the world at large into one's life.

The idea that thought rules one's Life is not new. Buddha is quoted to have said:

> "What we are today comes from thoughts of yesterday, and our present thoughts build our Life of tomorrow: our Life is the creation of our mind."

About a half century later the idea emerges in Judaism. In chapter 23, verse 7, of Proverbs of the King James Bible we find the phrase:

> "As a man thinketh in his heart, so is he."

In Christian theology this phrase alters the notion of predestination (and in Eastern mysticism that of karma), which points to the belief in personal responsibility for one's life as the means to transform from a robotic life to intentionally being.

Many "New Age" books that came forth during the twentieth century, particularly during the first half, are rooted in these ideas. One that directly relates to the proverb is James Allen's essay As a Man Thinketh, published in 1902. The essay centers on the theme:

> "A man is literally what he thinks, his character being the complete sum of all his thoughts" (7).

Perhaps only indirectly so, do we find mention in these works of the principle of resonance, the means by which the attraction occurs. It is rooted in the ancient idea that likeness attracts what is similar to it. It was known to the premodern world as "sympathetic magic" (see Frazer 1906).

However, in 1906 William Walker Atkinson not only used the phrase "law of attraction" but also added a vibratory element to his transformative thesis, found in the title of his book *Thought Vibration or the Law of Attraction in the Thought World*, stating within that "like attracts like."

While there were other writings that used the phrase "law of attraction," it seems that Atkinson's work ushered in a more scientific ethos of what it would mean to the future inhabitants of a world becoming progressively more technical. The phrase "like attracts like" implicates the scientific phenomena of resonance, synchrony, and frequency-and-phase relationships governed by the principle of wave interference.

As has been thus far presented, the idea of resonance is a central concept in modern science. In this frame of reference it describes the vibratory state of a system at rest.

It also reveals the fact that each system, albeit a particle or a galaxy, is characterized by a signature vibration known in physics as an eigenfrequency.

We can deduce why HI finds it advantageous to endow each system with a unique vibration. If everything were vibrating with identical frequencies there would be no separateness or differences. Everything would be clumped into one gargantuan mass. How cozy we'd all be!

One of the reasons for so much skepticism that thoughts have such attractive and transformative power comes from the anecdotal-and-unscientific nature to substantiate its claims.

Another more specific area of doubt comes from scant scientific information existing at that time about the nature of attention, since most, if not all of the anecdotal accounts are based on the ability of the mind to focus or concentrate on desired outcomes. No one then knew anything substantive about the nature of attention.

As you can see the law of attraction requires a highly focused mind to be successful. Hence, we might as well dub the law of accident the law of distraction (LOD) and keep the acronym LOA for the law of attraction, since the former moves the mind away from the target while the latter moves it towards it. The idea of steering the mind towards a goal brings the principle of feedback into the mind-control picture.

Now we have two scientific principles, which are the basis of transformative psiology, allowing greater understanding of how transformative power is related to volition. The holistic feedback loop takes it to another level.

Achievement Anxiety

Perhaps one of the most pernicious blockages to attaining higher states of being and consciousness is that of impatience and the belief that the process needs to be micromanaged. This uptightness prevents the psycho-organism from relaxing and allowing the conjunctive process to proceed on its own accord. We need to remember that transformative progress relies on tapping one's inherent potentials, not creating them from scratch.

Lying

Lying is the basis of much psycho-fragmentation. It reflects not telling or denying the truth about oneself and/or about others to oneself and others, insofar as one is privy to what is "true" about anything within one's capacity to perceive and understand.

The root of lying emanates from MCs, usually acquired during one's formative years, that cause the person to feel unworthy about themselves. It has many ramifications in regards to personality traits, like that of vanity that needs to "downgrade" another in order to feel worthy by comparison.

Repression

In the FCCP there are memory constructs that are programmed to inhibit or repress inputs that contradict one's acquired psychophysiological status quo (a.k.a., the ego, and belief system). It is proposed by neuroscience that this takes place in a structure in the limbic system of each brain hemisphere known as the amygdalae.

Lack of Forgiveness

The keeping of accounts on people that have in one way or another done us wrong, or are perceived to have done so, is a major energy and psychic divider, enormously contributing to psycho-fragmentation and disempowerment.

Complexes

The term "complex" was developed by Jung's school of depth psychology. It indicates a related group of emotionally significant ideas that are completely or partly repressed and that cause psychic conflict leading to abnormal mental states or behavior. Like the Z Complex, they subliminally affect every aspect of the psychophysiology, and play a significant role in one's psycho-fragmentation, leading to degrees of disempowerment, simply because the energy is tied up in these configurations.

Entrainment of Mirror Neurons

The concept could be explained technically by "mass entrainment" in which the size of a group involved in similar repetitive behavior would, through a predominance of positive feedback, have an entraining effect on its constituents.

Experiments with laboratory animals suggest that such mimicry is mediated by "mirror neurons."

> "A mirror neuron is a neuron that fires both when an animal acts and when the animal observes the same action performed by another. Thus, the neuron 'mirrors' the behavior of the other, as though the observer were itself acting. ... In humans, brain activity consistent with that of mirror neurons has been found in [various areas in the brain]" (Wikipedia: Mirror Neuron).

Therefore, "monkey see, monkey do" is not exclusive to that species.

The Unprotected Mind

On the Symbol, the ovoid in which the infinity loop is established represents a vibratory protective border. It acts as an aura or field that is based on the signature-golden-ratio vibrations of each individual. Based on this frame of reference it also acts as a sensorium/transmitter where it absorbs energy-and-information it needs and expels what it cannot use.

Being formulated with an irrational number, it maintains its "separation" from other influences by dint of its degree of dissonance and not being able to readily resonate with other constructs. It constitutes a kind of "friendly dissonance." Its purpose is to allow what's developing inside to do so with minimal interference from the "outer" world.

In symbolic terms it is often referred to as a "womb" in which a gestational or evolutionary process takes place that unfurls the information "involuted" in the form of fractal vortical solitons (i.e., Psilotons) of intrinsic information.

The vibratory border is sustained by the vigor of the self-referential infinity loop. Should the loop lose its metabolic robustness (i.e., its unresisted flow), it will lose some vibratory momentum and its protective border will to some extent be compromised and exposed to outer influence. As a consequence, the mental system becomes corrupted and weakened, and the physiology will be prone to disease and/or psychic invasion.

This is not any different in principle to the protective magnetic shield (aura) and the ozone layer of the Earth.

Therefore, the psychophysiological health of a system depends on the degree to which its parts contribute to overall synchrony, which constitutes the system's level of health and wholeness. As with a magnet, the strength of the field depends not only on the alignment of the atomic spins but also of the quantity of spinning units. For example, combine two small magnets together and they produce a stronger and larger field. Put two similar Psilotons together and they create a larger one by dint of constructive interference.

Unfriendly Other-Directed Agendas

We must not forget that the transformative principles being revealed in this book can (and to some extent have) also been used by those who do not support or respect the individual's Essence-right to develop their highest potentials. They are either motivated in selling something you don't need or are propagating some hidden "world order" as a power trip to subjugate humanity to their undisclosed ends.

TRANSFORMATIVE METHODOLOGIES

In this section we need to apply what has been thus far presented about the wave dynamics and its relationship to a cognitive cycle (also coined as a "psicle"), to explain the various methodologies used by the transformative tradition, particularly to:

1. Generate will.
2. Expand states of being and consciousness.
3. To enact our inherent potential to develop higher being bodies (OBEs of varying vibrations) that resonate with particular vibratory milieus in the FCCP/PUF.
4. To improve our relationship to our Selves, others, and the environment in which we are cosmically embedded.

The methods of transformative psiology aim to change the vibratory rate of the entire psychophysiology into an energetic state of synchrony, which is none other than the state of being holy, and affixing it ("crystallization") into a higher being body that can travel and vibrationally commune with other aspects of the cosmos. In these higher states of being the F-F system acquires a rate of processing fast enough to be conscious of and control a slower system in a more effective and intelligent manner.

Once properly developed, these higher bodies can co-exist with the physical one (as sheaths or embedded auras) or separate from it (but never from the cosmic medium, or the Pluripotent Unified Field (PUF) in which its modifications are embedded).

Throughout the centuries transformative systems have evolved a set of principles that formed the basis of their methodologies and practice.

PRINCIPLES

From the foregoing, we glean 4 Principles of particular significance to creating the WR. Because they are integrally related there will be some overlap in describing their psiological meanings and the rationales of why and how they contribute to the formation of methodologies. We begin with perhaps the most important principle in the transformative process: motivation.

1: MOTIVATION

Motivation is conventionally thought to involve a reason or reasons one has for acting or behaving in a particular way. It may involve conscious, non-conscious, and the mechanisms inherent in the satisfying of physical and psychological needs—e.g., food, safety, sex, etc. In a pathological sense, motives may take on algorithmic-like expressions of aberrant conditions, like obsessive-compulsive behavior.

In regards to psycho-transformative issues, in which the Self is in search of its Self, the question arises of why an individual would want to pursue a teaching and practice that does not at its core exclusively seek some kind of a conventional payoff, considering that the desired outcomes, such as the acquiring of higher being and consciousness, is not so definitively conceived.

It is thought that what leads one to be motivated to seek transformative knowledge is typically based on:

1. Fear of losing one's personal sense of identity, and seeking "salvation."
2. Dissatisfaction with one's physical and/or social environments.
3. Dissatisfaction with oneself, perhaps by realizing one's psycho-fragmentation.
4. Boredom: losing or never having had enchantment with life.
5. Experiencing a huge discrepancy between one's Essence and personality.
6. Fascination with the literature on the lives and feats of highly-evolved beings.
7. Following an ingrained family belief and practice.
8. A deep interest in existential issues.
9. The quest for personal and/or social empowerment.
10. The realization through personal experiences that life is more interesting and complex than what is described by conventional thinking.
11. Experiencing "paranormal" events that lead one to believe in the attributes of "spirit" and/or in an "after life."
12. An inclination towards subjective orientation rather than objective orientation.
13. All or any combination of the above.

While the list is not exhaustive it is very suggestive of the nature of what may motivate individuals to pursue a transformative teaching. These play a very important role in helping a student to understand him/herself ("Know thyself").

However, whatever the impetus may be, the biggest obstacle to motivating a person to pursue or stay on the course of a spiritual development program is the nature of the illusive goal, which is not conducive to forming an attentional F-F loop needed for efficient learning to take place.

Therefore, it would be quite difficult to focus the mind on something it cannot conceive. It is for this reason that the use of symbols, perhaps in the form of a prayer, mantra, icon, living master, belief system, and so on, can *temporarily* serve that purpose. One of the problems with these devices is that they often become the goal rather than the means towards attaining it.

One needs to stay with the "object" or tool until the WR has been engendered and the Presence has come into existence. Then one drops the techniques and focuses on the experience of being present to one's Self. In the meantime there will be all kinds of logical contradictions, which is inevitable as the MMC is being developed and the PSQ still rules the psycho-soma. During this phase it might be useful to adopt the following verse from Walt Whitman's "Song of Myself."

> "Do I contradict myself?
> Very well then I contradict myself;
> (I am large, I contain multitudes.)"

Belief

Belief has always played a key transformative role in human life. It constitutes the most powerful transformative agent in human psychology. Yet it is not an easy term to define. However, we can assume with some assurance that belief seems to engender a psychosomatic alignment with some psychic mass, and that its contextual coherence is the source of its transformative power, which is never simply just thought but a quantic effect, so to speak.

Not all beliefs are powerful. It all depends on the degree to which they form large, widely-distributed assemblies throughout the psycho-organism. An influential "believer" would then be one whose belief system has a high level of conceptual, emotional, and sensory integrality, such that a wide variety of concepts have been imbued with associative meanings.

Since the psycho-organism accepts what the mind considers to be "true" it doesn't really matter how logical it all is (though for certain intellectually-centered types this may be the ruling factor), but how subjectively "connected" it actually is. The belief system of instinctively-centered types would rely on what makes "sense" to them, since sensation is the language of this center. Emotionally-centered types would rely on how they "feel" about a situation.

This is precisely what allows the placebo or nocebo effects to take hold of the system, for in the former one would have a trusting belief in how, say, a medical pill or treatment would positively affect them, and, in the latter, how it would negatively affect them.

Therefore, just thinking about something does not of itself produce the desired psychosomatic effect; it requires the existence or creation of an inner volume of collective power, which as a quantity of Psiloton units produce emotional intensity with in-phase or out-of-phase wave interference patterns.

Belief doesn't happen overnight. It has been developing since infancy and continues to do so within the social network in which one is embedded. However, it is subject to being changed, but not as easily as one might assume. The key factor in modifying a belief system is to alter its modus of coherence or to bypass the editing effects of the PSQ. Here are some examples of some belief-altering modalities.

1. Subjecting the mind-body complex to torture, such that it causes the whole system to break down while, at the same time building another belief system in its place. This is the basis of "brain washing," often found in military systems and in prison camps.
2. Embedding a subject in the social-gravitas of a particular culture or subculture with the goal of being entrained by it. This modality relies on the contextual coherence of its doctrines and practices.
3. The use of fear to render subjects to go into survival mode and change their minds.
4. The use of entheogens to produce mind-altering effects that reveal to the subject that there's more to "reality" than what has been taught by the dominant culture.

5. The experience of "paranormal" events that reveal to the subject that there's more to "reality" than what has been taught by the dominant culture.

6. The use of logical arguments and experiments, especially in cultures strongly affected by the technological beliefs of modern times.

In any case, belief intensity is developed by the self-referential F-F looping of the cognitive cycle. If iterated beyond a certain threshold it acquires enough psychic mass able to entrain other MCs into its frame of reference. It then subjectively imbues meaning by the mind that something is not just true or real but that it is incontrovertible. This incontrovertibility forms the nucleus of the PSQ. However, it can also be an attribute of an intentionally created MMC.

Belief and Hypnosis

Hypnosis is a technique that bypasses the PSQ in order to introduce a weighted MC that can affect behavior on non-conscious as well as behavioral levels.

The minister of Divine Science Dr. Joseph Murphy (1898-1981) writes in his book *The Power of Your Subconscious Mind* a key statement about belief as a transformative law.

> "This law of *belief* is operating in all religions of the world and is the reason why they are psychologically true. The Buddhist, the Christian, the Muslim and the Hebrew all may get answers to their prayers, not because of the particular creed, religion, affiliation, ritual, ceremony, formula, liturgy, incantation, sacrifices or offerings, but solely because of belief or mental acceptance and receptivity about that for which they pray" (1963. 12, pdf. Italics added).

In terms of what is being proposed in this book, the "law of belief" has such powerful affects because it has over time, through numerous repetitions, acquired the *psychic mass* (a.k.a. reinforcement), so to speak, to entrain other memory constructs into its particular theme, which could be life limiting or enhancing.

To imbue it with power, the memory construct harboring the belief system is strengthened by vibratory predominance of delta during infancy and early childhood. The vibratory threshold is encoded at the fast heart rate, which in infancy reaches about 240 bpm (see Lad 1994).

The heartbeat of a human embryo begins at approximately 21 days after conception, or five weeks after the last normal menstrual period (LMP), which is the date normally used to date pregnancy in the medical community.

The context or message of the construct "defines" its reach and power.

In this quantitative regard the belief could be summed up briefly as a thought in your mind that has been subliminally or willingly incontrovertibly accepted as being "true" in one's quest for meaning. It then becomes the habituated "measuring rod" against which one accepts or rejects incoming information.

The process typically starts quite early in life, as modern research has revealed, during the time spent in the womb. Because it has psychophysiological consequences it may block proper development of any organ system, leading to aberrant health and behavior during infancy and childhood, which if not addressed may last a life time.

2: INTENTION

Intention fosters the development of a MC that has sufficient psychic mass to entrain habitual mental habits and behaviors to its frame of reference. It uses existing MCs in novel or unusual ways. However, intention implicates the notion of *will*, which must, to become effective, be iterated with the cognitive cycle into a functional habit.

From a developmental perspective, intention requires inputs to the cognitive cycle that are, as has been revealed, paradoxical in context in order to engender the WR. Obviously, this involves the use of creative imagination.

Creative Imagination In Fostering the Wholistic Response

As mentioned, research has revealed that imagination affects the same neuromuscular pathways involved in behavior (Decety 1996). The psiological rationale behind this is that imagery requires a greater number of MCs in its make up and, consequently, imagining the entire Self in its various modes of being evokes the polar opposites in a certain ratio (Phi::phi—1::.618…) and magnitude needed for the WR.

Paradox

Without paradox and mystery (amongst other conundrums) attention cannot be evoked or sustained at the levels needed for transformative effects. Recall that a cognitive system is designed to come into play as a kind of "watch dog" to keep track of OOPEs and to resolve contradictions.

To maintain proper vigilance the Self separates the super-fast operations of the non-conscious from the slower needs of the conditioned consciousness. This instantiates "two" minds, which systems of transformative psychology aim to merge into a state of oneness.

To do so they have discovered a way to equilibrate the oppositional aspects of the Self in order to draw on energies and wisdom from HI. Hence, they turn oppositional forces into complementary relationships, a methodology modern psychology has yet to discover.

In keeping with the thesis being propounded in this book, an intentional input needs to be a conjunction of incommensurables that equally evoke the polarities of the psycho-soma into a state of "stillness."

An effective way to accomplish this is with the use of wholistic symbols (e.g., the Symbol, yin-yang, the cross, etc.) in which subject and object, or opposite geometric signs, are given equal status.

Such symbols can be used as the nucleus around which all lines of transformative information in regards to one's MMC converge. In this case one uses the symbol as a mnemonic hub, the connecting agent to be remembered. By just thinking of it the WR can be triggered.

Turning Contradictions into Benefits

It now becomes clear why systems of Transformative Psiology (TPsi) generate contradictions in the form of physical and intellectual challenges, such as paradoxes (e.g., puzzles, koans, complex rituals, theological and ontological complexities, etc.), to keep the attentional system of the Self engaged; otherwise it would become bored and go to sleep.

A paradox is important not only for sustaining an alert mind; it also, at some point in the developmental process, is conducive to equilibrating the polarities of the organism and the mind into a neutral state of "stillness," quite likely due to the self-cancelling effect of destructive wave interference simultaneously affecting many modes of the Self.

When faced with the overwhelming experience of "stillness", the mental system interprets it as a vulnerable dysfunction. To "solve" the conundrum the Self automatically shifts into a higher mode in order to speed up its mode of operation. In Scripture this is referred to as the "quickening of the spirit."

The Interface of Two Incommensurable Orders

Where the commensurable and irreconcilable meet, the conditions for the attraction of certain transformative forces come into play. Devices such as "double mandalas" have been used to create this conjunction of opposites.

> "In the special class of figures that Jungian analyst Marie-Louise von Franz calls the 'double mandala' this STRUCTURAL feature is brought even more clearly into relief. *Double mandalas, because they focus on the interface between the two incommensurable orders - where form becomes emptiness and emptiness becomes form -* were throughout history also treated as magical devices. They were considered instruments of divination and prophecy, and utilized as spiritual guides. Meaning was attributed to randomized events made to occur 'outside' of the 'causal order' in such a way that the user of the device could obtain information about outcomes taking place in corresponding processes WITHIN the causal order" (Ibid. Italics added).

In terms being developed here of transformative psiology, "form" is an objectification in positive time-space of "emptiness" in negative space-time, where information has not been processed to be perceived by the senses and interpreted by the threshold of conditioned consciousness. When the combined imagery of the two "incommensurable" orders are contemplated they over time affect their intended paradoxical effect on the psychophysiology.

On the Symbol, these incommensurable orders are broadly intimated by the Subject-Object signs, but more forcibly shown to merge at the Star, where they reconcile into the standing waves of phenomenal "reality."

Esoteric Symbols as Unifying Archetypes

In their very thought provoking article "The Enneagram as Classic 'Double Mandala: Part I - The 'I Ching' and other 'Divination Machines',"

(1999b) John Fudjack and Patricia Dinkelaker introduce the subject of symbols that express a conjunction of opposites when viewed as mandalas.

A Mandala is a geometric figure representing the universe in Hindu and Buddhist symbolism. In psychoanalysis such a symbol presents in dreams, representing the dreamer's search for completeness and self-unity. In both cases the symbols incorporate in their designs a visual effect that has confusing yet ameliorative aspects. In Jung's depth approach to psychic contexts, mandalas have archetypal influence of the person's psychology.

When mandalas emerge from the Self they have a more unitary meaning, for the oppositional categories (e.g., subject and object in the Symbol) blend large swaths of the psyche that become organized in a new way.

> "When viewed as a mandala the Enneagram can be understood as a representation of the 'Archetype of the Self'. It is thus a symbol of an advanced stage of the individual's development in which consciousness as a whole (what Jung called 'the psyche') is *fundamentally organized in a new way that makes possible a reconciliation between two 'orders of existence' that are usually treated as 'incommensurable' and irreconcilable.* The two are sometimes described as 'the sacred and the profane' (Eliade), sometimes as 'the eternal and the temporal' (Von Franz), and sometimes simply as 'emptiness and form' (Buddhism)" (Fudjack/Dinkelaker 1999b. Italics added).

Because of their out-of-pattern conjunction they not only produce alertness and draw attention to them but they also generate memory. We don't need a major study to support this contention because it is verifiable common sense that strangeness and criticality robustly implant themselves in the mind. That is why esoteric symbols, which are characterized by the conjunction of opposites, are rarely if ever forgotten.

Essence's Signature as an Auto-stimulatory Imperative

At the individual level, signature is the basis by which an individual provides the necessary paradox for general awareness to exist.

This principle lets the practitioner know that he/she is a unique being sporting a finger-voice-molecular print, and a personal mode that needs to be honored and incorporated in the process.

This is especially important given that positive feedback is more easily enacted when the input and the output have high degrees of similarity. Therefore, auto-stimulatory exercises, like mantra, prayer, and such, are best performed when the expressive patterns—e.g., voice, vision, thought, movement—of the Self are performed on itself.

The principle also applies to highly cohesive groups acting in higher degrees of synchrony, as in, for example, intercessory prayer (see Dossey 1989). Studies done on the effects of intercessory prayer have not taken this variable of group synchrony into consideration, assuming that it already exists. This would account for variations in the often conflicting results.

Self Cancelling: The Eraser or Neutralizer of "Lower" Mind

The principle of destructive interference helps us to understand how wave systems interact with themselves and one another. Paradox, contradiction, etc., not only keeps the Self alert to itself and its environment but it also has other benefits in "purifying" the system of encumbering MCs that block energy from flowing as freely as it should.

A key psiological revelation is that this is the mechanism by which the Self erases or neutralizes the contexts of MCs by allowing contradictions to share the same mental space. When two or more opposing MCs are "given" or acquire same valuation, the PSQ cannot process them according to its binary frame of reference.

If held in the 1-to-7 cycle for a sufficient period of time they will clash and neutralize themselves in such a way that their contexts will be diminished/destroyed, and the once-configured energy will be released to be expended or used in another context with psychic mass, such as the MMC.

This effect is behind the "science" of why transformative teachings place much value in viewing the world, particularly its social environments, from these three evolving perspectives:

1. Tolerance: the attitude of allowing, given that the mind in its lower vibratory states is not capable of fully knowing why people act as they do.
2. Understanding: the attitude of approaching conundrums with patience and concern, which evokes higher processing rates of intelligence to help bring greater clarity about them.
3. Unconditional Love: the attitude that by having negative emotions about other people, regardless of what they do, not

only wastes valuable transformative energy for oneself, but further divides and disempowers one from attaining higher states of being and consciousness. Love produces the largest and most auspicious assemblies.

Paradoxical Unity and the Gospel of Thomas

Knowledge of using paradox to get the mind to self-cancel and open up to higher energy and intelligence is found in this tractate discovered in the Naq Hammadi scrolls. In the 22nd verse of the Gnostic Gospel of Thomas we find a more informative description of the transformative paradox:

> "Jesus saw infants being suckled. He said to his disciples, 'These infants being suckled are like those who enter the Kingdom.' They said to him, 'Shall we then, as children, enter the Kingdom?' Jesus said to them, 'When you make the two one, and when you make the inside like the outside and the outside like the inside, and the above like the below, and when you make male and female one and the same, so that the male not be male nor the female female; and when you fashion eyes in place of an eye, and a hand in place of a hand, and a foot in place of a foot, and a likeness in place of a likeness; then you will enter the Kingdom" (in Barnstone 1984, 302).

I also thought it could be described more scientifically in terms of the EEG parameters of the sleep-wake cycle in which the Hypo-metabolic and Hyper-metabolic ends of the spectrum conjunctively converge.

"The Thunder, Perfect Mind"

This strange title belongs to a tractate with ancient origins from the apocalyptic literature of the Gnostic tradition. It exemplifies like no other piece of writing the paradoxical nature of contexts used by transformative systems to clear the mind and prime it for higher reception. This is quite similar in principle to how Milton Erickson induced his patients into a deep susceptible state. Here's a link to an excerpt from an interview with Elaine Pagels discussing the poem: https://www.pbs.org/wgbh/pages/frontline/shows/religion/maps/primary/thunder.html

This paradoxical teasing of the mind is quite similar in principle to how Milton Erickson induced his patients into deep susceptible states, using confusing contexts. The tractate/poem appears to serve this purpose.

Self Cancelling: An Attractor of Higher Intelligence

In wiping out or neutralizing lower-and-slower MCs, the Self opens a low-pressure zone that will attract the workings of higher-and-faster cognitive systems. In the parlance of the "New Age" this is often referred to as "channeling," but how the channeling/downloading occurs has never been fully explained, as has been revealed to the author of this book.

In Scripture and other holy books, this notion has been alluded to as humility, meekness, surrender, peace, and the like. It has been symbolized as the "lamb of God," a "still lake," and as an "open vessel," and other imagery portraying calm, receptive, neutral states.

In the Christian Gospels, the image of a "virgin birth" symbolizes in biological-contradiction the impregnation of a lower principle by a higher one. When the symbolic aspect is lost there is a tendency to see it as a biological absurdity.

Recall that a coherent faster-operating system of HI has the advantage of "entraining" a slower one to its context and pace (see Bentov 1981, 37-38, for the technical rationale). In this rather technical regard it is the source of "impregnation."

The notion of a "virgin" is found in almost all transformative doctrines, preceding Christianity by centuries; it is one of those symbols that points to that of protecting an acquired value while it is in its nascent stages. To the objective-compulsive mindset this conundrum presents as a biological absurdity, leading them astray from knowing (gnosis) the effects symbolic images have on the potentials inherent in the physiology.

Love, the Great Transformer

The term "love" is used so loosely and indiscriminately in our times that for many it loses the gravitas of its all-embracing, even "terrifying," attributes.

There's something about love that intimates an enduring attentional process, which Shakespeare (perhaps Bacon or Marlowe) seems to have expressed perfectly in his 116th sonnet.

"Let me not to the marriage of true minds
Admit impediments. Love is not love
Which alters when it alteration finds,
Or bends with the remover to remove.
O no! it is an ever-fixed mark
That looks on tempests and is never shaken;
It is the star to every wand'ring bark,
Whose worth's unknown, although his height be taken.
Love's not Time's fool, though rosy lips and cheeks
Within his bending sickle's compass come;
Love alters not with his brief hours and weeks,
But bears it out even to the edge of doom.
If this be error and upon me prov'd,
I never writ, nor no man ever lov'd" (in Harbage 1969).

Psiological teachings embrace the state of love because it is the evoker of the largest assemblies, given that it tolerates, understands, and embraces all of what we are. This line of thought would be part of one's MMC.

Love does not reject the body, nor the mind, and by loving them aims to "see" them as they are, and by simply "seeing" without judgment it forms the basis of the most powerful conjunctive F-F cycles one can create.

It grows and develops the Self exponentially when it includes others and the world we live in. By itself, it is more than enough to transform the Self to attain its highest mental and spiritual potentials in which the "thing" aspect of the spirit seems to be absent.

Because it is all about paying unconditional attention to our Selves and the universe we live in, we find its physical substrate in the executive function of the frontal lobes, just at the juncture where the hemispheres meet and divide into sympathetic and parasympathetic neural pathways. Focusing on the central spot activates the paradoxical state of being in which the Presence emerges as an inner "stillness" open to higher influence.

Simulation of a Death-Like Event

If the Self in the physical mode of being "thinks" its going to die, due to being (fortuitously or intentionally) entropically equilibrated, it automatically accelerates the 1-to-7 cycle to speed up its transition to recur or continue its evolutionary development in another plane in negative space-time.

Realizing that this response ensures the continued existence of the Self in a spiritual mode of being, transformative teachings have discovered long ago that a death-like event (DLE) can be simulated to produce the same effect.

Like a Near Death Experience

The DLE is quite similar to a near-death experience (NDE), but it doesn't fulfill the vital-sign indicators of the NDE, where heartbeat, oxygen, and brain parameters do not appear "normal." Yet, conscious OBEs appear in both situations.

In both instances experience of the ingoing vortex is extraordinarily viewed as a tunnel with brightness at the apex, and being drawn towards it.

In both cases, realization that the Self can exist as a higher being body outside of the physical body is possible, leading to the verification that the Self continues its existence in another mode of being. It provides supporting evidence that there is no such thing as "death" as the cessation of signature identity or existence.

When seers and saints speak of dying they do not necessarily mean physical death, but a transformation of embodied potentials that are released to attract the life force. Consider this saying of St. Paul in Corinthians I of the King James Bible:

> "I protest by your rejoicing which I have in Christ Jesus our Lord, I die daily. If after the manner of men I have fought with beasts at Ephesus, what advantageth it me, if the dead rise not? let us eat and drink; for tomorrow we die" (15:31-32).

By saying that he dies daily means that he dissociates and is raised up from the physical body and its senses in the 4th state of consciousness.

However, the dissociation, according to the 1-to-7 attentional process, must be quickened (i.e., speeded up). This quickening includes what the person has "sown" and conserved in memory, for in the final analysis the strongest MC, with the created psychic mass, will prevail in the projected embodiment.

> "But some [man] will say, How are the dead raised up? and with what body do they come? [Thou] fool, that which thou sowest is not quickened, except it die: And that which thou sowest, thou sowest not that body that shall be, but bare

grain, it may chance of wheat, or of some other [grain:] But God giveth it a body as it hath pleased him, and to every seed his own body" (Ibid 15:35-38).

Baptism By Water to Generate a DLE

On contemplating the Symbol I have been informed that in ancient times the use of baptism by water was a method in which the body was held under water until it felt it was at the point of dying. In that instant, it would accelerate the cognitive cycle to produce an OBE, which would reveal the "truth" that would set one free: that death of the physical mode is not the end of our conscious existence.

This happens to many people upon facing moments of great danger, or in putting themselves in highly risky situations. Mountain climbers often report being outside of their bodies and seeing the body go through the climbing motions. There's no doubt that this is what took place with the OBEs that projected during the VW-Mack-Truck incident.

If we analyze the transformative methodologies we find that they progressively lead the system to a general equilibration that intimates entropic balance. Since the psychophysiology reacts to imagery (thought and/or spatially suggested) as if it were "real", we can understand why there can be such an "ultimate" dissociative-transformation when suggested by such symbolic contexts via all the centers.

Suggestions such as "stillness," "silence," and "paradoxical conjunction of opposites" (e.g., yin-yang symbol) require in a very general way the equilibration of active and passive forces of all the centers. While the ideal goal is one of UNITY, the practical one is that of psychophysiological conjunction as the precursor to that special state.

"Mansions" in the PUF/FCCP

Recall that at the time of death the pulse rises to the rate it had in the womb of 160 BPM (Lad 1993). Scientists have discovered a cognitive quickening in laboratory rats that is more coherent than is found during the ordinary waking state (Borjigin, et al, 2013). All this suggests that the Self is transitioning from the resistance of the physical mode of the Self into a less condensed vessel in which the cognitive cycle is much faster.

The inevitable quickening also amplifies the signature nature of each individual to "gravitate" towards a resonant-receptor milieu in the cosmic dimension of the FCCP. In the Christian Gospels these heavenly sites are referred to as the "mansions."

> "Let not your heart be troubled: ye believe in God, believe also in me. In my Father's house are many mansions: if [it were] not so, I would have told you. I go to prepare a place for you" (KJHB: John 14:1-2).

In Hindu philosophy the spiritual places in the universe are known as "*lokas.*"

Complementarity: A New Way of Thinking

In the title of this chapter we read "Psiology," which is a term that has been suggested to the writer as a way to not displace the conventional usage of "psychology," and all that scientifically pertains to it, but to rephrase its very useful findings in a holistic and counterintuitive frame of reference, which will suggest for us a moral and ethical frame of reference that can guide our personal as well as our social interactions.

This initiative does not exclude the unconventional insights of the researchers of old known as "seers," but seeks to dissolve incommensurables like subject-object, mind-body, matter-spirit, psychology-parapsychology, and the like that plague our way of thinking, and instead view them as complementarities.

These "incommensurables" have imposed the either/or way of logic that forms the basis of a historically revered dualistic cognition, which views the Self and the universe at large in schismatic terms.

While this kind of extreme-comparative thinking has its place in analytic and binary functions, it does not represent the Objective and Subjective potentials revealed to the author as being the creative basis of our existence.

It produces the rationale for fear-based psychology and collective behavior, and is summarily blind to the potentials of a love-and-understanding-based psychology and societies.

We can find a compelling example of this more advanced psiological thinking with knowledge about the inherent metabolic nature of the body. Metabolism both breaks tissue down, known as *catabolism*, and builds them up, known as *anabolism*. While these are oppositional functions they do not exclude one another in the holistic operations of maintaining a healthy equilibrium of function, known as *homeostasis*.

We don't think of either as being "good" or "bad" but only as having potential dysfunction by dint that their balanced relationship is by some means thwarted. A psiologist would therefore be one who would have a different kind of "measure" for what is "good," which is the cooperative functionality of opposites. The notion of "good" and "evil" is not a complementary relationship inherent in the psychosocial potentials of human life.

Moreover, psiology does not aim to replace spiritual or religious practice but to enrich it with a new understanding that has scientific as well as subjective implications.

A Moral Principle

From a psiological perspective "complementarity" is considered to be a moral principle from "Divine" origin. It implicates a kind of dynamic equilibration that seeks "stability" by their combined activity.

What imbues the principle with Divinity is strongly implicated in how the universe is structured to ensure objective stability with subjective freedom. It is deduced from the notion that the Creator wished to maintain objectivity (i.e., materiality, order) and at the same time, provide diversity, intelligence, and creativity.

To not have one overpower the other HI appears to have merged them in a dynamic way, such that the iteration of one leads to the other. For example, recursion equations of an "object" will, at some point, fractal into self-similar thematics. The mathematics and science of chaos has revealed this complementary nature in which order (objectivity) turns into chaos (subjectivity rich with information) and vice versa.

Different Kinds and Levels of Intelligence

In this book HI has a specific meaning. It points to the ability of a system to process energy and information at faster rates when compared to another system.

From this rationale, we may assume that there are other intelligences in the universe that have cognitive ability and creativity greater than the ones we are accustomed to witnessing on our plane of existence. It is in this regard that we can conceive of a HI to precede ours in terms of cognition and creative causality.

While we cannot know the precise nature of the origins of the universe, perhaps because we are "It" trying to "see" and understand itself, we are endowed with the ability to make ontological assumptions about our cosmic origins, and it is not unworthy of us to think of HI in the category of God or with some secular nomenclature.

Again, this kind of schismic thinking is psychologically and socially dysfunctional, and as we have seen throughout history, it produces the most heinous kinds of violence to humans and to the ecology of the planet.

Objectivity and Subjectivity of Equal Importance

What imbues the psiological approach with special significance is that it takes into consideration the need for objective as well as subjective potentials as being at the center of cognition on all levels of existence.

Therefore, "goodness" is what is fostered and what emerges from individuals and groups in developing and expressing their signature identities in the spirit of complementarities.

Love-and-understanding Based Existence

A love-and-understanding-based society is necessary for this to happen. Currently, the world we live in is predominantly ruled by fear-based rationales that define "reality" in harsh and lunatic terms, like the "real world out there" as if there were two places for reality, one being loving and peaceful and the other impersonal and rudely competitive.

Good and Evil

The notion of "good" and "evil" is simply a deviation from what maintains this stability, which would be the extremes on the positive or negative side of the contextual equation. It has been known throughout history as the "golden mean," which here is envisioned with the principle of complementarity, studied by how its constituent parts contribute to higher states of being and consciousness on all levels of existence.

3: ATTENTION

Throughout this book, attention has been portrayed as an accelerant of the cognitive cycle. As such, it accelerates the process by sustaining an impression mediated by a MC in the 1-to-7 cognitive cycle, which contributes to the development of a positive F-F cycle. When the cycle repeats for a certain period of time it engenders short-term memory, and if it repeats beyond this threshold it engenders the wherewithal for long-term memory.

While we are not aware that we use this process continuously as part of our subliminal cognitive endowment, we do have a rudimentary sense of what it is when we repeat to ourselves a number or a phrase we want to remember. We are then intentionally auto-stimulating ourselves with an input, and we know that the more we repeat it the greater the reinforcement. We intuitively know that repetition is what reinforces the memory we attempt to create.

The cycle will automatically begin whenever we attend to something in a general and/or a specific sense. So it is an inherent potential that replicates a progenitor system, any MC with some degree of predominance or from an ontological perspective (a.k.a., the Creator, God, the Divine, the Source, the Universe, etc.) in a self-similar fractal format.

Whenever the mental system encounters a contradiction, a contrasting discrepancy to existing MCs, or any kind of OOPE, it will automatically begin to accelerate in order to better understand it.

Iterating the cycle beyond a certain point generates fractal bifurcations that contribute to basic cognitive functions, especially to thinking. In this regard, attention is a psycho-accelerant that can acquire (by dint of its quantity of cycling) a force sufficient to override the pull of other MCs that have become habituated as powerful "needs," such as hunger, sex, sleep, healing time, and so on.

Moreover, because it can direct cognitive activity to different sites of the body-mind-spirit complex to enact specific behaviors, it also acts as a spatial directive capable of receiving from and projecting to anywhere within the FCCP of the Psiloton sphere of the "localized" personal Self or beyond into the broader "nonlocal" reaches of the PUF. Such "nonlocal" interactions may occur at transluminal speeds that appear to be "instantaneous" and "superconducting."

The Honing of Attention Span

Without exception, all systems in the transformative tradition employ and seek to hone attention in the pursuit of their goals. Over time, the span and scope of attention must include all the centers, including their positive and negative halves, in order to ignite the Wholistic Response.

Once again:

> "Positive feedback entrains a resonant context while negative feedback "holds" it in place. Hence, the attentional process is sustained by a complementary ratio of negative and positive feedback (attention: NF=PF) that generates a "steady state" (input at 1 equals output at 5). We experience the equivalence as recurring moments of "stillness." The extent to which we can sustain the balance, as a steady state, constitutes our attention span" (from chapter Mind is a Control System).

The following chart illustrates with some Sanskrit terms the different levels of attention span on a particular object (represented by the letter C of the alphabet).

The objective is to focus on C. The letters X, Y, X, W, V, and T are distractions. The letters in parentheses represent mental self awareness while the letters without parentheses represent "pure awareness" or, in our terminology, unfettered Presence.

ORDINARY MIND	(X)	(Y)	(X)	(W)	(V)	(T)
DHARANA	(C)	(C)	(Y)	(C)	(C)	(T)
DHYANA	(C)	(C)	(C)	(C)	(C)	(C)
SAMADHI	C	C	C	C	C	C

Since the goals involve (regardless of the system in question) bringing the whole body, mind, and spirit into a state of dynamic equilibrium that instantiates a paradoxical "holy" state in which the oppositional forces are equilibrated into an auspicious state of quiescence ($>0<$), the "object" needs to be the Self in its various modes of being.

It is important to point out that as a selective psycho-accelerant, attention attains a steady perceptual state by simultaneously incorporating two oppositional forces at the imagined input to do so: negative and positive feedback. As was mentioned, negative feedback is the deselective factor (at 4) while positive feedback is the selective one (at 3a) in the 1-to-7 cycle; they each immediately affect one another by default.

A Virtual-Attentional Feedback Device

One of the basic principles for improving any learning endeavor is through the use of feedback. As was previously discussed, feedback is almost always accompanied by feedforward behavior to alter oneself or the object in order to adjust behavior to meet the goal. It cannot be overstated that without a F-F loop the learning process is critically stunted, and as a consequence, so will be attempts at intentionally fostering higher states of being and consciousness.

Feedback Between Goal & Behavior

In Figure 4 a model of the attentional process was introduced to show the particulars of a 1-to-7 cycle that repeats in millisecond+ time frames like a movie. If the input at 1 resonantly matches the output at 5 the system experiences a rise in amplitude, telling it that the goal has been met. If not, the cycle is apt to continue until it achieves a match. It is important to remember that without feedback no efficient amount of learning can occur.

Consciousness, an Impossible Goal to Conceptualize or Imagine

One of the problems with many transformative exercises is that the goal (e.g., higher level of consciousness) cannot be clearly defined. Therefore, one must resort to something more palpable, like the use of a mantra or visual aid, as an "object" to inform one if the mental system is iterating (the 1-to-7 cycle) correctly and stabilizing a steady state.

The F-F loop occurs when one is distracted (by thoughts, etc.,) from focusing on the object, and one compares the output with the goal (the input at 1). Oftentimes the observer is absorbed (i.e., in phase with) the distractions, and it may take awhile to realize that the mark is being missed. Hence, the F-F process loses its ideal contiguous temporality (i.e., connected by a window of time). By replacing the distractions with the desired goal using the feedforward component, the cycle is refreshed and the process repeats.

This is done until the goal becomes a perceptual habit, such that whatever one is observing, the iterative activity will come of itself. However, there's a problem in that the habit is limited to a context, even though the ability to focus the mind has to some extent been strengthened.

Research has shown that this ability can be transferred to other situations when meditation is the modality to improve attention span.

> "Recent behavioral research shows that practicing meditation trains various aspects of attention. Studies show that meditation training not only improves working memory and fluid intelligence, but even standardized test scores" (Hasenkamp 2013).

The Necker Cube

A better method is to use objects that can serve as tools to accelerate the F-F process, and thereby increase attention span and learning time without dealing with too much context. A number of visual illusions developed by the Gestalt school of psychology appear to serve this purpose. One such object is the Necker cube (Figure 25).

The Necker cube is a symmetrical 2D object that the mind interprets in a 3D format. Its "illusory" nature was discovered in 1832 by the Swiss crystallographer Louis Necker, who noticed that crystals present cubic shapes that cause the mind to change perspective. It was noticed that when observed continuously the background-foreground appear to "exchange" places in an oscillating manner. Eventually it was used quite extensively by Gestalt psychology during the early part of the twentieth century to study its effects on the figure-ground nature of the perception.

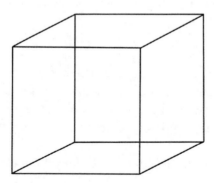

Figure 25: Necker Cube.

The Cube is Not Moving

However, in all cases the oscillation is seemingly difficult to stop. Since the cube is not actually moving it must be the dualistic mind. One compelling reason is that the mind processes information in either-or formats, and the rate of oscillation is a reflection of this. Because of the symmetry of the cube, the mind searches for an either/or resolution, but dithers in between the two states; hence, the oscillation. Since then, the Necker cube phenomenon has been the subject of numerous studies.

A Study of the Necker Cube with Long-Term Meditators

A recent study with the Necker cube by Jannis Wernery (2013) involved the ability of certain subjects to speed up or slow down the perceived inversions. Nevertheless:

> "These reports on voluntary control of bistable perception also showed that while reversals could be slowed down or sped up by participants, it was not possible for them to prevent reversals altogether" (Ibid, 83).

Yet, one study of long-term Buddhist meditators found that they could slow the reciprocal movement of binocular rivalry (another kind of bistable testing) and extremely prolong the observation of one of its phases better than short-term meditators or non-meditating subjects (Carter 2005, in Wernery, 83). This interval was called "dwell time."

Dwell Times

Dwell times indicate how long one percept is seen at a time, before perception changes to the other percept. A few other terms are used synonymously to dwell times in the literature (e.g. stability durations, reversal times, switching times, perceptual durations).

Necker Cube Provides Feedback of Mental Duality

Therefore, we can deduce that this visual device (also can be done with other senses) can provide a reference to how long the mind can dwell in the moment, using top-down intention.

322

Control Center at the Frontal Lobes

From the foregoing we've learned that attention is launched from the executive function in the frontal lobes of the brain. This constitutes the executive office or "throne" in Scripture, in which a "king" sits to rule the "kingdom within."

However, neuroscience reveals that the frontal lobes are not fully developed until about twenty-one years of age (this may vary in cultures that introduce attention-honing systems, such as mantra or prayer, at a very early age). Hence, since birth the child has been on instinctive pilot control—it basically eats, sleeps, and moves its bowels. During infancy the baby is in the grips of delta brainwaves, and, as a consequence, sleeps most of the time.

Gradually it starts to learn to move about, and stimulates the growth of its bones and muscles. Little by little the instinctive center is transferring its potential and power to the child's ability to move about and grasp things with its hands, bringing objects of exploration to its mouth.

Prior to that stage, the instinctive center has called all the shots, but as the movement center develops it begins to acquire a power it did not have before: the exploration and conquering of positive time-space, involving tasting, touching, grasping, throwing, walking, etc, and remembering what it has learned in negative space-time.

As the movement center continues to develop the intellectual center begins to abstract these movements in terms of objects in space separated by time.

Through such embodiments (see Thelen 2001) the child begins to understand basic concepts that will evolve over time to denote the emotional rhythms of language, especially the tones and facial expressions of "yes," "maybe," and "no," or absolute confusion.

Quite likely, the most excruciating lesson the child experiences during this period is related to bowel and bladder control. This begins to break the instinctive center's free reign by voluntary control of the anal and urinary sphincter muscles, which eventually becomes a habit. Mastery of the movement center over the instinctive center forms the roots of a value system.

This value system is recorded in the amygdalae as a limbic device that will sculpt the child's moral and ethical behavior for a long time to come (unless, as current research reveals, the development of the amygdalae is stunted, which is apt to lead to sociopathology). Its basic egocentric drives are now tempered by emotional control of the movement and instinctive centers.

As the intellectual center develops new conceptual models it begins to affect the values it has acquired, especially in terms of contradictions. At this stage the child's forefinger is busy pointing and asking questions that often leave the parents at wits end to find analogies that the child can understand.

Also, there are children who are born with numerous learning moments in "previous" life-times in positive time-space, as well as in negative space-time. These children may display unusual levels of intelligence and talent.

4: MEMORY

Memory can be viewed as the result of combinatorial potentials in the FCCP. In this regard each MC is a potential, like a letter in the alphabet, that contributes to a greater-than-itself effect. This appears to be the most effective way for nature to conserve energy, for otherwise each remembered event would need to be recorded separately in its entirety, requiring a much larger neural substrate. Neuroscience has quite convincingly shown that perception involves the whole brain, even the whole organism, in the formation of synchrony, the rhythm that binds disparate MCs, acting like atomic building blocks of the mind.

Is Memory Epiphenomenal?

In conventional forms of psychology memory is often thought of as some epiphenomenal figment of cognition that may disappear under certain pathological conditions (e.g., dementia, Alzheimers). Memory may also be "forgotten," which means that while it is harbored in the FCCP there's something in its context that encumbers its retrieval.

However, from a psiological perspective memory is "real" by dint that it, like all other condensates in the volume of the PUF, are fashioned from the same primal "substance" we conventionally call "energy," which is here conceived of as "potential energy" (PE) that converts upon being condensed into "kinetic energy," manifesting as fundamental waves and vortices.

Moreover, MCs are thought to be vortically configured energy waves (e.g., encoded coherent domains in the watery matrix of cells) that endure in the field as rolled up "continuous fractals" that form the levels/zones of cognition of the Psiloton volume.

At the subtlest levels memory is preserved and never "forgotten." By accessing these levels, for example through hypnosis, much of the history of the individual can be retrieved, or reconstructed (a more likely scenario).

Hence, memory is information conserved into self-referential vortical formations that configure the energy in a certain pattern, and transmit the information upon being unfurled.

Perceptual and Behavioral Fragments

Strangely enough, the "self" typically doesn't ordinarily remember its greater Self in its entirety, but in perceptual and behavioral fragments that for the most part occur on autopilot. While in these fleeting moments of existence, there is not enough processing time to evoke the whole as a unified and open system of perception, such that whenever and whatever we are involved with is witnessed integrally in body-mind-spirit.

Raymond Lull's Mnemonics

Raymond Lull (1232-1315), was a Catalan mystic, logician, and poet whose writings helped to develop the Romance Catalan language and widely influenced Neoplatonic mysticism and Christian philosophy throughout medieval and 17th century Europe.

Though there's much to know about Lull's Christian philosophy and personal life, what is of keen interest to me is how he blended a wide variety of complex ideas using a system of logically crafted symbols in the form of cartwheels and charts as mnemonic devices.

These devices were linked by an "alphabet" of categories, which the practitioner of his system needed to learn by heart. This provided the "code" by which these commensurable as well as incommensurable ideas could be bound together in a kind of subjective rationale that would have a powerful affect on the belief system and behavior of the practitioner.

Before leaving Lull, I must not fail to mention his impact on our modern times. Lull had a great influence on Gottfried Leibniz, the

German equivalent to Newton, both discovering the infinitesimal calculus at about the same time. Perhaps the most significant is the recognition of being a pioneer of computation theory, especially due to Lull's systems of organizing concepts using devices such as trees, ladders, and wheels. These have been analyzed as symbolic classification systems that had import to modern computation systems (see Welch 1990; also Abbott/Damberg 2003, for a computerized version of the Lullian Art on a PC).

Lull's mnemonic graphics also had a far reaching influence on contemporary art, architecture, and culture, as demonstrated by Daniel Libeskind's architectural construction of the Studio Weil in Port d'Andrax in Majorca, Spain, completed in 2003. Referring to the Lullian wheels, Libeskind states:

> "Studio Weil, a development of the virtuality of these mnemonic wheels which ever center and de-center the universal and the personal, is built to open these circular islands which float like all artwork in the oceans of memory" (Libeskind, 2011 in Wikipedia: Ramon Llull).

Giordano Bruno's Memory Devices

Giordano Bruno (1548-1600) was an Italian Dominican friar, philosopher, and cosmological theorist, whose radical views disfavored him with the leaders of the Catholic Church, which eventually led to his being burned at the stake.

He is known for his cosmological theories, which conceptually extended the revolutionary Copernican model. He proposed quite presciently that the stars were just distant suns surrounded by their own planets outside of the solar system, existing in an infinite universe with no celestial body at its center. He also raised the possibility that these planets could even foster life. He was subsequently honored as a martyr who gave his life for the preponderance of science over religion.

> "Bruno's infinite universe was filled with a substance—a 'pure air,' aether, or spiritus—that offered no resistance to the heavenly bodies which, in Bruno's view, rather than being fixed, moved under their own impetus (momentum). Most dramatically, he completely abandoned the idea of a hierarchical universe" (Wikipedia: Giordano Bruno).

Like Lull, he was also known for his works on the improvement of memory using mnemonic graphics and symbols.

The efficacy of these devices was to connect various fields of information at the different nodes and to study their linkages as combinatorial possibilities.

Reinforcement is the Magnitude of the Assembly

Without going into the details of these mnemonics techniques, on first sight they hold a strong affinity with the hypothesis that they foster neural assemblies of large size and scope, which goes towards supporting the research that memory is reinforced by how it is widely distributed throughout the nervous system with each iteration of the 1-to-7 cognitive cycle.

Repetitive Process as Ritual

In keeping with all that has been presented, it behooves us to keep in mind that reinforcement is essentially the repetition of a process, which, according to the thesis being propounded here, is the 1-to-7 cognitive cycling of a percept.

Hence, when repeating a prayer, mantra, or affirmation we are creating an enduring memory that needs to be "stronger" than the PSQ. This is the purpose of all kinds of ritual.

One must keep practicing the transformative process until it acquires the status of a habituated, yet evolving MMC that is at least equal to or greater in magnitude to the PSQ, and thereby be able to neutralize and entrain it into its nature. Afterwards, intentional practice is not necessary since it would be taken up by "long-term potentiation."

Using OOPEs to Reinforce Memory

It is interesting to note that there is a tendency to remember strange events or OOPEs, given that by their strangeness they have an accelerating effect on the cognitive cycle, which in turn propagates a plethora of fractals on a particular thematic.

This quantitative connectivity factor is held here to be the rationale for facilitating efficient memory creation and recall by dint of the scope and breadth of the resulting assembly.

For example, if we wanted to remember a particular event we could couch it in exaggerated metaphors involving scale and descriptives. Simply exaggerating the size of all that is perceived in that moment would render it more accessible to recall at some future time.

Exaggerated Icons and Architecture

Though rarely recognized as such, the building of temples and spiritual objects to enormous size suggests that this reinforcement technique extends back to remote antiquity. Many temples and cathedrals were built with these "intuited" principles in mind.

Upon entering these huge structures one is immediately immersed in the symbolic images, statues, and icons within the enormous space. For the peoples of those times, these sacred spaces generated a strange-mystical beauty that echoed the oral transmission of their personal and transcendental beliefs.

Many of these structures display golden-ratio proportionality (see Doczi 1981) that couples and resonates with "The golden mean as clock cycle of brain waves" (Weiss, et al, 2003), which implicates how the large and the smaller couple energetically according to this "Divine proportion."

Imaginary Exaggeration

Also, using the imagination to exaggerate the features of goals symbolized by objects, including that of the body, utilizes larger swaths of memory, and therefore increases assembly size and the potential for manifestations.

Great Beauty

Whenever we experience something that elicits the sense of beauty in us we are apt to have an enduring memory of it. There's something in the composition that connects with our sensorium in ways that cannot easily be explained. However, we know by experience that it pervades our entire being, fulfilling that quantitative factor being espoused thus far.

This would be true, regardless of the kind of beauty. It could involve for example: the visual arts, music, literature, poetry, architecture, culinary arts, dance, sports, mathematics, nature, and even sunrises and sunsets (which, by the way never repeat the same pattern).

Pavlovian Connectivity & the Surreal

Memory is also facilitated by temporal connectivity. This was proven by the Russian physiologist and Nobel laureate Ivan Pavlov (1849–1936), who is best known for his studies on the conditioned reflex. His experiments with dogs showed how the secretion of saliva can be stimulated not only by food, but also by the sound of a bell temporally associated with food.

Because the objects that present themselves in highly charged emotional moments are bound by temporal connectivity we remember the entire scenario, mainly at the non-conscious level, which often emerges in our dreams in strange surrealistic patterns. They can also be included in deep symbolic expressions that emerge from the 3rd and 4th states of conjunctive consciousness.

The Physical Mode of the Self and Symbolism

Though not expressed in such terms, throughout the history of the transformative tradition the human Self was viewed as a fractal, a microcosmos (e.g., in the "image of God"), of the Divine, the intelligence behind the macrocosmic order.

Trust

It behooves the practitioner to have a sense of unflinching certainty or trust that the intended purpose is manifesting NOW, that is, already formulated as an evolving "seed" in the FCCP; not necessarily as behavior at 6, even though the image can be construed as behavioral.

It is important to keep in mind that we are embedded in the Psiloton, the fractal equivalent of the PUF and its condensations, and the present moment is everywhere throughout its volume, regardless of whether it has entered our conditioned threshold of consciousness.

In advanced practitioners the temporal "interval" between the fractal "seed" stage and the behavioral manifestation is much shorter than in non-advanced practitioners. This often appears as "miracles" and "magic" to slower mind systems.

Anticipation of all levels of being and consciousness has its roots in the limbic system, where all of our expectations are founded (see Freeman 2000). The physical organ of the subtlest MCs would be the pineal gland, operating in the ultra-violet blue-shift-part of the EM spectrum (see Bosman 2000).

Magnitude & Formation of a Master Memory Construct

This integrality is the essence of the Wholistic Response and the quantitative source of all power. As research has revealed (Melloni, *et al*, 2007), it is the size, scope, and depth of the neural assembly (as the collective physical aspect of all MCs) distributed throughout the FCCP that determines the quantity and quality of conscious events.

While the research focuses on "ordinary" states of consciousness, the psiological perspective expands on the findings to show that size, scope, and depth can be increased to incorporate the whole of the Self, including its polar attributes, in all of its modes of being.

The developmental process is the "mechanism" which facilitates the attentional process that needs to be prolonged beyond a certain rate of cycling, in order for it to frequency double and begin to fractal and bifurcate, thereby inwardly increasing the magnitude of MC connectivity in the FCCP. These extended micro-constructs are then amplified by the outgoing vortex into larger conscious synchronous assemblies that are referred to as "states."

Therefore, magnitude of MC connectivity throughout the FCCP is a very important factor that directly correlates to changes in being and consciousness. When the magnitude of a MC is sufficient to create changes it is called a master memory construct (MMC).

Essence's Signature

We are all born with signature, given, according to the revelation, that we are fractals with nuances of difference and processing speed of the primal progenitor system (a.k.a. "God" or other deific or secular terms). Based on the principle of constructive interference we can accelerate the nature of this essential nature (a.k.a. Essence) by repeating it to itself, that is by turning the image of the Self, in any or all of its modes of being, back into its Self (signified in Scripture as "I am that I am"). This Self-referential activity amplifies the Self to operate at a greater level of coherence and amplitude, forming the basis of bliss (and similar emotional states), and with a faster ability to process energy and information.

Therefore, silently talking with one's Self may have a similar effect.

State-dependent Memory/Learning

State-dependent memory or learning is the phenomenon through which memory retrieval is most efficient when an individual is in a similar psychophysiological state as they were when the memory was formed. Studies have found that it has practical implications for memory retrieval.

> "The term is often used to describe memory retrieval while in states of consciousness produced by psychoactive drugs—most commonly alcohol—but has implications for mood or non-substance induced states of consciousness as well" (Wikipedia: State Dependent Memory).

While much of what is imagined occurs fortuitously in response to the outer environment (around 85% of the time), the ability to intentionally imagine is key, as to which the voluminous "self help" literature attests (e.g., Miller 1978), in goal-directed behaviors, of which spiritual development is one. Interestingly, improvement of athletic performance has been the main use of this technique.

CREATION OF A MASTER MEMORY CONSTRUCT

The PSQ is the functional MMC for ordinary states of being and consciousness. Its psychic mass has been built over many years of social conditioning, and its range in the FCCP is typically localized to the default mode network (DMN). Though it is not conducive to producing the WR, it fluctuates from moment to moment in response to outer events. Each fluctuation is, for the most part, too weak to generate a wholistic effect, but in some instances it can, but with little or no enduring benefits. In many instances these mini-convergences are about negative emotions.

The PSQ maintains a level of stability by vibrationally separating the non-conscious domain from a predominance of habit patterns through which the psycho-organism perceives and reacts to the world.

The aim in transformative psiology is to create a new mediating MMC that would support Essence and the transformative goal, and yet maintain functionality in the positive time-space universe.

Built in Secret

In this regard, the new MMC would need to be gradually built outside of the influence of the PSQ until it acquires enough psychic mass to entrain/transform the old MMC into its context. This needs to be done gradually and, to some considerable extent, in "secret" so as to not allow it to be usurped by the PSQ.

Transformation, Not Destruction

Also, loss or critical disruption of the PSQ without having a suitable replacement would leave one in a critical state of confusion that may lead to not being able to function effectively in positive time-space. Therefore, it is not recommended that one try to destroy the PSQ, bypass it, or, as some purport, to transcend it without firstly establishing the new MMC.

Enough Gravitas to Entrain the PSQ

Therefore, the first step is the creation of a master memory construct (MMC), that at some stage of development will have enough "gravitas" to entrain the non-conscious domain to generate will, permanent 'I' (sense of Self as a Presence), and an expanded level of consciousness.

A practical method for creating an evolving MMC is to formulate it in personal terms using four of the questions that were asked about the transformative tradition.

1. What is my goal or purpose?
2. What prevents me from attaining it?
3. What methodologies help to overcome the obstacles?
4. How do I know if I'm moving towards the goal?

In this way you'll have a framework that organizes the information in a succinct and more coherent manner. This makes it easier to remember, and, of equal importance, it contributes associatively to the formation of larger and more coherently-powerful neural assemblies.

The guiding maxim for this section is "United we stand, divided we fall."

As explained, for the Self to transition into a higher vibratory state of being and consciousness, it needs to produce a Wholistic Response in which all of its aspects are vibrating in a synchronous manner, as an oscillation between the positive and negative values that have been "given" equal status (via acceptance, love, etc.) in psychic space.

For example, when two memory constructs of equal value move towards each other they enact a cancelling effect. If they move in concert they will merge and amplify, and morph into a third context known as "concatenation" involving the golden ratio (see Roopun, et al, 2008).

How Valid Must the Model Be?

As I have been working with the Symbol to organize my research into a cognitive model of the universe, I kept wondering how it would impact my Self and its developmental process.

Throughout the expository process I would fall into moments of "objective compulsivity" and would wonder if the model would pass some kind of "peer review" to validate its assumptions. Even though I knew that it was a lopsided way of thinking, at times it would overwhelm me and make me feel that all this theorizing was but a bunch of rubbish.

In those moments the sense of Presence would be eclipsed and I would need to struggle to come out of the miasmic cloud. No matter how I contemplated the Symbol I could not get out of this smoke screen. I would take a break and do something else with my mind, and in those side-tracking moments I could "hear" the whisper of the Presence come to the rescue with variations of this reinvigorating theme:

> "It doesn't matter what model of the universe you concoct or accept just as long as it can relate to your Self, especially in its physical mode of being, and cause it to vibrate in synchrony. Remember, the organism enacts your belief, regardless of how logical or not, in your interpretation of reality."

This brought to mind Ouspensky's phrase "we need to shake the machine." Shaking the machine means to get the Self, in its physical as well as in its psychological modes of being, to vibrate as a whole in synchrony. The term "shake" doesn't quite convey this meaning, but that's what it suggests.

In this regard, it is not just a group of neurons connected in synchrony throughout the brain to produce the phenomenal prerequisites of awareness, but the entire psycho-soma functioning in alignment as one. This would involve the watery matrix of memory throughout the body in which the electromagnetic signature (EMS) of the DNA is embedded, the fascia, and the enveloping heart wave.

Therefore, the symbols one uses to model the universe and/or the Self need to have associative connectivity via the imagination with one's entire sense of being, which initially is brought forth by the physiology in the form of sensations. If, for the sake of illustration, we equate the physiology to a gong, the symbol would equate to the baton used to make it vibrate.

Of course the gong needs to be clear of obstructions for the energy to move efficiently through it. If done properly, the sound we hear consists of the fractal harmonics of a subtler kind of processing, which we refer to as the "mind."

To begin this transformative process we begin with enough overtone activity in the normal waking state to auto-stimulate ourselves into the standing properties of TS waves. In doing so we form a self-referential F-F loop with our Selves. If we keep iterating the percept that we have intentionally created, the looping becomes a robust positive feedback cycle that will continue subliminally as "long-term potentiation."

Protecting the MMC in Virgin Psychic Space

An important activity the School encouraged was for students to go to art museums to take in fine impressions. While at the Detroit center, a group of students decided one day to take a trip to the National Gallery in Washington, DC, to view the collection of works by the Italian Renaissance painter Raffaello Sanzio da Urbino (1483-1520), popularly known as Rafael, who was considered to be a "conscious being" by the Teacher. At the Gallery I spent a bit more time looking at The Alba Madonna in its circular frame.

After contemplating it for a while, the esoteric meaning of the painting suddenly occurred to me. The Madonna has the Christ child on her lap and to her right is St. John the Baptist. In this beautiful symbolic scenario, the Madonna represents the unused parts of the centers, the virgin, that harbors and protects the developing Christ child, who is the "seed" or MMC of the transformative gospel that is to revolutionize the common mind into a new way of thinking and emotionalizing.

The new way of thinking is based on being aware of oneself so as not to judge, and the new kind of emotionalizing was that of unconditional Love (i.e., agape), for oneself and others. Love could only occur in a mind that was at peace with itself. A peaceful mind is a balanced one in which the negative and positive halves are equal (as represented by the triangular envelope). I remained before Rafael's masterpiece for a long time, held by the insight and subtle energy rippling throughout my body.

The impression and its message penetrated my entire being. For the first time I experienced something quite explicitly with all of my centers, and in doing so a very powerful state was formed. Even after so many years have passed, the mere thought of the Alba Madonna recreates the state it produced when I first saw it.

The beautifully arranged colors speak to the emotional center while the triangular spatial composition that frames the characters within the circular frame speaks to the moving center. The theme of mother and child connects with the regenerative and nurturing aspect of the sex center. The message alluding to the Gospel stories of Jesus evokes the intellectual center. In higher centers it all comes together in an energetic holistic convergence that is spiritual.

Repetition: Accelerating the F-F Loop of Wholistic Causation

We need to come back to the basic model of the attentional process, illustrated in Fig. 4, to appreciate the iterative 1-to-7 cognitive cycle that needs to be incorporated in order to create a powerful MMC that can entrain relevant contexts in the FCCP that would satisfy the intended goal.

The F-F loop of wholistic causation was more explicitly modeled in Fig. 15 to show the interactions of expectations, related to the conditioned mind, and out-of-pattern events (OOPEs). As was explained, the ratio between such novel events (of which the transformative goal is certainly one) and the habits of mind, determines if they will be neutralized to allow for HI to come into the scene. Hence, in the symbolic dimension of the Gospel of John we read:

> "In the beginning was the ratio, and the ratio was with God,
> and the ratio was God" (Seife 2000, 26).

If the OOPE does not equate in terms of psychic mass to that of the conditioned mind, the PSQ would continue to rule one's psychology without any resistance.

As is well known, this pattern has been developing since early childhood, and as some systems aver, as a memory construct recurring over "life times." Therefore, it has without question the psychic mass to entrain the psychophysiology into an enduring persona, or when fragmented into multiple personas. Note, that when it doesn't develop properly it can lead to psychophysical aberrations, especially when it diverges radically from the needs of Essence. Therefore, care needs to be taken when we tamper with it.

When Sharing the Same Psychic Space

Initially, the iterative practice will be fraught with conflicts, now that opposites are sharing the same psychic space. This is expected, and the adept needs to be made aware that this is an inevitable phase, requiring compassionate guidance and support. Over time, the iterative process will reinforce the MMC to the point where it will assimilate and take over the role of the PSQ, allowing for continued development in challenging social contexts. At this point the MMC becomes a "force."

The Wholistic Response In Yoga & Other Systems

While the symbolism in Yoga is different from its Western expressions, the principle of unifying the seven chakras is essentially the same. The process aims at aligning the various neural plexus (which also implicates endocrine centers) distributed vertically throughout the organism into an equilibrated resonant system.

Once the balance is established it is able to draw "spiritual" resources from systems more intelligently, processing energy and information at much faster rates, with greater access to the cosmic realms of the FCCP.

The Neutralizing Effect of "Balance"

It is important to understand that the aim is to temporally align (i.e., rhythmically entrain) the five lower centers in a state of paradoxical alignment (e.g., sympathetic = parasympathetic) in order for them to equilibrate, neutralize, and conjunctively attract the integrating and amplifying (i.e., the psycho-accelerative) effects of the life force entering through the spiritual centers from higher levels of the FCCP.

Triggered at Will

The sensation and physical attributes of the Wholistic Response are connected to the Symbol (the image of the physical mode of the Self, or any other "object" associatively related to it), and remembered at the attentional site in the frontal lobes (i.e., the higher emotional, ajna, etc.). This is where the Wholistic Response can be triggered at will.

Specific Gestures to Trigger the MMC

Specific bodily gestures can be used as conditioned triggers, referred to in Hindu and Buddhist practice as mudras (e.g., a symbolic hand gesture used in ceremonies and statuary, and in Indian dance). Because of their peculiarity these gestures and/or postures are connected to "virgin" areas in the psychophysiology in which information can be "isolated" and conserved away from the PSQ while the MMC is developing.

Once established there as a "habit", the auspicious response can be temporarily evoked or permanently established by just connecting to that spot. Recall that humans are microcosmic fractals of the whole universe, and in this regard higher centers in the head are where the inner Divine rules its universe writ small.

MMC Must Initially Develop in "Secret"

However, while it is being developed the MMC needs to be kept out of the hands of the PSQ to avoid being usurped and distorted. Historically, this has best been accomplished with the use of esoteric symbols in the form of myths, rituals, and allegories. The alchemists brought this technique to a highly developed art form, displaying the virtues of the subjective potential.

In Scripture it is alluded to with the phrase "let not thy left hand know what thy right hand doeth."

> "Take heed that ye do not your alms before men, to be seen of them: otherwise ye have no reward of your Father which is in heaven. Therefore when thou doest [thine] alms, do not sound a trumpet before thee, as the hypocrites do in the synagogues and in the streets, that they may have glory of men. Verily I say unto you, They have their reward. But when thou doest alms, let not thy left hand know what thy right

hand doeth: That thine alms may be in secret: and thy Father which seeth in secret himself shall reward thee openly" (KJHB: Matthew 6: 1-4).

MMC Must be Valued as a Treasure

In the following verse I will annotate in parentheses the psiological meaning of the common terms with their transformative meanings.

> "Lay not up for yourselves treasures upon earth, where moth and rust doth corrupt (in positive time-space), and where thieves break through and steal (the habitual response of the PSQ): But lay up for yourselves treasures in heaven [in the executive function of the frontal lobes], where neither moth nor rust doth corrupt, and where thieves do not break through nor steal: For where your treasure is, there will your heart be also (higher emotional center). The light of the body is the eye (the "cosmic egg" light body of the pineal gland): if therefore thine eye be single [sustained attention], thy whole body shall be full of light (quickened by positive F-F loop from sound to light). But if thine eye be evil (fragmented), thy whole body shall be full of darkness. If therefore the light that is in thee be darkness, how great [is] that darkness!" (KJHB: Matthew 6: 19-23. Parentheses added).

All myths, parables, and allegories are rendered in ordinary, yet unsuspecting, language that to the informed reader connects to a deeper meaning, referred to in Scripture as the "word," which, by the way also translates from the ancient Greek as the "ratio" (see Seife 2000).

METHODOLOGIES FOSTERING THE WHOLISTIC RESPONSE

A methodology is defined as "a system of methods used in a particular area of study or activity" (NOAD). In regards to transformative psiology (TPsi), methodologies are studied in terms of how they are conducive to forming the Wholistic Response (WR).

In other words, they can be viewed as modalities that have demonstrated through experiments and experience the ability to cause the psychophysiology, which is also referred to as the system, to vibrate in synchrony.

From the foregoing it has been concluded that for this to occur, four principles need to be included in the transformative process: motivation, intention, attention, and memory.

Accordingly, a person must have the desire that is intentionally enacted by focusing attention to reinforce a MMC; this will set the stage for engendering the WR that will cause the system to release the Presence, which in turn will become the key percept of the F-F loop.

Therefore, there are essentially two methodological categories that need to be considered:

1. Modalities for total system activation; and
2. Amplification of the Presence via a continuous F-F looping.

Let's elaborate on each category.

MODALITIES TO ACTIVATE THE WHOLE SYSTEM

Everything in the body-mind-spirit complex is implicitly connected. However, under ordinary conditions these interconnections are divided by what and what is not selected at 3a by the ingoing vortex. The potential of connectivity in the FCCP is infinite if one considers all the possible permutations in the volume of the Psiloton, not to mention the connectivity with other Psiloton systems.

Therefore, any modality that is to be used to evoke the largest degree possible of connectivity (i.e, assembly size) must avoid having specificity in terms of context, such as would occur with affirmations desiring a particular result.

That is why, for example, certain nondescript sounds, referred to as mantras, are used in Yoga, and why making affirmations to manifest powers of all kinds (siddhis, magic, telecommunication, channeling, etc.), including those related to healing, is not initially encouraged.

Nevertheless, this would be allowed to one who has fully developed the existence of the Presence to the level of an enduring habit (a.k.a. "crystallization") in which a high level of conjunctive consciousness endures.

With this rule of thumb, we can understand how any center—instinctive, movement, emotional, intellectual—in the organism can be activated in a conjunctive manner to produce the WR, which in turn paves the way for the emergence of the Presence.

The key cognitive function that determines how, when, and how much of these centers can be activated thusly is the imagination, especially when it is mediated through the heart wave so as to reach the system-wide analog system of the fascia in which a very-generalized memory can be formed.

It is, therefore, in this rather technical sense that the specific and general purpose of all kinds of modalities may be understood. This quest for equilibration needs to begin with the physicality of the organism, since it is the repository of all of our inherent potentials.

Tapping the Inherent Potentials of the Organism

In biological terms the organism is the ground from which the mind emerges, primarily as a kind of "watch dog" to the ever-changing outer world.

In order for it to do so it must project from its inherent nature a cognitive system that is able to process energy and information at a rate that is faster than the cellular matrix. That, of course, would be a molecular one, which in turn, when confronted with issues the molecular matrix cannot resolve, projects an electromagnetic field, and so on.

I was given the opportunity to experience the difference between these two states when I projected out of the organism in molecular as well as in electromagnetic bodies. Each one was cognitive and alive. The former, as you may recall, got stuck in a corner of the ceiling while the latter one went right through the walls like a radio wave.

Yet before all this took place, the Self immersed in the organism did not presumably foresee its own potential, that it could develop into a cognitive force that could have the wherewithal to rule the organism through the F-F loop of holistic causation. This is symbolically stated in the Gospels as:

> "And the light [Divine potential] shineth in darkness; and the
> darkness comprehended it not" (St. John 1:5. Bracket added).

In many cases the mind remains tethered to the will of the organism, but should it, for some reason, buck the instinctive thresholds the body will actually fuel the development of the mind to accelerate beyond its rate to process energy and information in order to "protect" itself.

Size and Scope of the Assembly

While projecting accelerative fields to resolve issues, slower ones cannot resolve what has been postulated to be the modus operandi for all cognitive events, since it is limited to what is selected by the input being processed by the ingoing vortex. This leaves behind an enormous repertoire of energy and information by default. Hence, ordinary cognition is not conducive to generating the WR.

Since assembly size and scope are key variables to cognition, the question arises: how much of the Self is one able to select, and how is one able to do it? Obviously, it would be necessary to include the enormous amount of energy left behind in the FCCP by default.

A key transformative benefit comes from the fact that the organism believes what the mind perceives, whether it be "actual" or created from the imagination. According to what has been presented thus far, the notion of "mind" needs to be understood as the emergent or projected system used to process energy and information at rates the slower system cannot produce at its current rate.

Therefore, body, mind, and spirit are progressions of human potentials defined by processing speeds. From this qualifier we note that the quickening of the organism becomes the mind, and the quickening of the mind becomes the spirit.

Systems of spiritual development simply tap into this natural potential of progressive emergence and take it to another level. As has been explained, this is done by auto-stimulating oneself, using the imagination, with an intentional positive F-F loop to accelerate the process into a higher level of being and consciousness. If one is able to repeat the conjunctive states at will it becomes an identity-defining habit.

Wanting to See and/or Know Accelerates the Mind

The transformative process begins with the desire (motivation) to want to know one's Self. It begins by wanting to experience the most palpable object of the Self, the organic mode of being.

Obviously, there's much of the biology that we cannot see, but can only sense somewhat. Interestingly, however, the moment we turn attention to its solidity, we find that we have to some degree energized it with a positive F-F cycle, and, if sufficiently dwelled upon, the "mind" inherent in the potential of the physiology (especially the neural network) begins to manifest to the observing mind, which by the way is going a bit faster than the physical mode of being in order to "see" a modicum of its nature.

Simply wanting to see the body in its entirety automatically evokes the attentional potential to speed up its processing rate of the physical image, because this is not something the PSQ is used to doing. Once the image is established it feeds back to the body, which in turn feeds forward to the mind. However, this wholistic observing requires an extra amount of energy, which the organism is not so happy in providing.

Under ordinary conditions we may be able to "tell" it to move a finger or walk in a certain way, but we'd have difficulty having it give up enough energy for the mind to become aware of as much as is possible of the entire system for an extended period of time. At some point it will revert to habitual modes of thinking and behavior.

If, for example, hunger crops up, the project embarked upon would most likely be postponed as "you" head for the cookie jar. If this isn't the case you'd be swamped by imagery and thoughts of food. The key thing that may have been learned is that the instinctive center is a formidable force, and that not only is it powerful but it is also quite stingy with its resources.

The only strategy to overcome its force is with an equally powerful MMC. If the MMC is to be effective it must have already established its frame of reference to as many of the modes of the Self as possible, using the pulsing broad magnetic field of the heart to establish the principle of magnitude.

INSTINCTIVE MODALITIES

Instinctive modalities involve connecting with those functions of the organism that are normally considered to not be under conscious control and to operate autonomically. Not being aware of them makes it very difficult, or close to impossible, to develop a F-F connection to establish a learning cycle in which the function could be intentionally modified. Nevertheless, the transformative tradition has found ways, with specific psiological understanding, to bridge this gap.

Food

Perhaps, the most widely used method to affect instinctive tone and health is through the ingestion of foods and liquids. For example, in Ayurveda, the ancient medicinal science of India, the balance of three *"doshas"* (a.k.a. humors) known as *Vata* (related to air), *Pitta* (related to fire), and *Kapha* (related to earthiness) is connected to the predominance of these attributes in foods and how they are ingested to create equilibrium to compensate for one's body type (based on a predominance of one or two of the *doshas*). A similar system is found in Chinese medicine, and in other systems. Many Hindu transformative teachings recommend forms of vegetarianism in order to lighten the metabolic load on the body. However, there are transformative systems that have no such dietary restrictions. Yet they all have produced examples of beings with high levels of consciousness.

Fasting

Fasting is a common practice amongst many transformative systems. It relates to the withholding of food (i.e., solid food) while praying and making humble supplications to a Higher Power (God, etc.). Its consistent mention in Holy Scripture strongly suggests that the authors/believers of these texts (whatever the historicity of how they came about) were privy to the wholistic affects it had on the entire psycho-soma.

Nowadays, the discoveries of synchrony in neuroscience, and its scope throughout the system, help explain how the dearth of resources, in this case food (also: oxygen, and impressions, as in sensory deprivation), has a conjoining effect on individual units, which in this case are the cells and MCs distributed throughout the system. This is one more example of how commonly felt threats bring people, as well as cellular and memory constructs, into cooperative (in some instances as mob behavior) unity to solve a perceived (actual or imagined) problem.

A classic biological example of this phenomenon is found in experiments with colonies of the slime mold.

> "Slime mold or slime mould is an informal name given to several kinds of unrelated eukaryotic organisms that can live freely as single cells, but can aggregate together to form multicellular reproductive structures. Slime molds were

formerly classified as fungi but are no longer considered part of that kingdom" (Wikipedia: Slime Mold).

What makes the behavior of the slime mold such an interesting area of study is how they convert from a life of individual cells into an aggregate of cells that act as one unit. When their food sources are plentiful the slime mold morphs into individual cells that roam their habitat in diverse directions, sort of doing their "own thing." However, once the food supply runs short they send out molecular signals to one another and return to morph back into a singular composition acting in unison as an individual cell. It seems that the slime-mold effect is the modus operandi of many, if not all kinds of cells, including the diverse cells of the human body.

Yogis have known since ancient times that by slowing the breath, and thereby reducing the oxygen to the brain, the mental system slows down its hyper responsiveness to stimuli. Meditators experience this effect as they watch the breath and relax the musculature. As the breathing slows down it appears to prolong the intervals between the in breath and the out breath. At some point it appears to take a long pause, seemingly ceasing altogether. During this cessation the brain cells (as well as the other cells of the organism) appear to go into synchrony, forming a pleasant sense of Presence. From the example of the slime mold it would seem that the lessening supply of oxygen triggers a unifying effect as a mechanism of collective survival.

Could this also be what's behind experiments involving sensory-deprivation? Conventionally, it is not thought that impressions entering the organism have any role in sustaining the function of the sensory system (the sensorium). However, as Gurdjieff proclaimed, from a psycho-transformative perspective, impressions—the smallest units of thoughts, feelings, and sensations—are a subtle form of "food" for the organism, with particular emphasis on the development of the soul.

Hence, fasting could be psiologically defined as the withholding of any kind of life-sustaining substance while praying and making humble supplications to a Higher Power (God, etc.) as the means to unify the psycho-organism into an integral state. The following quotes are examples involving fasting taken from the King James Bible.

> "Then I proclaimed a *fast* there, at the river of Ahava, that we might afflict ourselves before our God, to seek of Him a right way for us, and for our little ones, and for all our

substance. For I was ashamed to require of the king a band of soldiers and horsemen to help us against the enemy in the way: because we had spoken unto the king, saying, The hand of our God [is] upon all them for good that *seek* him; but His power and His wrath [is] against all them that *forsake* Him. *So we fasted and besought our God for this: and He was intreated of us*" (Ezra:21-23. Italics added).

We glean from this passage that to seek the will of God requires that it be sought with all of one's body, mind, and soul, that is via the Wholistic Response. Moreover, that this wholeness is the "measure" by which the seeker does not "forsake Him." In the following passage we glean another aspect involved in the practice of fasting.

"But as for me, when they were sick, my clothing [was] sackcloth: *I humbled my soul with fasting; and my prayer returned into mine own bosom*" (Psalms 35:13. Italics added).

It implores the practitioner that fasting be accompanied by humbleness, and, furthermore, that it be directed to the broadcasting heart wave centered in the bosom. In the next passage we learn that fasting has its place in the transformative process.

"Pharisees used to fast: and they come and say unto him, Why do the disciples of John and of the Pharisees fast, but thy disciples fast not? And Jesus said unto them, Can the children of the bride chamber fast, while the bridegroom is with them? as long as they have the bridegroom with them, they cannot fast. But the days will come, when the bridegroom shall be taken away from them, and then shall they fast in those days" (Mark 2:19-20).

In the above quote from Mark, it is made clear that fasting is what brings the "bridegroom," that is, the Presence, who as long as "he" remains fasting is irrelevant, but once "he" is gone, fasting is required to bring "him" forth again.

Moreover, fasting needs to be done personally in secret, simply because it appears to be intimately involved in protecting the MMC during its formative stages.

"Moreover when ye fast, be not, as the hypocrites, of a sad countenance: for they disfigure their faces, that they may appear unto men to fast. Verily I say unto you, They have their reward. But thou, when thou fastest, anoint thine head, and wash thy face; That thou appear not unto men to fast, but unto thy Father which is in secret: and thy Father, which seeth in secret, shall reward thee openly" (Matthew 6:16-18).

Sensorial

The sensorial system—involving the auditory, optic, taste, olfactory, tactile, and proprioceptive systems as well as the meridians, the electromagnetic field of the heart, and the biochemical receptors— is utilized in transformative psiology as the means to equilibrate the centers into a functional balance. Note that each of these modalities has a spectrum of sorts used to differentiate information impacting the sensorium in the form of sound, photons, biochemicals, and pressure waves. The transformative aim is to integrate the spectral order into a state of harmony. In this regard, the proprioceptive system is of special importance given that it is naturally designed to maintain balance via continuous feedback loops with the Self and the outer environment; though this happens most of the time in autopilot. Paying attention to one's movements can create an auspicious connection to that feedback phenomenon.

Entheogens

Biochemical approaches use natural (e.g., mushrooms, ayahuasca yielding psilocybin) and synthetic substances (e.g., LSD, DMT, etc.) that appear to chemical analysis to have affinity for various neural receptors with blocking (antagonistic) or enhancing (agonistic) effects on one's mood and perceptual states. Perhaps blocking effects produce a predominance of negative feedback while enhancing effects have a predominance of positive feedback; either of which may induce positive or negative experiences. Entheogens that prevent the cognitive system from being contained in the default mode network (DMN) are conducive to opening the flood gates of perception, leading to a psychedelic smorgasbord of lucid dream-like and symbolic contexts. It is interesting to note that many of these biochemicals are produced by the organism, such as DMT (N, N-Dimethyltryptamine), which is naturally produced in the pineal gland and is thought to be involved in spiritual seeing and OBEs. While some of these substances open the gates of perception they don't necessarily bring the psycho-soma into the desired equilibrium.

Psychic Mass & Topdown Alterations of DNA

When a mode of the Self gains enough "psychic mass," so to speak, it acquires control over the system "below" it. In this regard, the saying "the higher can see and control the lower, but not the inverse" quite literally applies.

In such advanced cases, one can imagine that a physiological effect will occur that can alter expression of the DNA molecule. A powerful example of such control is found in the case of Swami Rama by the doctors at the Menninger Clinic (Green/Green 1997). He quite dramatically demonstrated that he could use his imagination to generate a lump on his thigh, an indication that he could affect the physiology encoded in the DNA molecule, and just as easily remove it, a feat that goes toward supporting the epigenetic hypothesis that thought and emotions affect the nucleus of our physicality.

Great Danger

While we don't ordinarily think of it in this way, danger, especially to the organism, has the propensity to generate huge neural assemblies. It is as if the entire psycho-organism becomes focused on what could be of considerable threat to its existence. Of course, danger presents different kinds and levels of criticality to different people.

Different systems bring the notion of danger into their methodologies by warning their students of the dire consequences of not pursuing the prescribed developmental process with seriousness. In almost all transformative teachings the sense of danger subtly looms in the background, given that the evolutionary path is a "movement" from being "less" to being "more," with some developmental chaos in between.

Some teachings aim to neutralize this potential trace of anxiety by indicating in their beliefs that we are already what we are seeking to become. This seems like a gentler approach, which of necessity needs to include other MC aggregators to acquire magnitude.

All allusions to the "end of days," "end of the universe," "Hell, fire, and brimstone," and the like are based on creating a sense of danger (whether being "true" or not) in the believer. Their anxiety-producing effects make them poor choices to accelerate the cognitive cycle (quickening). Love and compassion are far superior by many orders of magnitude in terms of time and their transformative effects.

MOVEMENT MODALITIES

The movement center is characterized as those functions that are under voluntary control, such as the ambulatory musculature (involving the somatosensory strip in the brain). Because the instinctive and movement centers are held to operate with the same grade of energy (Gurdjieff) working with movement is a key way to affect processes and functions in the rather inaccessible instinctive center. Also, because the voluntary musculature is so well distributed it is useful as a conduit in activating specific as well as greater portions of the system in a simultaneous manner.

Postures

Asanas are the classic example of the use of postures that contribute to manifesting the WR. Asanas constitute a series of simple to complex postures adopted in performing Hatha yoga. When observed with a clinical eye, asanas require balancing the body in a wide variety of challenging ways. In order to maintain an "effortless" state of equilibrium reflected throughout the body the mind needs to go faster to enact such a generalized control.

Mudras

Mudras are symbolic hand gestures or poses used in Hindu and Buddhist ceremonies and statuary, and in Indian dance. Since these gestures or poses may differ from one cultural setting to another it is assumed that they have been subjectively used to enact certain behaviors and meanings. Once they have been learned they can be used to elicit objective outcomes (quite likely by Pavlovian conditioning). Hence, the left and right sides of the body can be manipulated to evoke the WR and its paradoxical effect by triggering the brain hemispheres to work in synchrony.

Movements

All sorts of movements are used by the transformative tradition to simultaneously engage both the negative and positive halves of the movement center and also to increase the metabolic rate, and thereby distribute the information conserved in the MMC to all of the tissues in the system. In this way the cellular matrix, mainly consisting of water, can operate on the same page and spread transformative effects via the molecular as well as the electromagnetic field. This typically happens when the movement in question brings the organism to a level of exhaustion and then exceeds it to tap an extra source of energy. This often leads to being in the zone. Movement exercises may be quite simple, though not easy to do, for example the extremely-long-barefoot walks performed by Australian aboriginals (see Morgan). They may also be quite complex, like the Movements performed by Gurdjieff's pupils around the symbol of the enneagram.

Under the category of the movement center we find a vast and varied repertoire of: dances (both sacred and secular); running (such as is exemplified in the *Zen of Running* by Rohé 1975); kneeling and prostration to humble oneself and open oneself to more fully accepting the MMC, and complex rituals that merge instinctive with intellectual and emotional modalities. Of course, any prolonged physical activity and being present to what one is doing serves the same purpose.

Relaxation

Relaxation of the whole body requires that we envision it, which of cognitive necessity instantiates a positive feedback loop that if held for a sufficient period of time can elicit the WR. Moreover, the scope of relaxation when conjoined with alertness expands access to the deepest levels of the FCCP.

Breath Control

Because breathing is both an involuntary as well as a voluntary function, the transformative tradition uses it to gain some degree of control over the non-conscious domain. This would include the metabolic rhythm of the entire system and the auspicious heart wave by accessing the sympathetic and parasympathetic branches of the autonomic nervous system. By simply slowing the breath the parasympathetic system can be evoked, and by speeding it up the sympathetic one can be brought into action. Since simultaneous slow breathing and fast breathing is not physically possible, the transformative practice of conjoining the two is by combining movement and slowing the breath at the same time. One of the best modalities for this conjunctive result is that of T'ai Chi. Another technique involves exceeding the respiratory thresholds by using hyperventilation while remaining very calm.

Laying on of Hands

Because the hands play such an important role in almost all activities performed by humans, the organism dedicates large amounts of enervation and brain space to them. We can say that they are both receptive as well as transmissive "organs" that have physical and symbolic powers. Since much energy flows to and through them, they not only act as "antennas" but also as "receivers" of "subtle" energies, which since time immemorial have been used thusly by all aspects of the transformative tradition.

Conjunction of Hands

Another interesting aspect of the psiology of hands is how they are typically clasped together in prayer and well-wishing. This conjunction instantiates not only symbolic but also psychophysiological equilibrium.

EMOTIONAL MODALITIES

Though it has been postulated that emotions are the vibratory "side effects" of intellectual and imaginary functions they are, nonetheless, remembered as MCs in their own right. In this regard, they are able to modify mood states by being activated in a resonant manner. Often emotional states are confused with sensations, which are biological "side effects" based on the fact that the cellular matrix is vibratory in nature. What is of particular interest in this vibratory nature of emotional as well as biological activity is that it appears to have the "purpose" of indicating to the Self signals of "right" and "wrong" behavior by the language of wave harmonics.

Beauty

Though this hasn't been proposed as a theory of why beauty in all of its forms affects us, it cannot be denied that when it does, which may be different for each individual, it may very well be because it has (however brief) a wholistic effect on our being. An important aspect of this response is quite likely due to the harmony the piece or moment being experienced possesses, in which seemingly disparate elements come into some kind of synchrony that is transferred, perhaps by entrainment, to us and "makes" us whole.

Therefore, beauty, in whatever context it may be found—Nature, architecture, art, music, dance, etc.—is wonderful and inspiring to dwell upon in order to not only help focus the mind but also to allow its composition to flow through us to elicit the Presence.

Music's Affect on DNA

Gariaev's mediation of information via the "acoustic field" to affect DNA has affinity to research performed by the French musician and physicist Joel Sternheimer. He discovered a quantum method for the regulation of DNA protein biosynthesis in plants with specifically designed sounds. Hence, he used sounds that we hear at the classical level (as ordinary sounds) to impact biological processes at the quantum level.

Sternheimer composed musical-note sequences based on these quantum parameters, which help plants grow and develop in specific ways. For example, some melodies increase the size of tomatoes while others enhance a particular flavor.

A US patent application describes in detail the extent of his findings (Sternheimer 2002). The patent also includes melodies for cytochrome oxidase and cytochrome C, which are two proteins involved in human respiration. It also includes sound sequences for troponin C, which regulates calcium uptake in muscles.

This may very well be the basis behind the behavior of talking to plants to enchant them and make them grow and flourish. However, it also implicates the effects specific sounds and music can have on our genetic blueprint.

Love, Compassion

As has been repeatedly mentioned throughout this book, the emotions of love and compassion towards one's Self and others produce, from a technical perspective the largest system-wide assemblies and synchrony. It doesn't take a rocket scientist to understand that these, for the most part learned, states create not only the positive emotions of joy and euphoria but also can spread to others by sympathetic entrainment. From a psiological view they are both rationally objective in how they may be subjectively expressed in a great number of ways.

Clearing of Psychic Space Through Atonement

The term atonement has religious overtones, but it captures in one fell swoop the need to clear and open one's psychic space, which here is the various modes of the Self.

> "Atonement (also atoning, to atone) is the concept of a person taking action to correct previous wrongdoing on their part, either through direct action to undo the consequences of that act, equivalent action to do good for others, or some other expression of feelings of remorse" (Wikipedia: Atonement).

The term has a close association to that of forgiveness, reconciliation, being sorry, remorse, conscience, and repentance. From a technical perspective, the clearing of psychic space makes it more conducive to flow and interact more freely with the FCCP, and facilitate a greater connectivity of the psycho-soma.

There are a number of ways in which atonement can be put into practice. For example:

1. Personal apologizing to others.
2. Self acceptance and personal forgiveness.
3. Rituals in which the past is forgiven and the clarity of the present is embraced.
4. The practice of tolerance, understanding, and love.

In all cases, atonement allows for contradictions to share the same psychic space, which instantiates wave formats seeking to enter into phase but undergoing transitional clashes one experiences as the "warring factions within" or the "dark night of the soul." At some point all this will naturally clear up since wave intelligence always seeks the most harmonious manner in order to save energy.

An excellent spiritual program for working with atonement issues is provided by Amy Jo Ellis's *Healing Your Family Tree Through the Court of Atonement* (see Bibliography).

> "The Court of Atonement is a spiritual principal that solves conflicts at the soul level. With less than 20 minutes of reading, you will understand how to use this process and begin healing yourself, your loved ones and even your pets.
>
> This simple process clears energy in a way we have not really experienced before.
>
> We have seen problems that have plagued families for decades, simply disappear after only a few days, while others were miraculously gone in hours!" (Ellis 2018)"

INTELLECTUAL MODALITIES

The intellectual center works with thoughts in the form of ideas and concepts, derived from cultural settings by default and/or specialized educational settings in which one learns a logical system. Both contribute to what one is apt to believe to be "true," and in this regard it affects one's behavior and relationship to other people.

Because the DNA molecule has syntactical function, as Gariaev has experimentally shown (1992), the thinking process is linked to it, which may very well be how hypnotic, self-induced, or other directed suggestions can have physiological effects (see Swami Rama in Green/ Green 1997) and more implicitly with the placebo/nocebo effects noted in clinical testing.

Doctrinal Texts

All written documents, however simple or complex—bibles, holy books, codices, tablets, written in various media—address the intellectual center. They are typically the basis of the MMC of the system in question, and may be difficult to interpret and/or decipher.

Confusing Contexts & Assembly Size

Contextualized assemblies form resonant associations around the thematic or idea being thought of or imagined. They remain within this "border," so to speak. On the other hand, shocks of some intensity bypass this constraint and are more generalized. However, it's possible to use context to achieve a more general shock-like (OOPE) effect.

Zen Koan

The Zen koan is a perfect example of how a confusing context is used to invoke more of the Self via the FCCP. The query enters the mental system at the input (1), it is then crunched by the ingoing vortex (2) but when it interfaces with the FCCP it is confronted by the limbic editor and becomes "confused." This increases the processing time as the vortex spins in search of a contextually resonant connection, which is impossible to find.

As it does, it expands the search, which increases the scope and size of the outgoing assembly. If not able to find a resolution it dithers between possible solutions, which are often wisely rejected by the Roshi (Zen Master). When the conditioned mind becomes frustrated and surrenders, it equilibrates and draws in a finer level of the FCCP operating at a faster speed. Only then is the "right answer" more likely to come forth.

A koan is a question expressed as riddle—like "What is the sound of one hand clapping?"—it's used in Zen Buddhism to demonstrate the inadequacy of logical reasoning and to provoke enlightenment. It uses context, the question, in order to frustrate the context generator of the conditioned "mind." The quest to answer the question becomes very emotional, for it has been imbued with developmental importance.

This very emotional quest reminds me of the affects the assignment to write a short story had on me. Searching for the elusive answer had a dissolving or bypassing effect on the conditioned ego (PSQ), which would then release more arousing energy from the RAS (reticular activating system) to facilitate more total system involvement and the formation of larger assemblies from the FCCP.

IMAGINATION: THE KEY FUNCTION TO ALL PSYCHO-TRANSFORMATIVE MODALITIES

A study mentioned earlier, done by Jean Decety in 1996, revealed that imagination and executed actions share the same neurological pathways. This is an exciting piece of information because it has so many implications for the field of conventional psychology as well as transformative psiology.

It informs us that what we think and imagine not only neuroglialy precedes our actions but also instigates global patterns of embodiment below the threshold of awareness. This strongly suggests that non-conscious mentation secretly primes us to act in certain ways.

It also modifies somewhat the findings of Melloni previously quoted: "unconscious perception" not only involves "local coordination" of neuroglial (she didn't use this term) activity but also global pathways as well. Therefore, "mass action" is not only occurring in the brain but also throughout the whole organism, via the "holistic media."

This, as previously explained, especially involves the fascia, interstitial and cellular water, the magnetic field of the heart (i.e., the heart wave) that also reaches near or beyond the speed of light, the entire psycho-organism and the extended field that surrounds it.

Knowledge of the transmitting and receiving sensitivity of the heart wave is found in Scripture. In Proverbs of the King James Bible we find the phrase:

"As a man thinketh in his heart, so is he" (23:7).

This also goes towards explaining why the numerous subliminal variables occurring during state dependent learning are holographically included in every memory-forming event.

Imagination Evokes Larger Assemblies

The transformative advantage of imagination over other modes of mind is that it evokes much larger, broadly distributed assemblies. Therefore, in using more neuroglial assets throughout the nervous system it, as described by the research, would generate a more expanded state of consciousness involving greater portions of the whole psychophysiological system.

Keep in mind that as the size and scope of the assembly increases, so does the need of the mental system to accelerate the rate of information processing to handle the scope of information. Hence, the need to "handle" complexity "makes" the mind go faster.

As a matter of fact (i.e., a recurring mode of objectivity) the whole mind-body-spirit complex is always involved in every cognitive moment, but what keeps one from being privy to the totality of it all is our conditioned threshold of awareness, mediated by the size of the assembly, and the editing function of the "emotional brain" centered in the limbic system.

Limiting the assembly-forming response is also how the system conserves energy, by restricting how much of this resource is used to engage in the mundane activities of everyday life. We don't need cosmic consciousness to peel carrots and drive cars, although these simple acts are not as simple as they seem when viewed with an expanded consciousness.

At the non-conscious level there is an enormous orchestration of interconnected movements (estimated—by Schmid (2016)—to occur at 320 Gigabytes per second) that involve about 50 trillion cells, of which the neuronal network is but a small part (only 100 billion neurons, excluding the glia and their holistic connectivity). The neuroglial assemblies that form the vibratory threshold of conditioned consciousness are but the tip of the psychophysiological iceberg.

While synchrony goes towards explaining how memory constructs resonantly connect via feedback to form contextual assemblies, characterized by fast frequency and amplitude modulations, this in itself does not explain how the the different contexts are integrated into a seamless percept. Freeman's model of cognitive vortices does, especially how multiple vortices merge into a singular output cycling with the 1-to-7 cycle in millisecond+ speeds. As they merge there is a concomitant computational processing of the MCs based on wave interference.

What We Image In Our Heart, So Shall We Be

The Psiloton Self is a self-referential cognitive system. We become what we cogitate, especially when it is done in sympathy with the heart. Keep in mind that there are instinctive, movement, intellectual, and emotional kinds of cogitation. If our cogitation is intentional and sufficiently repeated via the 1-to-7 cycle and connected to the heart wave, the Self will have the power to change its Self in a godlike manner.

Not acknowledging the broadcasting power of the heart, modern psychology assumes that all cognitive events take place in the brain. While it does have information of broad neural, immunological and endocrine effects on thinking and behavior, it assumes the distribution is facilitated linearly via nerves and biochemical channels. The moment-to-moment holistic involvement is not only local but also socially and cosmically informed.

Image in the Heart is Worth A Million Words

Earlier, we elaborated on the heart wave that envelops the organism in the form of a toroidal soliton, which in a cognitive sense is dubbed the Psiloton.

On the Psiloton model the heart wave flows outwardly over its surface and reenters at the top, where the ingoing vortex transforms its space and time into the subtlety of fractal zones. As it does so it makes a "selection" from the repertoire of the FCCP, which is then taken up by the outgoing vortex to produce a percept and/or a behavioral response. The cycle repeats in millisecond+ time frames like a quantum movie.

What we think or imagine is iterated thusly, but the percept is challenged by a tussle of habits and OOPEs, and the next round of cognition is altered by the outcome of these events, as modeled by the F-F loop of holistic causation (Fig. 15).

Envisioning the Universe in One's Self

An important and empowering aspect of practice is to connect everything in the universe to one's image of Self. This is a wisdom derived from ancient teachings.

Self observation creates the condition for a positive F-F loop to activate the cellular matrix to transfer its intrinsic information and potential into a subtler body that can transform, if sufficiently reinforced, into a "mental body" able to dissociate from the physical one into the appropriate medium in the surround.

Astrotheology

From a common perspective the notion of astrotheology takes on the following meaning.

> "The term astro-theology is used in the context of 18th-to-19th-century scholarship aiming at the discovery of the original religion, particularly primitive monotheism. Unlike astrolatry, which usually implies polytheism, frowned upon as idolatrous by Christian authors since Eusebius, astrotheology is any "religious system founded upon the observation of the heavens", and in particular, may be monotheistic. Gods, goddesses, and demons may also be considered personifications of astronomical phenomena such as lunar eclipses, planetary alignments, and apparent interactions of planetary bodies with stars. Astro-theology is used by Jan Irvin, Jordan Maxwell and Andrew Rutajit (2006) in reference to "the earliest known forms of religion and nature worship", advocating the entheogen theory of the origin of religion" (Wikipedia: Astrolatry).

From a Psiological Point of View

From a psiological perspective astrotheology is an ancient science that studied the dynamic patterns of celestial objects to determine their influence on Earthly and human behavior. It equated its dynamism to the astrological scheme as the macrocosmic order and its affects on the microcosmic ones.

Monotheism

Its tendency towards monotheism suggests a deep psychological understanding of the transformative effects of a MMC and its polytheistic fractal extensions.

The Cyclic Process of the Ancient World

It served as a guidance system for agricultural and organizational purposes and as, perhaps, the first fully expressed study of a self-referential-cyclic *process*, which can be viewed as macro-fundamentals and their micro-fractal harmonics.

Involution, Devolution, Evolution

It preceded in scope the twenty-four hour sleep-wake cycle by many orders of cyclic magnitude extending into the thousands of years that formed the characteristics of eras or ages in terms of involutionary, devolutionary, and evolutionary stages of development linked to auspicious moments of conjunctions and alignments.

Expansive Neural-Assemblies

Astrotheology connected human life to the dynamics of the cosmos at large. There wasn't a moment in the life of astrologically-imbued peoples that they were not experiencing immersion in the neural-expansive assemblies of an enormous cosmic drama that opened the powers of the human spirit to the miraculous enchantment of the times, that seems so fictitious to the modern and postmodern objective-compulsive mindsets.

Objective-Subjective Potentials

In Gurdjieff's transformative system it is held that these symbols form an "objective language" through which knowledge of human development is transmitted. He referred to this development in quantitative terms, by "how much" of the individual, in terms of the seven psychic centers (similar to the chakras in Yoga), are fully integrated into a harmonious system.

However, because subjective imagination can creatively modify how the physiology behaves, it must be included in the transformative equation. It therefore follows that objectivity and subjectivity are two sides of the cognitive coin that cannot be excluded. They both affect each other in a co-creative manner: Objectivity⇔Subjectivity.

The aim is to use the subjective dimension to elicit a large assembly able to produce the Wholistic Response, the objective result.

THE ENTRAINING FORCE OF HIGHER CENTERS

Modern physiology and psychology have not yet come to recognize the existence of higher centers in the brain. In the transformative tradition these centers are known by various names, of which the "*Ajna*" and the "*Sahasrara*" are the more commonly known Sanskrit terms used in Yoga.

> "Ajna, or third-eye chakra, is the sixth primary chakra in the body according to Hindu tradition. It is supposedly a part of the brain which can be made more powerful through meditation, yoga and other spiritual practices just as a muscle is. In Hindu tradition, it signifies the subconscious mind, the direct link to the brahman. While a person's two eyes see the physical world, the third eye is believed to reveal insights about the future. The third eye chakra is said to connect people to their intuition, give them the ability to communicate with the world, or help them receive messages from the past and the future" (Wikipedia: Chakra).

In the Fourth Way literature it is referred to as the "higher emotional center." It is technically exemplified in the Psiloton model to be in a fractal zone next to FZ0, which is the site with affinity to the Sahasrara, the "higher mental center" in the Fourth Way.

> "The Sahasrara is described as a lotus flower with 1,000 petals of different colors. These petals are arranged in 20 layers, each layer with approximately 50 petals" (Wikipedia: Chakra).

It is located above or at the crown of the head, and is psiologically interpreted to mean a full activation of the cellular matrix and their energetic output of the brain hemispheres (signified by the lotus flower); again the quantitative factor. In other traditions this experience, which is typical of the high 3rd and 4th states of conjunctive consciousness, is described as a "halo" around the head or as the "crown" worn by a "king" ruling over the "kingdom within." This is the broad field that develops due to the self-cancelling effect. For those who have not experienced the neural activation, it comes with the roaring sound of "rushing waters."

What makes these two centers of keen interest to those who study the transformative process, especially those involved with depth psychology, is their symbolic mode of communication (Gurdjieff, Jung). Because of their high degree of conjunctive operation, these centers are constantly compressing dualities (multiplicities) into a symbolic language that will make immediate sense to those who are in that comparable state.

According to Gurdjieff, these centers are already functioning in every human being, but they are by dint of the PSQ ("false personality") separated, which is here interpreted to mean that they don't share the same frequency and can't enter into synchrony. Therefore, all transformative efforts need to be directed to the lower centers in order to bring them into phase and the right frequency to connect with the higher centers and be entrained into their speed and context.

Esoteric Symbols as Psycho-transformative Tools

Our lives are ruled by symbols. We use them to communicate with one another and to keep records of important events. The quality of our lives would surely be compromised without them.

> "Man, it has been said, is a symbolizing animal; it is evident that at no stage in the development of civilization has man been able to dispense with symbols. Science and technology have not freed man from his dependence on symbols: indeed, it might be argued that they have increased his need for them. In any case, symbology itself is now a science, and this volume (Cirlot's Dictionary of Symbols 1971) is a necessary instrument in its study." (Herbert Read in Cirlot 1983).

Relationships in the natural world are skewed towards the objective while those in the spiritual world have higher subjective degrees of freedom, which often defies the tendency to find objective meanings in them.

> "Given that every symbol 'echoes' throughout every plane of reality and that the spiritual ambience of a person is essentially one of these planes because of the relationship traditionally established between the macrocosm and the microcosm, a relationship which philosophy has verified by presenting Man as the 'messenger of being' (Heidegger): given this, then it follows that every symbol can be interpreted psychologically".

In a psycho-symbolic sense "room" may indicate a mental enclosure and "dead" doesn't necessarily mean "deceased" but "forgotten." So, for example:

"[T]he secret room of Bluebeard, which he forbids his wife to enter, is his mind. The dead wives which she encounters in defying his orders are the wives whom he has once loved, that is, who are now dead to his love.

This subjective extension of the symbolic has opened up the psychological interpretation of dreams, which Jung and the depth psychologists used to develop a way to understand the language of the unconscious.

"Jung emphasizes the twofold value of psychological interpretation; it has thrown new light upon dreams, daydreams, fantasies and works of art, while on the other hand it provides confirmation of the collective character of myths and legends. He also points out that there are two aspects of the interpretation of the unconscious: what the symbol represents in itself (objective interpretation), and what it signifies as a projection or as an individualized 'case' (subjective interpretation). For our part, objective interpretation is nothing more nor less than understanding. Subjective interpretation is true interpretation: it makes the widest and profoundest meaning of a symbol in any one given moment and applies it to certain given examples." (Cirlot 1983, xivi-xlvii).

What makes symbols, however, so vital for transformative psiology is their effects on the nature of the organism. If we consider again the compelling neurological findings that show that imagination evokes the same neuronal pathways as does actual behavior (Decety 1996), we now have a quantitative biological reference that correlates with qualitative, subjective, psychological experience.

It also correlates with those who support the radical epigenetic view that DNA does not dictate how and what we are (Lipton 2005); we encounter a scientific basis for understanding how the use of symbols and the meanings we attribute to them can have transformative effects on all aspects of the Self.

These variable attributes constitute the impact the subjective potential has on the objective nature of our physical reality.

From all that has been presented, we find the theme of using imagery as the best method with which to develop magnitude (also as "psychic mass"), given that there are many more subjective links to the

MC repertoire than there are objective ones. It therefore constitutes the "best method" by dint that it facilitates the development of large assemblies by the power of associative suggestion, which in the case of transformative evolution of the Self must elicit the general state of polar equilibration of all of the centers.

From a learning perspective, the physical mode is the most auspicious to initially dwell upon, simply because it is objectively available to form a F-F cognitive cycle in order to accelerate the mental mode to a higher and controlling frequency.

By turning attention inward we initially encounter the body, not only in its biological manifestation but also in its gravitas as to what controls the mind to satisfy its needs.

Symbols Are Not Artifacts

The study of symbols is often overlooked in modern psychology (the exception being Jungian depth psychology) because symbolizing appears to be an artifact of human psychology, not something that emerges naturally to compensate for the limited thresholds of practical cognition.

According to the scholar of esotericism Manly Hall, the ignorance of symbols can lead to gross misunderstanding of Scripture.

> "While the greatest minds of the Jewish and Christian worlds have realized that the Bible is a book of allegories, few seem to have taken the trouble to investigate its symbols and parables" (Hall 1928, 410).

Yet when properly explained, these allegories would reveal another hidden level of mystical meaning and perception.

> "When Moses instituted his Mysteries, he is said to have given to a chosen few initiates certain oral teachings which could never be written but were to be preserved from one generation to the next by word-of-mouth transmission. Those instructions were in the form of philosophical keys, by means of which the allegories were made to reveal their hidden significance. These mystic keys to their sacred writings were called by the Jews the Qabbalah (Cabala, Kaballah)" (Ibid).

From a psiological perspective the Cabala is a MMC incorporating many associative lines of thought that as a symbol invokes the Whole of the Self in its various modes of being. The symbol is a necessary device that a higher system of being and intelligence needs to communicate a multidimensional inclusiveness in a practical manner for a "slower" mind.

According to this revealed rationale (i.e., logos) the "As above, so below" of Hermetic lore means that they are alike in function but not in processing speed. Otherwise, if they were identical there'd be no need for a merger with the "above."

Relevance of Symbolic Obscurity

When it comes to contextual coherence as a progenitor of psychophysiological unity we're bound to encounter paradoxes. In one scenario the notion of secrecy to protect the MMC from the ravages of the PSQ and vulgar distortions is instantiated by symbols and semantic contexts that remain obscure to the conditioned "logic" of the practical mind. Therefore, in Scriptural terms:

> "[W]hen thou doest [thine] alms, do not sound a trumpet before thee, as the hypocrites do in the synagogues and in the streets, that they may have glory of men. Verily I say unto you, They have their reward. But when thou doest alms, *let not thy left hand know what thy right hand doeth*: That thine alms may be in secret: and thy Father which seeth in secret himself shall reward thee openly" (St. Matthew 6:2-4. Italics added).

In this genre we find, for example, the rationale of the diverse alchemical graphics and symbols bordering on the bizarre.

The strange beauty of these symbols is that it encompasses the essence of transformative psiology, and is contextually sufficient unto itself. Each figure encompasses a vast synthesis that when integrated brings forth multiple levels of meaning and a rich psychophysiological effect to the observant practitioner. It is a pictorial approach to forming the MMC. Symbolic pictorial approaches are useful, considering that they evoke large-all-embracing assemblies. Consider the Alchemical symbol below (Figure 26) in regards to the conjunction of opposites to evoke a neutralizing effect that draws higher "spiritual" potentials.

While this practice of hiding from the PSQ "isolates" the MMC during its developmental stage, it creates a kind of "babel" that could not be understood by other systems. Hence, we find the esoteric traditions by default divorced from sharing information. If occurring in different cultures, the differences of language are yet another alienating factor. Also, by sustaining a degree of uniqueness, the system in question could help its adepts from wandering, and losing contextual continuity, by claiming doctrinal exclusivity.

Nevertheless, if viewed by anyone while in the third or fourth state of being and consciousness, the meaning is quite effortlessly revealed.

Figure 26: Alchemical marriage of king and queen, sun and moon in which a third principle (as "spiritual mercury" in the form of a bird and star) enters from "above." From the "Philosopher's Rosegarden" of Arnaldus von Villanova (c. 1240-1311).

Over Emphasis on the Historicity of Holy Texts

Scholars searching for the historicity of holy persons (e.g. Moses, Krishna, Buddha, Jesus) mentioned in religious as well as in secular texts, have found it difficult to verify the objectivity of their existence (see Murdock, and Bonnacci 2012, for the alleged astrotheological origins of them). On inquiring of the Presence about these findings I was given rather interesting answers.

Q: Is it true that many, if not all, of the key figures alleged to have lived according to the scripture and doctrinal texts of the religious and secular systems of the world did not actually exist?

A: The "texts" that you are asking about were not descriptions of reality or historical events; they were manuals of spiritual development expressed in the language of symbols, particular to that culture, of which astrology seems to have been the most appropriate at the time. The personages mentioned in them are actually human potentials that some people at the time may have brought into fruition, and in these cases, could have been historical beings.

Q: Why then did they not simply directly transform people according to this knowledge and create a highly-evolved race?

A: Doing so would disrupt the necessary principle of freedom, based on subtle fractal dissimilarities, in the cosmic order, ordained by the Creative Intelligence, and that would by the law of resonance turn every individual into a cookie-cutter puppet, extinguishing thereby the rich drama of existence and the process of creative thinking. We need to also understand the fundamental necessity of the principle of freedom, which requires that variety must exist in the universe. Otherwise, as I have pointed out before, the universe would gel into one ginormous clump. Those who by surreptitious wielding of power seek to create strict uniformity in organic life, which of course includes humanity, will at some point fail, for the principle is ingrained in everything. Think of the problems derived from over-inbreeding which, for example, led to hemophilia.

Q: So what was the essence of these symbolic teachings?

A: If you study what history has presented you'll notice that the message, despite all the political and self-serving meddling that went on and continues to do so, throughout history you'll quite likely notice that in every instance the message has been the introduction of a developmental process that, above all, is directly related to learning and creativity. Hence, the legends and the process, once deciphered and understood, appear quite the same across the board, since the methodology is quite universal, and it is for this reason that scholars believe that they have been forged or copied from one another. Whether that occurred or not is not the most important issue.

Q: What about the often quoted miracles performed to prove their powers?

A: Magic and miracles are actually the manifestation of laws pertaining to a higher plane on a lower one. Recall that each higher level of being and consciousness is progressively endowed with greater degrees of freedom. These beings use these attributes to get people on a certain level to believe in their own inherent potentials, and to make efforts to develop them.

Q: Are they punished if they refuse to do so?

A: No, their level of being and consciousness is sufficient to "decide" their future. There is, however, always a choice.

Q: What is the process?

A: It is learning how to initially modify the body, simply because the body is, as I said, an energetic wave form subject to modification. What makes the body such an auspicious tool for transformation is that it is bristling with potential in a very condensed manner, which it will release given that what is presented to it by the mind is believable in terms of its language.

Q: What is its language?

A: Purely symbolic, mediated by sensations, which are vibrations reflecting geometric variables. However, it is unsurprisingly relational to the astro-geometry of the Earth within the planetary system of the sun. Since remote antiquity this model has dominated how the universe was viewed, forming the basis of a developmental system reflected in the sleep-wake cycle of organic life on Earth. Today its effects are spoken of, for example, as "circadian cycles" that have external as well as internal influence.

Q: How do these cycles relate to human psychology?

A: It is from these cycles that our brain waves are fashioned, according to the Schumann resonances, of about 7-8 Hz, surrounding the Earth. Of course, the ancient model was more inclusive, such as is found in the Zodiac with its calendric modulations occurring in a kind of temporal sine wave, which is what cycles are all about. So, in this respect, we have a science with objective as well as subjective aspects. Something that is of little significance to modern science in general and psychology in particular.

Q: Is the symbolic language the realm of the subconscious?

A: Not as something separate from the physiology but intrinsic to how it processes information based on its needs identified by their vibratory signatures. For example, the color red means many things to the instinctive intelligence. It may signify the loss of blood, food in the form of meat, and also warmth as in the wavelength of infrared. In relation to the loss of blood it may also signify danger, war, and violence. However, in relationship to warmth it may also signify acceptance and love. So, you see, every moment in life can only be understood in a very broad, yet evolving, subjective context.

Q: What were the symbols used to describe the process?

A: All processes have a beginning and end that form a cycle of sorts. For the ancient world it was, as was mentioned, in the form of the Zodiac, representing the relationship of larger cycles to their smaller harmonics. All the major religions of the time used it in the development of their principles and methodologies, and even used it to develop quite sophisticated methods of health and healing. The process describes how the universe on all scales of existence devolves into matter beginning in Aries, and progressively "falling" to Taurus, Gemini, Cancer, Leo, and Virgo, the bottom of the *involutionary* part of the cycle. At Libra the cycle reverses into an *evolutionary* one and progressively rises up out of materiality through Libra, Sagittarius, Capricorn, Aquarius, and culminates in Pisces. At its point of return the beginning and the end converge to create a novel dispensation (a.k.a. the "alpha and omega").

Q: Recurrence?

A: Yes, but not in the absolute sense. In keeping with our many exchanges, the cycles are not circular per se but in vortical spins that are subject to morphing in unsuspected ways due to the sensitivity of initial conditions; again, fulfilling the principle of freedom. That is why history never actually repeats, and the so-called notions of karma and reincarnation have much novelty to be taken advantage of than is conventionally attributed to them. So recurrence is not a "bad" thing but something which is completely misunderstood.

Q: Can we then say that the Zodiac is the precursor of modern science and its study of processes?

A: It certainly contributed to this mindset, but we need to keep in mind that upon abandoning the subjective aspect of reality the intellectual enlightenment that began from the 15th century onward set the Western world on an objective-compulsive trajectory. This led to being progressively mesmerized by the objective aspect as the source of all things and, at the same time, the dismantling of trust in the subjective-magic-like potentials inherent in all people, particularly on the whole notion of a Creator. Notice how terms like "magic," "spirit," "miracles," and the like have been turned on their heads to currently mean something extraordinary or simply, when to some extent displayed, as some kind of trick by this currently well-established collective sense of doubt.

Q: Isn't doubt the basis of modern science?

A: That's precisely the nature of the problem. By causing people to doubt, according to the scientific method, they over time lost their subjective identity which is the basis by which the spirit works to produce "magic" and "miracles." For example, the power of telecommunication is inherent in every human being, such as is still evident with the Australian aboriginals, but when it was put into doubt in the cosmopolitan world, the materiality of the cell phone replaced it. The same calamity occurred with the ability to imagine, the true source of many of our human powers, with the advent of the TV. These inventions are not the problem per se but become so when they are taken over and controlled by a few. As this technology advanced, human imagination, that is, in terms of it being used as a psycho-transformative tool, diminished. This had powerful consequences on individual freedom, simply because now it was possible to program huge swaths of the human population with all kinds of stupid-self-serving images, mainly to sell products and develop the predictable behavior of consumer robots.

Q: Are we in an inevitable downward spiral?

A: It can only begin to come to a halt when a group of people come to realize the extent to which their subject-potentials have been taken over by minds drunk with power and excessive materiality. They need to abandon the obsessive-compulsive worship that has infiltrated their psychology and begin to learn how to clear themselves with psiological principles that they can master and share with others. We don't evolve by relying on a machine to take over our potential to "advance" what is inherent in our nature.

Q: Is there such a place as "heaven"?

A: Heaven is the place where one's vibrational signature is drawn to, which, by the way is where we are now and to where we might be tending, based on how we alter our being and consciousness.

ASSESSMENT

Spiritual development, as is any other kind, is a learning process, and as such, functions more efficiently when it is accompanied with a feedforward-feedback (F-F) loop that informs the learner if the intended goal is being met. It also informs what needs to be altered to the goal via a feedforward effect and/or to the learner in order to match the intent with the objective (6 with 1) in order to instantiate a positive feedback loop to sustain and amplify it.

In this regard, it was pointed out that attempting to use consciousness as the goal poses many insurmountable problems simply because it is such a featureless objective. However, we can experience a certain field-like aspect about its "nature," which has been hypothesized to be how its isotropic smoothness is modified by the MCs that develop within its "volume." If we, for example, think of "nothing" it offers no contrast to the uniformity of the field and in this respect we'd experience, by comparative thinking, a kind of "void."

It is for this reason that instead of attempting to go for the ultimate goal, it is best to understand that its potential resides in every mode of the Self, and that it can be evoked by a more palpable modality to activate the whole system to vibrate in synchrony.

Yet, even this vibration is a byproduct of its existence, which can be used to accelerate the 1-to-7 cognitive cycle. This vibration is a sign of its Presence, which is what remains, and can only continue to expand when the system equilibrates and allows a finer energy to enter and quicken the Presence itself.

Since we are, quintessentially the sum and "substance" of the Presence, we can only know in those glorified moments what it has "transcended" but not precisely what it is.

In this regard, we can experience levels of phenomena normally not available to the conditioned consciousness, but not the source of consciousness itself; yet strangely we know that we are in its grips.

Therefore, the F-F process is best initially done with the modalities and then with the sensory and feeling effects that emerge from working with them. In any case, therefore, one can assess one's level of development by determining the scope of what is being experienced and the amount of time one is able to experience it.

The scope can be personally assessed by how much of one's Self one is simultaneously aware of in conjunction with the depth of the memory field (i.e., the FCCP). Note that the more one is aware of the fullness of the Presence in one's Self, more of the field will be encompassed, simply because the speed of mental processing must increase to encompass greater regions of the field (signifying the nature of conjunctive states of being and consciousness).

Hence, the most important variable to determine is the temporal one, which then becomes: how long can one focus on one thought, image, feeling, and/or sensation, or all the preceding while being very relaxed (in order to access the greatest portion of the FCCP). It is OK to use a watch/clock, and record results in a journal, with comments and insights, if desired. Set up a scale, using 10 minutes as the goal, like so:

Degree of Focus Based on Minutes: LO 1-2-3-4-**5**-6-7-8-9-10 HIGH

Over time, the range can be increased to encompass more time. At some point the Presence will linger as a subtle vibration permeating all of one's being, and it will be experienced more fully throughout the day. From that phase onward one will be able to trigger it at will by simply thinking of it.

A METAPHOR FOR COMPASSIONATE EXISTENCE

There is a dynamic spiritual presence at work throughout the universe, which is both within and outside of us. Throughout history it has been given many names and explanations of its existence. I simply refer to its mystery and magnificence as the Presence. However, no one truly knows its true "nature," and for this reason no one cannot claim to own it. Nevertheless, each person has been endowed to directly experience it without relying on a particular genetic make up, clergy, or any form of ritual, set of codes, liturgy or creed, simply because the creativity of the Presence is infinite. While there are many paths to help one experience and worship it, none are exclusive.

Because no two individuals are identical they will have unique experiences of its existence, and each would express these moments of revelation in unique ways. Therefore, go forward, and may the Presence and the Logos be with you.

ABBREVIATIONS
& SYMBOLS

> - a vortical construct moving in the right or clockwise direction

< - a vortical construct moving in the left or counterclockwise direction

>< - self cancelling

>> or << in phase mergers

>0< - signifies destructive interference and self-cancelling

⊤ - entropic balance

1-to-7 - cognitive cycle

a.k.a. - also known as

AK - auto-kinetic

AM - amplitude modulation

ANG - autonomic neuro-glial system

BPM - beats per minute

c - speed of light in a "vacuum" (299,792 km per second)

CNGS - central neuro-glial system

CSBC - conjunctive states of being and consciousness

D - dimension (e.g., 1D, 2D, 3D, 4D, etc.) as scale invariance in self-similar formats

DLE - death-like event

DMN - default mode network, conditioned sense of Self ('I')

DNA - deoxyribonucleic acid, carrier of genetic information

EDA - electrical dermal activity

EEG - electroencephalograph

e.g., - for example

EMS - electromagnetic signal

FCCP - field of creative and cognitive potentials

FM - frequency modulation

fMRI - functional magnetic resonance imaging, reveals brain metabolism

FP - false personality

Gb - gigabyte(s): a unit of information equal to one billion (10^9) or, strictly, 2^{30} bytes

GSR - galvanic skin response

HH - Hodgkin-Huxley

Hz - Hertz: cycles per second

'I' - the sense and/or feeling of Self; also as a predominant memory construct (MC)

ibid - from the same previously cited source

i.e., - that is to say

KE - kinetic energy

KJHB - King James Holy Bible

LOA - law of attraction

LOD - law of distraction

LPCS- laser photon correlation spectrometer

LTP - long-term potentiation

MC - memory construct; also as a network of connectivity

MRI - magnetic resonance imaging

ms - milliseconds: one thousandth of a second

NDE - near death experience

NF - negative feedback

NIB - not in bibliography

NOAM - New Oxford American Dictionary

O - objective

OAM - orbital angular momentum

OBE - out-of-body event(s) or experience(s)

OCD - obsessive-compulsive disorder

OOPE - out-of-pattern event(s)

PE - potential energy

PF - positive feedback

Phi - golden ratio: 1.618… major section

phi - golden ratio 0.618... minor section

PNGS - peripheral neuro-glial system

PSQ - psychophysiological status quo; a.k.a. ego

PUF - Pluripotent Unified Field: first level of "created" order

S - subjective

SAM - spin angular momentum

SCD - subjective-compulsive disorder

SQUID - superconducting quantum interference device

TP - true personality

Tpsi - transformative psiology

WR - Wholistic Response

X - chromosome

Y - chromosome

CREDITS FOR ILLUSTRATIONS

Front cover: Adaptation of William Blake's "The Ancient of Days" from series Europe: A Prophecy, 1794. Public domain. The Symbol is added.

Figure

1. Marriage and the Vortex (left), and Cosmic Egg (right). Author.
2. The Symbol. Author.
3. Lateral view of the brain. From Gray's Anatomy (1918). Labels were added. Public domain.
4. Interpretation Freeman's Vortices involved in cognition. Author.
5. Psiological interpretation of the development of a Bose-Einstein concentrate. This image, created on Dec. 31, 1994, is in the public domain in the United States because it is a work of the United States Federal Government, specifically an employee of the National Institute of Standards and Technology, under the terms of Title 17, Chapter 1, Section 105 of the US Code.
6. Curling and unfurling of HEM waves towards opposing scales. Author.
7. EEG Cognitive vortex (not to scale) exemplifying contractive and expansive transitions of scale. Author.
8. Anatomy of the toroidal-soliton attributes of the Psiloton, that more integrally defines the entire 1-to-7 cognitive cycle as a singular self-referential dynamic. Author.
9. Typical soliton shapes. Author.
10. Illustrating the concept of light trapped in vortex loops and superposed on the self-referential dynamics of a soliton, yielding flower-like patterns. A: Derived from "A table of prime knots up to seven crossings." I, the copyright holder of this work, Jkasd, release this work into the public domain. Wikipedia: https://commons.wikimedia.org/wiki/File:Knot_table.svg B: Soliton, Author. C: derived from "A poster with twelve flowers of different families." Permission is granted by the copyright holder, Alvesgaspaar, to copy, distribute and/or modify this document

under the terms of the HYPERLINK "https://en.wikipedia.org/wiki/en:GNU_Free_Documentation_License"GNU Free Documentation License, Version 1.2 or any later version published by the HYPERLINK "https://en.wikipedia.org/wiki/en:Free_Software_Foundation"Free Software Foundation; with no Invariant Sections, no Front-Cover Texts, and no Back-Cover Texts. Wikipedia: https://commons.wikimedia.org/wiki/File:Flower_poster_2.jpg

11. Constructive Interference. Adapted from Haade, open use under GNU Free Documentation Lic. (Wikipedia).

12. Destructive Interference. Adapted from Haade, open use under GNU Free Documentation Lic. (Wikipedia).

13. Beat Frequency. Left side (constructive +/- destructive interference) is Adapted from Haade, open use under GNU Free Documentation Lic. (Wikipedia). Right side is adapted from file licensed by its author, Ansgar Hellwig, under the Creative Commons Attribution 2.5 Generic license in Wikipedia: https://commons.wikimedia.org/wiki/File:Beating_Frequency.svg.

14. The Limbic System (lines and sites are approximate). The ACC & PCC roughly point to the anterior cingulated cortex and the posterior cingulated cortex, respectively. Adapted from: OpenStax College, the author and copyright holder of this work—Limbic Lobe.jpg—grants permission of use under the following Creative Commons Attribution Share-Alike 3.0 Unported License.

15. Feed-Forward-Feedback Loop of holistic causation. Author.

16. Dynamics of positive feedback loop. Author.

17. Fractal developments in the thinking process based on the parameters of the EEG. Author.

18. Cornstarch and water solution under the influence of sine wave vibration. I, Collin Cunningham, the copyright holder of this work, hereby publish it under the Creative Commons Attribution-Share Alike 3.0 Unported license. Wikipedia: https://commons.wikimedia.org/wiki/File:CornstarchCymatics_cc.jpg

19. Transmission of DNA genetic information into water through electromagnetic waves. Created from data gleaned from Montagnier, *et al*, 2010.

20. llustrated summary of the process performed by the Laser Photon Correlation Spectrometer. Author.

21. Communication between brain networks...Communication between brain networks in people given psilocybin (right) or a non-psychedelic compound (left). This work has been released into the public domain by its author, David Nutt. This applies worldwide (Wikipedia).
22. Conjunctive states of Being & Consciousness based on the parameters of the EEG. Author.
23. Staff of Osiris. This pine cone staff in the Egyptian Museum Turino, Italy is a symbol of the solar god Osiris (left). Public domain. Modern depiction of Caduceus (right). This work has been released into the public domain by its author, Rama. This applies worldwide (Wikipedia).
24. EEG Correlates to Locality/Nonlocality of the FCCP. Author.
25. Necker cube. Author.
26. Alchemical marriage of king and queen, sun and moon under the influence of spiritual mercury. From the "Philosopher's Rosegarden" of Arnaldus von Villanova (c. 1240-1311), manuscript in the Vadiana Library, St. Gallen. Public domain.

BIBLIOGRAPHY/ REFERENCES

Abbott, Steven; Dambergs, Yanis, (Sept. 20, 2003) "Blessed Raymond Lull's Memory Training Tools" - http://lullianarts.narpan.net/cont.htm.

Abraham, Ralph H., (1994) *Chaos, Gaia, Eros: A Chaos Pioneer Uncovers the Three Great Streams of History.* New York, NY: HarperCollins Publishers.

Adler, Jerry, (May 2012) "Erasing Painful Memories," *Scientific American,* 56-61.

Aihara, Herman, (1980) *Acid and Alkaline.* Oroville, CA: George Ohsawa Macrobiotic Foundation.

Alfinito, Eleonora, *et al,* (June 14, 2011) "The dissipative quantum model of brain: how does memory localize in correlated neuronal domains," *Quantum Physics:* arXiv:quant-ph/00060661.

Alfred, Jay, (2005) *Our Invisible Bodies.* Victoria, BC, Canada: Trafford Publishing.

Amador, Vega; Trans. James Heisig, (2003) *Raymond Lull and the Secret of Life* New York, NY: The Crossroad Publishing Company.

Amzica, Florin, *et al,* (2002) "Glial and Neuronal Interactions during Slow Wave and Paroxysmal Activities in the Neocortex," *Cerebral Cortex* 12 (10): 1101-1113.

Anderson, Jeffrey S., *et al,* (Nov. 16, 2010) "Topographic maps of multisensory attention," *Proceedings of the National Academy of Science of the U S A,*107(46): 20110–20114.

Anthony, Sebastian, (Jun. 25, 2012) "Infinite-capacity wireless vortex beams carry 2.5 terabits per second," *Extreme Tech* (https://www.extremetech.com/?s =orbital+angular+momentum).

Appali, R.; Petersen, S.; Rienen, U. Van, (2010) "A comparison of Hodgkin-Huxley and soliton neural theories," *Advances in Radio Science,* 8, 75–79.

Apple, Inc. (2005-2009) *New Oxford American Dictionary.*

Āranya, Hariharānanda, (Translated by Mukerji, P. N.), (1983) *Yoga Philosophy of Patanjali.* Albany, NY: State University of New York Press.

Aratyn, Henrik, *et al,* (June 1999) "Toroidal solitons in 3+1 dimensional integrable theories," *Elsevier Science,* Vol. 456, Issues 2-4, 162-170.

Arenander, A. T., (2004) "Brain Patterns of Self-awareness," (paper, pp. 1-12, on the internet from: B. Beitman and J. Nair (eds.), *Deficits of Self-Awareness in Psychiatric Disorders*, New York, in press, 2004).

Ash, David, (July 2012) *Vortex of Energy: A Scientific Theory*. United Kingdom: Puja Power Publications (pdf format).

Austin, James. (2000) *Zen and the Brain: Towards an Understanding of Meditation and Consciousness*. Cambridge, MA: The MIT Press.

Axmacher, Nikolai, *et al*, (2006) "Memory formation by neuronal synchronization," *Brain Research Reviews*, 52 170-182. (C) 2006 Elsevier B.V. All rights reserved.

Axmacher, Nikoilai, *et al*, (Feb. 25, 2010) "Intracranial EEG Correlates of Expectancy and Memory Formation in the Human Hippocampus and Nucleus Accumbens," *Neuron*, Vol. 65, Issue 4, 541-549. (Copyright © 2010 Elsevier Inc. All rights reserved.)

Backster, Cleve, (Winter, 1968) "Evidence of a Primary Perception in Plant Life," *International Journal of Parapsychology*, Vol X, No 4.

Bai, Yu, *et al*, (Feb. 28, 2011) "Review of Evidence Suggesting That the Fascia Network Could Be the Anatomical Basis for Acupoints and Meridians in the Human Body," *Evidence-Based Complementary and Alternative Medicine* Volume 2011, Article ID 260510, 1-to-6.

Baker, Gregory L., (1992) *Religion and Science: From Swedenborg to Chaotic Dynamics*. Jamaica, NY: The Solomon Press.

Balli, Tugce, *et al*, (2007) "On the Linearity/Non-Linearity of Mental Activity EEG for Brain-Computer Interface Design," *Kuala Lumpur International Conference on Biomedical Engineering Proceedings*, Volume 15, 2007, 451-454.

Bamford, Christopher, (ed.), (1980) *Homage to Pythagoras: Rediscovering Sacred Science*. Hudson, NY: Lindisfarne Books.

Barnstone, Willis (ed), (1984) *The Other Bible*. San Francisco, CA: Harper San Francisco.

Bayne, Tim, (2007) "Hypnosis and the Unity of Consciousness," Originally published in "Hypnosis and Conscious States," Graham Jamieson (ed), Oxford, U. K.: Oxford University Press (93-109).

Beck, Robert C., (Mar. 10, 1978) "Preliminary Research Report, ElF Magnetic Fields and EEG Entrainment," http://www.elfis.net/elfol8/e8elfeeg2.htm.

Beck, Thomas E, *et al*, (Spring 2003) "A Quantum Biomechanical Basis for Near-Death Life Reviews," *Journal of Near-Death Studies*, 21(3).

Becker, Robert O.; Marino, Andrew A., (1982) *Electromagnetism and Life*. New York, NY: State University of New York Press.

Becker, Robert O.; Selden, Gary, (1987) *The Body Electric: Electromagnetism and the Foundation of Life*. New York, NY: William Morrow and Company.

Becker, Robert O. (1990) *Cross Currents: The Promise of Electromedicine, The Perils of Electropolution*. Los Angeles, CA: Jeremy P. Tarcher, Inc.

Becker, Robert O. (1991) "Evidence for a DC Primitive Electrical Analog System Controlling Brain Function," *Subtle Energies,* Vol 2, No 1, 71-88.

Becker, Robert O., (1992) "Modern Bioelectromagnetics & Functions of the Central Nervous System," *Subtle Energies,* Vol 3, No 1, 53-72.

Benford, M. Sue, (1999) "Spin Doctors: A New Paradigm Theorizing the Mechanism of Bioenergy Healing" *Journal of Theoretics,* June/July V. 1, No. 2.

Bennet, John G., (1990) *Enneagram Studies*. York Beach, ME: Samuel Weiser, Inc.

Benson, Herbert; Klipper, Miriam Z., (1975; 2000) *The Relaxation Response.* New York, NY: Harper Collins Publishers, Inc.

Bentov, Itzhak, (1981) *Stalking The Wild Pendulum: On the Mechanics of Consciousness*. New York, NY: Bantam Books.

Berman, Louis, (1928) *The Glands Regulating Personality: A Study of the Glands of Internal Secretion in Relation to the Types of Human Nature*. New York, NY: The Macmillan Company.

Berkowitz, Ari, (July 21, 2016) "Is Your Nervous System a Democracy or a Dictatorship?" *Scientific American.*

Besant, Annie; Leadbeater, Charles W., (1919) *Occult Chemistry: Clairvoyant Observations on the Chemical Elements*. London, England: Theosophical Publishing House (a Project Gutenberg eBook: www.gutenberg.net).

Bharati, Jnaneshvara, (2011) *Yoga Sutras of Patanjali, Narrative Style: A Translation From Sanskrit to English*. Self publication: SwamiJ.com.

Bharati, Jnaneshvara, (2011) "Mandukya Upanishad and Yoga: Twelve Verses on OM Mantra," Self publication: SwamiJ.com.

Bisiach, Edoardo, *et al,* (1999) "Unilateral neglect and disambiguation of the Necker cube," *Brain,* 122, 131–140.

Blanke, Olaf, *et al,* (Sept. 19, 2002) "Stimulating own-body perceptions," *Nature,* 419, 269-270.

Blavatsky, Helena, (1888) *The Secret Doctrine: the Synthesis of Science, Religion, and Philosophy*. England, London: The Theosophical Publishing Company.

Blavatsky, Helena, (1877) *Isis Unveiled: A Master Key To The Mysteries of Ancient And Modern Science And Theology*. England, London: The Theosophical Publishing Company. England, London: The Theosophical Publishing Company.

Block, Richard A., (1989) "Unilateral Nostril Breathing Influences Lateralized Cognitive Performance," *Brain and Cognition*, 181-190.

Bohm, David, (1980) *Wholeness and the Implicate Order*. London, England: Routledge.

Bohm, David, (1994) *Thought as a System*. New York, NY: Routledge.

Bohm, David; Peat, David, (1987, 2nd ed. 2000) *Science, Order, and Creativity*. London, England: Routledge.

Bolles, Edmund Blair, (1988) *Remembering and Forgetting: Inquiries into the Nature of Memory*. New York, NY: Walker and Company.

Brighton, Christine, (2005), *"The Resurgence of Pre-Modern Holistic Mind-Body Healing Concepts Into the Techno-Holism of Modern Times*. (A Dissertation Submitted to the Faculty of Trinity College of Natural Health in Candidacy for the Degree of Doctor of Naturopathy.)

Borjigin, Jimo, *et al*, (Aug. 27, 2013) "Surge of neurophysiological coherence and connectivity in the dying brain," *PNAS*, vol. 110, no. 35, 14432–14437.

Bonacci, Santos, (Aug. 14, 2012) "Secret of Secrets: The Elixir of Life, Hiding in the Bible Part 1," Presentations on *You Tube*: https://www.youtube.com/watch?v=cWjz9pB70vc.

Borreli, Lizette, (July 9, 2013) "Can An Organ Transplant Change A Recipient's Personality? Cell Memory Theory Affirms 'Yes'," *SCIENCE/TECH*.

Bosman, Saskia, (2000) "A holistic research project on the role of the pineal gland as an intermediary between the physical and metaphysical world of experience," *Private paper on the internet*.

brainMD: https://www.brainmdhealth.com/blog/how-your-brain-is-like-the-universe/

Breitman, Bernard D., (Apr. 24, 2009) "Brains Seek Patterns in Coincidences," *PsychiatricAnnalsOnline.com*, 1-9.

Briggs, John; Peat, David, (1989) *Turbulent Mirror: An Illustrated Guide to Chaos Theory and the Science of Wholeness*. New York, NY: Harper and Row Publishers.

Brighton, John, (Feb., 2000) "The Gate of Stillness Between the Finite and the Infinite," *The Messenger* (Newsletter, New York New Church), 21-22.

Brill, A. A., (Ed., translator), (1938) *The Basic Writings of Sigmund Freud*. New York, NY: Random House, Inc.

Buccheri, R., (2011) "Time and the dichotomy subjective/objective. An endophysical point of view." *Istituto di Astrofisica Spaziale e Fisica Cosmica - Sezione di Palermo Consiglio Nazionale delle Ricerche, Via Ugo La Malfa 153, 90146 Palermo, Italy.*

Buck, Richard Maurice (1901) *Cosmic Consciousness: A Study On The Evolution Of The Human Mind.* Bedford, MA: Applewood Books.

Budzynski, Thomas, (undated web paper) "The Clinical Guide to Sound and Light."

Bundle, Michael, (Jan. 26, 2013) "Physicists Find Evidence that the Universe is a Giant Brain," *Huffington Post UK.*

Burckhardt, Titus, (1986) *Alchemy: Science of the Cosmos, Science of the Soul.* (Translated from the German by William Stoddart.) Dorset, Great Britain: Element Books, Ltd.

Burr, Harold Saxton, (1972) *Blueprint for Immortality: the Electric Patterns of Life.* London, England: Northumberland Press, Ltd.

Buser, Steven; Cruz, Leonard, (2015) *DSM-5 Insanely Simplified: Unlocking the Spectrum within DSM-5 and ICD-10.* Asheville, NC: Chiron Publications.

Cade, Maxwell; Coxhead, Nona, (1979) *The Awakened Mind: Biofeedback and the Development of Higher States of Awareness.* New York, NY: Delacorte Press/Eleanor Friede.

Callahan, R.; Callahan, J., (1996) *Thought Field Therapy and Trauma: Treatment and Theory.* Indian Wells, CA: Thought Field Therapy Training Center.

Capolupo, Antonio, *et al*, (2013) "Dissipation of 'dark energy' by cortex in knowledge retrieval," *Physics of Life Reviews*, Vol 10, Issue 1, 2-9.

Capra, F., (1975; 1984) *The Tao of Physics: An Exploration of the Parallels Between Modern Physics and Eastern Mysticism.* New York, NY: Bantam Books.

Capra, F., (1996) *The Web of Life: A New Scientific Understanding of Living Systems.* New York, NY: Anchor Books, Doubleday.

Carey, George W.; Perry, Inez Eudora, (1920) *God Man: The Word Made Flesh.* Los Angeles, CA: The Chemistry of life Company.

Carhart-Harris, Robbin L., *et al*, (Feb. 7, 2012) "Neural correlates of the psychedelic state as determined by fMRI studies with psilocybin," *PNAS*, Vol. 109, No. 6.

Carhart-Harris, Robbin L., *et al*, (Feb. 3, 2014) "The entropic brain: a theory of conscious states informed by neuroimaging research with psychedelic drugs," *Frontiers in Human Neuroscience*, Vol. 8, Article 20.

Carhart-Harris, Robbin L., *et al*, (Mar. 1, 2016) "Neural correlates of the LSD experience revealed by multimodal neuroimaging," *PNAS*, 1-to-6.

Carlson, Dan, Sonic Bloom: http://dancarlsonsonicbloom.com/About. html#bio.

Carpenter, Patricia A; Davia, Christopher J., (2005) "Mind and Brain: A Catalytic Theory of Embodiment," *Department of Psychology, Carnegie Mellon Univ.*

Cathie, Bruce (downloaded 12/31/10) "Acoustic Levitation of Stones," in Website: *Antigravity and the World Grid*, of David Hatcher Childress: http://www.bibliotecapleyades.net/ciencia/antigravityworldgrid/ciencia_ antigravityworldgrid.htm#contents.

Celick, Murat, *et al*, (Jan. 2015) "The Golden Ratio of the Human Heart," *Gulhane Medical Journal* 57(1):1-4.

Chandler, Russell, (1993) *Understanding the New Age*. Grand Rapids, MI: Zondervan Publishing House.

Chariton of Valamo, Igumen (Compiler); Kadloubovsky, E.; Palmer, G. E. H., (Translators), (1978) *The Art of Prayer: An Orthodox Anthology*. London, England: Faber and Faber Limited.

Chevy, Frédéric, (2014) "Solitons with a Twist." *Physics* 7, 82.

Chopra, Deepak, (June 1, 2013) "Your Brain Is the Universe," *Huffpost*.

Church, Dawson, (2007) *The Genie in Your Genes: Epigenetic Medicine and the New Biology of Intention*. Santa Rosa, CA: Elite Books.

Church, George, *et al*, (Aug. 16, 2012) "Next-Generation Digital Information Storage in DNA," *Science*: 1226355.

Cirlot, J. E., (1983) *A Dictionary of Symbols*. New York, NY: Philosophical Library.

Cislenko, L. (1980) *Structure of Fauna and Flora With Regard to Body Size of Organisms*. Moscow, Russia: Lomonosov-University.

Colgin, Laura Lee, *et al*, (2009) "Frequency of gamma oscillations routes flow of information in the hippocampus." *Nature*, 2009; 462 (7271). Article reporting on this research: "How the Brain Filters out Distracting Thoughts to Focus On a Single Bit of Information," in *ScienceDaily*, Nov. 23, 2009.

Collin, Rodney, (1971) *Theory of Celestial Influence: Man, The Universe and Cosmic Mystery*. Newburyport, MA: Samuel Weiser.

Cook, Theodore A., (1979) *The Curves of life*. New York, NY: Dover Publications, Inc.

Cooper, Linn F.; Erickson, Milton H., (1959 Second Ed. 2006) *Time Distortion in Hypnosis: An Experimental and Clinical Investigation*. Wales, UK: Crown House Publishing, Ltd.

Coué, Émile, (1922) *Self Mastery Through Conscious Autosuggestion*. New York, NY: Malkan Publishing Co., Inc.

Cousto, Hans, (2000) *The Cosmic Octave: Origin of Harmony*. Mendocino, CA: LifeRhythm Publication.

Csikszentmihalyim, Mihaly, (1991) *Flow: The Psychology of Optimal Experience*. New York, NY: HarperCollins Publishers.

Cutting, James E., *et al*, (Feb. 5, 2010) "Attention and the Evolution of Hollywood Film," *Psychological Science Online First*—doi:10.1177/0956797610361679.

Dambergs, Yanis, (2003) *Mnemonic Arts of Blessed Raymond Lull*. Website: http://lullianarts.narpan.net/index.html.

Dardik, Irving I., (1996) "The Origin of Disease and Health Heart Waves: The Single Solution to Heart Rate Variability and Ischemic Preconditioning," *Cycles*, Vol. 46, No. 3.

Dartnell, Lewis, (May 1, 2005) "Chaos In The Brain," +Plus Magazine: http://plus.maths.org/content/chaos-brain.

Davia, Christopher J. (2005) "life, catalysis and excitable media: A dynamic systems approach to metabolism and cognition." In Jack A. Tuszynski (ed) (2006) *The Emerging Physics of Consciousness*. Heidelberg, Germany: Springer-Verlag, 229-260.

Davis, Daniel, (1842), *Manual of Magnetism*. Boston, MA: William S. Damhell, Printer.

Davis, Roy, (1974) *The Anatomy of Biomagnetism*. Green Cove Springs, FL: Self Publication.

Davis, Roy; Rawls, Walter C., (1975/1999) *The Magnetic Effect*. Metairie, LA: Acres, U.S.A., Inc.

Davis, Roy; Rawls, Walter, (2000) *Magnetism and Its Effects on the Living System*. Metairie, LA: Acres, U.S.A., Inc.

De Boer, Ocke, (2014) *Higher Being Bodies: A Non-Dualistic Approach to the Fourth Way, with Hope*. Mount Desert, Maine: Beech Hill Publishing Company.

Decety, Jean, (1996) "Do imagined and executed actions share the same neural substrate?" *Cognitive Brain Research*, 3: 87-93.

De Laszlo, Violet Staub (ed), (1959) *The Basic Writings of C. G. Jung*. New York, NY: Random House (The Modern Library).

Delude, Cathryn, (Mar. 22, 2012) "Researchers show that memories reside in specific brain cells," *MIT News* (Picower Institute for Learning and Memory), 1-2.

Desyatnikov, Anton S., *et al*, (Oct. 2012) "Spontaneous knotting of self-trapped waves," *Scientific Reports*, Vol. 2, 771.

Diamond, John, (1979) *Your Body Doesn't Lie: Unlock the Power of Your Natural Energy!* New York, NY: Harper & Row Publishers, Inc.

Diffen: https://www.diffen.com/.

Dillo, Clay, (Jan 13, 2011) "Can DNA Teleport Itself? One Researcher Thinks So," *Collaboratory*: U. S. Air Force.

Di Trapani, P., *et al*, (Aug. 29, 2003) "Spontaneously Generated X-shaped Light Bullets," *Physical Review Letters*, 91, 29.

Doczi, György., (1981) *The Power of Limits: Proportional Harmonics in Nature, Art and Architecture*. Boston, MA: Shambala Publications, Inc.

Doesburg, Sam M., (July 2009)"Rhythms of Consciousness: Binocular Rivalry Reveals Large-Scale Oscillatory Network Dynamics Mediating Visual Perception," *PLoS ONE*, Vol 4, Issue 7, 2-14.

Donavan, William, *et al*, (May 17, 2012) "Compressions, The Hydrogen Atom, and Phase Conjugation: New Golden Mathematics of Fusion/Implosion: Restoring Centripetal Forces," Web publication at: http://www.fractalfield. com/mathematicsoffusion/

Donders, F. C., (1869) "On the speed of mental processes." In W. G. Koster (ed), Attention and Performance II. Acta Psychologica, 30, 412-431. (Original work published in 1868.)

Douillard, J., (1994) *Body, Mind, And Sport: The Mind-Body Guide To Lifelong Fitness And Your Personal Best*. New York, NY: Crown Trade Paperbacks.

Duboc, Bruno, *et al*, "The Brain From Top to Bottom," Website: http:// thebrain.mcgill.ca/index.php.

Dresler, Martin, *et al*, (Nov. 8, 2011) "Dreamed Movement Elicits Activation in the Sensorimotor Cortex," Current Biology 21, 1833–1837 (© Elsevier).

Dyches, Preston, (April 7, 2015) "The Solar System and Beyond is Awash in Water," *NASA*.

Dyczkowski, Mark S. G., (1989) *The Doctrine of Vibration: An Analysis of the Doctrines and Practices of Kashmir Shaivism*. Dehli, India: Motilal Banarsidass.

Echel, John, (Apr. 5, 2018) "Superlight and Magnetricity - Cosmic Forces by John V. Milewski," *Aether force*: http://aetherforce.com/ superlight-and-magnetricity-cosmic-forces-by-john-milewski/

Eddington, Arthur S., (1927) *The Nature of the Physical World*. New York, NY: The MacMillan Company.

Egely, George, (1989) *Egely Wheel: Vitality Meter.* Budapest, Hungary: Dimenzió Ltd.

Egne, Hakan, (Oct. 9, 2012a) "Super Fluid Universe," You Tube: https://www.youtube.com/watch?v=eawL3WxzkUk

Egne, Hakan (2012b), "Fluid Dynamics and Universe," Web article: http://super-fluid-universe.8m.com/.

Ekeocha, Tracy C. (2015) *The Effects of Visualization & Guided Imagery In Sports Performance.* A thesis submitted to the Graduate Council of Texas State University in partial fulfillment of the requirements for the degree of Master of Arts with a Major in Health Psychology.

Ellis, Amy Jo, (2018) *Healing Your Family Tree Through The Court of Atonement.* Self publication as eBook: https://www.amyjosings.com/

El-Naschie, Mohamed, (2007) "The Fibonacci code behind super strings and P-Branes - An answer to M. Kaku's fundamental question," *Chaos, Solitons and Fractals* 31, 537–547.

Emoto, Masaru, (2004) *The Hidden Messages in Water.* (Translated by David A. Thayne), Hillsborough, OR: Beyond Words Publishing, Inc.

Encyclopedia of the Unusual and Unexplained: http://www.unexplainedstuff.com/index.html.

Epstein, D., (2000) *Healing Myths, Healing Magic: Breaking the Spell of Old Illusions; Reclaiming Our Power to Heal.* San Rafael, CA: Amber-Allen Publishing, Inc.

Erickson, Milton H. "Deep Hypnosis and Its Induction," (in LeCron 1968, 71-112).

Evans, John, (1986) *Mind, Body and Electromagnetism.* Worcester, GB: Element Books.

Fehmi, Jeffrey S., (Mar. 4, 2009) "Attention to Attention," (A version of this manuscript was published in *Applied Neurophysiology and EEG Biofeedback.* Publisher, Future Health, Inc. Editor, Joe Kamiya).

Fell, Juergen, *et al*, (2006) "Rhinal–hippocampal connectivity determines memory formation during sleep," *Brain* (2006), 129, 108–114.

Fideler D-R (ed), Guthrie, K-S (Compiler and Translator) (1988) *The Pythagorean Sourcebook and Library.* Grand Rapids, MI: Phanes Press.

Fields, R. Douglas, (Aug. 29, 2012) "Meet Your Glia," in *The Doctor Will See You Now: http://www.thedoctorwillseeyounow.com/content/mind/art3792.html.*

Fischer, Wolfgang, (Oct. 20, 08) "Wave Structure of Matter: Interconnectedness of Being, Full Spectrum Responsibility," ***GAIA*** *- PRO VITA ET FUTURA MUNDI*: http://emanzipationhumanum.de/english/human/wsm.html.

Fitzpatrick Jr., Daniel (2001) *A New Look At Dark Matter.* (Web ebook).

Fitzpatrick Jr., Daniel (2002) *Fitzpatrick's Theory of Everything.* (Web ebook).

Fitzpatrick Jr., Daniel (2013) *Phase Symmetry Makes Quantum Theory More Complete.* (Web ebook: http://rbduncan.com/phase.symmetry.pdf).

Florin, Esther, *et al*, (Feb. 2015) "The Brain's Resting-State Activity is Shaped by Synchronized Cross-Frequency Coupling of Oscillatory Neural Activity," *NeuroImage*, 111.

Frawley, David, (1990) *The Astrology of the Seers: A Guide to Vedic (Hindu) Astrology.* Salt Lake City, UT: Passage Press.

Frazer, James George, (1906) *The Golden Bough: a study of magic and religion.*

Freeman, Walter J., (1990) "Searching for signal and noise in the chaos of brain waves." In *The Ubiquity of Chaos,* ed. S. Krassner. Washington, DC., American Association for the Advancement of Science, 47–55.

Freeman, Walter J., (2000) *How Brains Make Up Their Minds*, New York, NY: Columbia University Press.

Freeman, Walter J., *et al*, (June 2006) "Nonlinear brain dynamics as macroscopic manifestation of underlying many-body field dynamics," *Physics of Life Reviews*, 1-32.

Freeman, Walter J., (2009) "Vortices in brain activity: Their mechanism and significance for perception," *Neural Networks* 22 (2009) 491–501.

Freeman, Walter J., *et al*, (14 Oct., 2011) "Cortical phase transitions, non-equilibrium thermodynamics and the time-dependent Ginzburg-Landau equation," arXiv:1110.3677v1 [physics.bio-ph].

Freeman, Walter J.; Vitiello, Giuseppe, (2010) "Vortices in brain waves," *International Journal of Modern Physics B, 24:17, 3269-3295.*

Freeman, Walter J. (2008) "Nonlinear Brain Dynamics and Intention According to Aquinas," *Mind & Matter* Vol. 6(2), pp. 207-234

Friedlander, Joel, (1993) *Body Types: The Enneagram of Essence Types.* Yorktown, NY: Inner Journey Books.

Fudjack, John; Dinkelaker, Patricia, (February, 1999a), "Enneagram as Mandala - Part I: Ego, Self, and Liminocentric Structures." Source: http://tap3x.net/EMBTI/j4selfb.html.

Fudjack, John; Dinkelaker, Patricia, (March, 1999b) "The Enneagram as Classic 'Double Mandala: Part I - The 'I Ching' and other 'Divination Machines'," Source: http://tap3x.net/EMBTI/j4selfc.html.

Fuller, Buckminster R., (1975-1979) *Synergetics.* Macmillan Publishing Co., Inc. (Source: http://www.rwgrayprojects.com/synergetics/synergetics.html).

Gariaev, Peter P., *et al,* (1992) "Investigation of the Fluctuation Dynamics of DNA Solutions by Laser Correlation Spectroscopy." *Bulletin of the Lebedev Physics Institute,* n. 11-12, 23-30.

Gariaev, Peter P., *et al,* (1994) "The DNA-wave Biocomputer," *Institute Control of Sciences Russian, Russian Academy of Sciences.*

Gariaev, Peter P., *et al,* (2006) "Crisis in Life Sciences: The Wave Genetics Response," *EmergentMind.org.*

Gaskell, George A., (1981) *Dictionary of All Scriptures and Myths.* Avenel, NJ: Random House Company.

Gegoriou, Georgia, G, *et al,* (2009) "Long-range neural coupling through synchronization with attention," Progress in Brain Research; 176: 35-45.

Gentet, Luc J., *et al,* (Feb. 11, 2010), "Membrane Potential Dynamics of GABAergic Neurons in the Barrel Cortex of Behaving Mice," *Neuron* 65, 422–435, (© Elsevier Inc.).

Georgiev, Danko Dimchev, (2004) "Electric and magnetic fields inside neurons and their impact upon the cytoskeletal microtubules," *Division Of Electron Microscopy, Medical University Of Varna, Bulgaria.*

Ghalay, Maurice, *et al,* (2004) "The Biological Effects of Grounding the Human Body During Sleep as Measured by Cortisol Levels and Subjective Reporting of Sleep, Pain, and Stress," *The Journal of Alternative and Complementary Medicine,* Vol. 10, No. 5, 767-776.

Gilden, David L., (2001) "Cognitive Emissions of 1/f Noise," *Psychological Review,* Vol. 108, No. 1, 33-56.

Gilsinan, Kathy, (July 4, 2015) "The Brains of the Buddhists: What Compassion Does to the Brain," *The Atlantic.*

Gleick, James, (1987) *Chaos: Making a New Science,* New York, NY: Viking Penguin.

Glassman, Robert B. (1999) "Hypothesized neural dynamics of working memory: Several chunks might be marked simultaneously by harmonic frequencies within an octave band of brain waves" *Brain Research Bulletin, Vol. 50, No. 2, pp. 77–93,* (Copyright (©) 1999 Elsevier Science Inc.)

Gonabadi, Nematollahi Sultan Ali Shahi, (2007) Sufi Order: http://www.sufism.ir/sufischool.php.

Goodenough, Ursula, (1998, 2000) *The Sacred Depths of Nature.* New York, NY: Oxford University Press.

Gourine, Alexander, *et al*, (July 30, 2010) "Astrocytes Control Breathing Through pH-dependent Release of ATP," *Science*, 329 (5991), 571-575.

Grebennikov, Viktor S. (1997) *My World* (*Moi Mir.* Novosibirsk, Russia: Sovetskaya Sibir).

Greber, Johannes, (1932; Translation 2007) *Communication With The Spirit World: Its Laws And Purpose.* (Translated & corrected from German To English by Joseph F. Greber and Elsa Lattey.) Teaneck, NJ: The Johannes Greber Memorial Foundation.

Greber, Johannes, (1937) *The New Testament: A New Translation Based on the Oldest Manuscripts.* Teaneck, NJ: The Johannes Greber Memorial Foundation.

Green, Elmer & Alyce, (1997) *Beyond Biofeedback.* Knoll Publishing Co., (Chapter II: Self-regulation: East and West, pp. 197 _ 218 - Source: http://www.sadhanamandir.org/BFB2.pdf).

Kleckner-Greenfield, Susan A., (1995), *Journey To The Centers Of The Mind: Towards a Science of Consciousness.* New York, NY: W. H. Freeman and Company.

Gresbæsbøll, Kaare, (2006) "Function of Nerves - Action of Anesthetics," *Gamma* 143, 27-36.

Guiley, Rosemary Ellen, (1991) *Harper's Encyclopedia of Mystical & Paranormal Experience.* San Francisco, CA: Harper San Francisco.

Gullette, Alan (Spring 1976) "Nothing is Sacred; Or, The Concept of Nothing in Zen," (Paper for course in *Religious Studies 3770: Zen Buddhism,* by Dr. Stan Lusby and Dr. Camp).

Gunkelman, Jay, (May 12, 2010) "Understanding Consciousness: An Emergent Property of Mind-Brain Interaction," Video on You Tube: http://www.scientificexploration.org/talks/28th_annual/28th_annual_gunkelman_consciousness_mind_brain_interaction.html.

Gurdjieff, George I., (1963) *Beelzebub's Tales to his Grandson or An Objectively Impartial Criticism of the Life of Man.* (pdf ebook format.) New York, NY: E.P. Dutton & Co., Inc.

Hall, Manly. (1928) *The Secret Teachings of All Ages: An Encyclopedic Outline of Masonic, Hermetic, Qabbalistic and Rosicrucian Symbolical Philosophy: Being an Interpretation of the Secret Teachings Concealed Within the Rituals, Allegories, and Mysteries of all Ages.* San Francisco, CA: H.S. Crocker Company, Inc.

Hall, Nina, (Editor), (1993) *Exploring Chaos: A Guide to the New Science of Disorder.* New York, NY: W. W. Norton and Company.

Hankey, Alex, (Jul. 31, 2006) "Studies of Advanced Stages of Meditation in the Tibetan Buddhist and Vedic Traditions. I: A Comparison of General Changes,"

Evidence-Based Complementary and Alternative Medicine, Vol 3, Issue 4, 513–521. (Advance Access Publication.)

Hanslmayr, Simon, *et al*, (2011) "The Relationship between Brain Oscillations and Bold Signal during Memory Formation: A Combined EEG–fMRI Study," *The Journal of Neuroscience*, 2 Nov., 2011, 31 (44): 15674-15680.

Hanslmayr, Simon, *et al*, (2012) "Prefrontally Driven Downregulation of Neural Synchrony Mediates Goal-Directed Forgetting," *The Journal of Neuroscience, October 17, 2012 • 32(42):14742–14751*

Harbage, Alfred (ed), (1969) *William Shakespeare: The Complete Works*. Baltimore, MD: Penguin Books.

Harms, John K., (2003) "Time-lapsed reality visual metabolic rate and quantum time and space," *Kybernetes*, Vol. 32 Issue 7/8, pp. 1113-1128.

Hasellhoff, Eltjo H., (2001) *The Deepening Complexity of Crop Circles: Scientific Research and Urban Legends*. Berkeley, CA : Frog, Ltd.

Hassenkamp, Wendy, (July 17, 2013) "How to Focus a Wondering Mind," *Greater Good Magazine* (Source: The Greater Good Science Center at the University of California, Berkeley).

Healy, Kevin, *et al*, (Oct. 2013) "Metabolic rate and body size are linked with perception of temporal information," *Animal Behavior* (in Science Direct, Elsevier), Vol 86, Issue 4, 685-696.

Hebb, Donald O., (1949) *The Organization of Behaviour*. New York, NY: John Wiley & Sons.

Heimburg, Thomas; Jackson, Andrew D., (July 12, 2005) "On soliton propagation in biomembranes and nerves." *Proceedings of The National Academy of Sciences*, vol. 102, no. 28, 9790-9795.

Hendricks, Luke, (2010) "The Healing Connection: EEG Harmonics, Entrainment, and Schumann's Resonances," Journal of Scientific Exploration, Vol. 24, No. 3, pp. 419–430.

Herrigel, Eugen, (Translated by R. F. C. Hull, 1971) *Zen In The Art Of Archery*. New York, NY: Vintage Books (a Division of Random House).

Highlen, Pamela S., *et al*, (1979) "Psychological Characteristics of Successful and Nonsuccessful Elite Wrestlers: An Exploratory Study," *Journal of Sport Psychology*, 1, 123-137.

Hitchcock, Ethan Allen, (1858) *Swedenborg, a Hermetic Philosopher*. New York, NY: D. Appleton & Company.

Ho, Mae-Wan, (Oct. 17, 2007) "Thermodynamics of Organisms and Sustainable Systems," (Invited lecture for conference on *Environment, Agriculture, Food, Health and Economy*, World Food Day, La Sapienza University, Rome, Italy).

Ho, Mae-Wan, (2013) *The Rainbow and the Worm: The Physics of Organisms*. Hackensack, NJ: World Scientific Publishing Company.

Ho, Mae-Wan, (2014), "Illuminating Water and Life," *Entropy*, 16, 4874-4891.

Ho, Mae-Wan, (Mar. 17, 2014) "Golden Music of the Brain" (Story of Phi Part 3), *Institute of Science In Society (ISIS)*.

Hobson, J. Allan, *et al*, (2010) "Transcendental consciousness wakes up in dreaming and deep sleep," *International Journal of Dream Research* Vol 3, No. 1, 28-32.

Hoffman, Erik, *et al*, (Spring 2001) "Effects of a Psychedelic, Tropical Tea, Ayahuasca, on the Electroencephalographic (EEG) Activity of the Human Brain During a Shamanistic Ritual," *maps*, volume XI number 1, 25-30.

Hoffmann, Erik, (May 12, 2010) "New Brain - New World," Web blog: http://www.newbrainnewworld.com/?New_Brain_-_New_World.

Hoffmann, Erik, (2012) *New Brain - New World: How the Evolution of a New Human Brain Can Transform Consciousness and Create a New World.*, England, London: Insights (Hay House UK, Ltd.).

Hofstadter, Douglas R., (1999) *Gödel, Escher, Bach: An Eternal Golden Braid*. New York, NY: Basic Books, Inc.

Hu, Hupping; Wu, Maoxin, (Feb 18, 2013) "Human Consciousness as Limited Version of Universal Consciousness," *Journal of Consciousness Exploration & Research*, 4(1): pp. 52-68.

Hu, Hupping; Wu, Maoxin, (2004) "Spin as Primordial Self-Referential Process Driving Quantum Mechanics, Spacetime Dynamics and Consciousness," *NeuroQuantology*, Issue 1, 41-49.

Hu, Hupping; Wu, Maoxin, (3/15/2003) *Spin Mediated Consciousness Theory*.

Huang, Tina L; Charyton, Christine (2008) "A comprehensive view of the psychological effects of brainwave entrainment," *Alternative Therapies, Sep/Oct 2008, Vol. 14, No. 5, 38-49*.

Huang, Xiaoying, *et al*, (Dec 9, 2010) "Spiral wave dynamics in neocortex," *Neuron* 68, 978–990.

Ivanov, B. A., (1995) "Two-Dimensional Magnetic Solitons and Thermodynamics of Quasi-Two-Dimensional Magnets," *Chaos, Solitons & Fractals*, (special issue "Solitons in Science and Engineering: Theory and Applications) 5, 2605.

Ives, Crystal (2005) "Human Beings as Chaotic Systems," Article on the web: http://www.imarkswebs.com/bk/classic+fifty+electrical+diagram/.

James, William, (1890) *The Principles of Psychology*. New York: Henry Holt, V.1.

James, William, (1902) *The Varieties of Religious Experience*. London, England: Longmans, Green, & Co. (Reprint by eBooks@Adelaide 2009: http://ebooks.adelaide.edu.au/j/james/william/varieties/complete.html.

Jaynes, Julian, (1976) *The Origins of Consciousness In The Breakdown Of The Bicameral Mind*. Boston, MA: Houghten Mifflen Company.

Jenny, Hans, (2000) *Cymatics: A Study of Wave Phenomena and Vibration*. New Market, NH: MACROmedia Publishing.

Jensen, Ole, *et al*, (May 17, 2007) "Human gamma-frequency oscillations associated with attention and memory," *TRENDS in Neurosciences*, Vol.30 No.7.

Johari, Harish, (1989) *Breath, Mind, and Consciousness*. Rochester, VT: Destiny Books.

Jung, Carl G., (and M.-L. von Franz; Joseph L. Henderson; Jolande Jacobi; Aniela Jaffe) (1964) *Man and his Symbols*. New York, NY: Anchor Press Doubleday.

Jutras, Michael J; Buffalo, Elizabeth A., (2010) "Synchronous neural activity and memory formation," Current Opinion in Neurobiology, 20:1-to-6.

Kadloubovsky, E.; Palmer, G. E. H., (Translators), (1977) *Writings From The Philokalia On Prayer Of The Heart*. London, England: Faber and Faber Limited.

Karagulla, Shafica, S., (1969) *Breakthough To Creativity: Your Higher Sense Perception*. Los Angeles, CA: DeVorss and Co., Inc.

Kardec, Alan, (1857; Translation to English by Anna Blackwell, 6th edition, 1996) *The Spirits' Book*. Brazil, Rio: FEDERAÇÃO ESPÍRITA BRASILEIRA.

Kent, James L., (2010) *Psychedelic Information Theory: Shamanism in the Age of Reason*. Seattle, WA: PIT Press/Supermassive, LLC.

Kieffer, Gene, (ed), (1998) *Kundalini for the New Age: Selected Writings of Gopi Krishna*. New York, NY: Bantam Books.

King, Chris, (8-9-14) "Entheogens, the Conscious Brain and Existential Reality,"

King, Chris (Aug. 20, 2014) "The Central Enigma of Consciousness," New Zealand, Auckland: Mathematics Department, Univ. of Auckland.

King, Moray B., (2002) *Tapping The Zero-Point Energy*. Kempton, IL: Adventures Unlimited Press.

Kirsch, Irving, (1999) "Hypnosis and Placebos: Response Expectancy as a Mediator of Suggestion Effects," *anales de psicología*, Vol. 15, No 1, 99-110.

Kleckner, Dustin, *et al*, (Sept. 5, 2012) "Creation and dynamics of knotted vortices," *Nature Physics* 9, 253–258.

Klee, Maurice M., (2014) "Biology's built-in Faraday cages," *American Journal of Physics* 82, 451.

Klimesch, Wolfgang, (April 1999) "EEG alpha and theta oscillations reflect cognitive and memory performance: a review and analysis," *Elsevier: Brain Research Reviews*, Vol 29, Issues 2-3, 169–195.

Klimesch, Wolfgang, (Nov. 12, 2013) "An algorithm for the EEG frequency architecture of consciousness and brain body coupling," *Frontiers In Human Neuroscience*: http://dx.doi.org/10.3389/fnhum.2013.00766.

Knight, Robert T., *et al*, (Jan. 21, 2018) "Cortico-limbic circuits and novelty: a review of EEG and blood flow data," *Rev. Neuroscience.* 9, 57-70.

Koestler, Arthur, (1964) *The Act of Creation*. England, London: Hutchinson and Company, Ltd.

Koestler, Arthur, (1967) *The Ghost in the Machine*. England, London: Hutchinson and Company, Ltd.

Komech, A.I., *et al*, (2004) "On Attraction to Solitons in Relativistic Nonlinear Wave Equations," *Russ. J. Math. Phys.* 11, no. 3, 289-307.

Koob, Andrew, (2009) *The Root of Thought: Unlocking Glia—The Brain Cell That Will Help Us Sharpen Our Wits, Heal Injury, and Treat Brain Disease*. Upper Saddle River, NJ: FT Press.

Kopell, Nancy (Course Organizer), (2009) *"Rhythms of the Neocortex: Where Do They Come From and What Are They Good For?" Short Course II*, Society for Neuroscience.

Korn, Henri, *et al*, (2003) "Is there chaos in the brain? II. Experimental evidence and related models," *Biologies* 326 787–840.

Kosevich, A. M., *et al*, (Oct. 1990) "Magnetic Solitons," *Physics Reports* (Elsevier) Vol. 194, Issues 3-4, 117-238.

Krieger, Kim, (2003) "Bullets of Light," *Physical Review Focus 12, 7 (2003)*.

Krishna, Gopi, (1970) *Kundalini: The Evolutionary Energy in Man*. London, England: Stuart & Watkins.

Krishna, Gopi, (1988) *Kundalini For the New Age: Selected Writings by Gopi Krishna*. Gene Kieffer, Editor. New York, NY: Bantam Books.

Kuittinen, Petri, (1999) "Noise in Man-Generated Images and Sound," Web article: http://mlab.uiah.fi/~eye/mediaculture/noise.html.

Lad, Vasant D. (1993) *Ayurveda: The Science of Self-Healing*. Delhi, India: Motilal Banarsidass Publishers.

Lad, Vasant D. (1996) *Secrets of the Pulse: The Ancient Art of Ayurvedic Pulse Diagnosis.* Albuquerque, NM: The Ayurvedic Press.

Lamb, Trisha, (Compiler), (2004) *Psychophysiological Effects of Yoga.* Prescott, AZ: International Association of Yoga Therapists (IAYT).

Lamsa, George M., (1968) *Holy Bible: [Translated] From The Ancient Eastern Text.* San Francisco, CA: Harper & Row Publishers.

Lanza, Robert; with Bob Berman, (2009) *Biocentrism: How Life and Consciousness are the Keys to Understanding the True Nature of the Universe.* Dallas, TX: Benbella Books, Inc.

Larson, Robin, *et al,* (Eds.), (1988) *Emanuel Swedenborg: A Continuing Vision, A Pictorial Biography and Anthology of Essays and Poetry.* New York, NY: Chameleon Books, Inc. (for the Swedenborg Foundation, Inc.).

Laurence, Aurieli, (Sept. 12, 2006) "What is a Scalar Wave?" *Electromagnetic Frequencies* (Blog): https://cellphonesafety.wordpress.com/2006/09/12/what-is-a-scalar-wave/

Laszlo, Ervin, (2004) *Science and the Akashic Field: An Integral Theory of Everything,* Rochester, VT: Inner Traditions.

Layne, Scott P. (Spring 1984) "A Possible Mechanism for General Anesthesia," *Los Alamos Science, 23-26.*

LeCron, Leslie M., (ed), (1968) *Experimental Hypnosis.* New York, NY: Citadel Press.

Lee, Richard H., (Ed. 1999) *Scientific Investigation into Chinese Qi-Gong.* San Clemente, CA: China Healthways Institute.

Leet, Leonora. (2003) *The Kabbalah of the Soul: The Transformative Psychology and Practices of Jewish Mysticism.* Rochester, VT: Inner Traditions.

Lehar, Stephen, (2004) *Boundaries of Human Knowledge.* Mahwah, NJ: Lawrence Erlbaum Associates.

Lele, Ramchandra, (1986) *Ayurveda and Modern Medicine.* Bombay, India: Bharatiya Vidya Bhavan.

Leon, James, (2008) "Principles of Theistic Psychology: The Scientific Knowledge of God Extracted from the Correspondential Sense of Sacred Scripture," *Web document, continuously updated and expanded: http://www.soc.hawaii.edu/leonj/theistic/.*

Leon, James, (2012) "Swedenborg Encyclopedia of Theistic Psychology: The Ideas of Emanuel Swedenborg (1668-1772) Expressed In Modern Scientific Psychology," *Web document, continuously updated and expanded: http://www.soc.hawaii.edu/leonj/leonj/leonpsy/instructor/gloss.htm.*

Lépinard, Denys (as of 2015): http://www.ontostat.com/anglais/index.htm.

Lesondak, David, (2014) "Fascia and the Mind/Body Connection." Lecture Ulm Univ., Gemany: https://www.youtube.com/watch?v=2Yo_f6U0WKo.

Lewin, Roger, (Dec. 1980) "Is Your Brain Really Necessary?" *Science* vol. 210.

Li, Jun Z., (2013) "Circadian rhythms and mood: Opportunities for multi-level analyses in genomics and neuroscience," *Bioessays* (published by WILEY Periodicals, Inc.) 36: 305–315.

Lilly, John C., (1972) *The Center of the Cyclone: An Autobiography of Inner Space.* New York, NY: Julian Press, Inc.

Lin, Zhicheng; Murray, Scott O., (Jan. 2014) "Unconscious Processing of an Abstract Concept," *Psychological Science*, vol. 25, no. 1, 296-298.

Liou, Chien-Hui, *et al*, (Nov. 15, 2007) "Correlation between Pineal Activation and Religious Meditation Observed by Functional Magnetic Resonance Imaging," Nature Precedings : 1328.1.

Lipton, Bruce, (2005) *The Biology of Belief: Unleashing the Power of Consciousness, Matter and Miracles.* Santa Rosa: CA: Mountain of Love/Elite Books.

Lomdahl, Peter S.; Bigio, Irving J. (Spring 1984) "Solitons in Biology," *Los Alamos Science*, *1-22.*

Lomdahl, Peter S., Spring 1984, "What is a Soliton?" *Los Alamos Science*, *27-31.*

Loye, David. (2000) *Arrow Through Chaos: How We See Into The Future.* Rochester, VT: Park Street Press.

Luce, Gay G., (1973) *Body Time: Physiological Rhythms and Social Stress.* New York, NY: Bantam Books.

Lugt, Hans J., (1983) *Vortex Flow in Nature and Technology.* Malabar, FL: Krieger Publishing Company.

Lull, Raymond, (1305) *Ars Magna: Ars Generalis Ultima* (The Great Art: The Ultimate General Art). Palma de Majorca, Kingdom of Majorca (now Spain).

Lutz, Antoine, *et al*, (Nov. 16, 2004) "Long-term meditators self-induce high-amplitude gamma synchrony during mental practice," *PNAS*, vol. 101, no. 46, 16369–16373.

Mac Cormac, Earl; Stamenov, Maxim, (Eds.), (1996) *Fractals of Brain, Fractals of Mind: In Search of a Symmetry Bond.* Philadelphia, PA: Benjamins Publishing Company.

MacFadden, Johnjoe, (2002) "Synchronous Firing and Its Influence on the Brain's Electromagnetic Field: Evidence for an Electromagnetic Field Theory of Consciousness," *Journal of Consciousness Studies, 9, No. 4, 2002, pp. 23–50.*

McGillion, Frank (1980) *The Opening Eye.* England, London: Coventure Ltd.

Maltz, Maxwell, (1969) *Psycho-Cybernetics*. New York, NY: Prentice-Hall, Inc.

Maltz, Maxwell, (2001) *The New Psycho-Cybernetics: The Original Science of Self-Improvement and Success That Has Changed the Lives of 30 Million People*. New York, NY: Prentice-Hall Press.

Mann, Edward W.; Hoffman, Edward, (1990) *Wilhelm Reich: The Man Who Dreamed of Tomorrow*. Northhamptonshire, England: Crucible, Aquarian Press.

Manning, Jeane, (1996) *The Coming Energy Revolution: The Search For Free Energy*. Garden City Park, NY: Avery Publishing Group.

Margulis, Elizabeth Hellmuth (Nov. 2, 2017) "Music is not for ears," *Aeon* (Mag.).

Marinelli, Ralph, *et al*, (Fall/Winter 1995) "The Heart is not a Pump: A Refutation of the Pressure Propulsion Premise of Heart Function.," *Frontier Perspectives* Vol. 5, No. 1, 15-24.

Martin, Andrew, *et al*, (2005) "Solitons and Vortices in Atomic Bose Einstein Condensates," *Durham BEC Project*, Durham University.

Matthews, Ronald E., (Undated) "Harold Burr's Biofields: Measuring the Electromagnetics of Life," *Subtle Energies & Energy Medicine*, Volume 18, Number 2, Page 55-61.

Maurer, Leon H., (July 10, 2010) "How Unconditioned Consciousness, Infinite Information, Potential Energy, and Time Created Our Universe: Proposing A New Scientific Paradigm," *Journal of Consciousness Exploration & Research*, VOL 1 – No. 5, pp 610–624.

McCraty, Rollin, *et al*, (2006) *The Coherent Heart: Heart–Brain Interactions, Psychophysiological Coherence, and the Emergence of System-Wide Order*. Boulder Creek, CA: HeartMath Research Center, Institute of HeartMath.

McGillion, Frank, (1980) *The Opening Eye*. London, UK: Coventure, Ltd.

McKnight, J. T., *et al*, (2001) "Attention and neurofeedback synchrony training: clinical results and their significance," *Journal of Neurotherapy*, Vol. 5(1-2), 45-61.

Meehan, Jodina, (2008...) *Journal of Cymatics* (http://cymatica.com/).

Melechi, Antonio, (4/19/17) "Every school of psychology has its own theory of the unconscious," *Aeon Essays* (aeon.co), 1-14.

Melloni, Lucia, *et al*, (March 14, 2007) "Synchronization of Neural Activity across Cortical Areas Correlates with Conscious Perception," *The Journal of Neuroscience*, 27(11): 2858-2865.

Merrick, Richard, (2011) *Interference: A Grand Scientific Musical Theory*. ISBN: 978-0-615-20599-1.

Meyl, Konstantin, (1996) *Scalar Waves: From an extended vortex and field theory to a technical, biological, and historical use of longitudinal waves.* (First edition of lecture series, published in pdf format on the web.)

Meyl, Konstantin, (2012a) "Task of the introns, cell communication explained by field physics," *Journal of Cell Communication and Signaling.* 6: 53–58.

Meyl, Konstantin, (2012b) "About Vortex Physics and Vortex Losses," *Journal of Vortex Science and Technology*, Vol. 1, Article ID 235563, 1-10.

Michelmann, Sebastian, *et al*, (Aug. 5, 2016) "The Temporal Signature of Memories: Identification of a General Mechanism for Dynamic Memory Replay in Humans," *PLOS Biology*, 1-27.

Mikeska, H. J. (1978) "Solitons in a one-dimensional magnet with an easy plane," *Journal of Solid State Physics*, 11.

Miller, Emmett E., (1978) *Self Imagery: Creating Your Own Good Health.* Berkeley, CA: Celestial Arts.

Miller, George A. (1956) "The magical number seven, plus or minus two: Some limits on our capacity for processing information," *Psychological Review* 63 (2), 81-87.

Miller, Mark: Neuro: brains are gorgeous at the right magnification - https://www.flickr.com/photos/neurollero/sets/366106/.

Miltner, Wolfgang H. R., *et al*, (1999) "Coherence of gamma-band EEG activity as a basis for associative learning," *Nature* 397, 434-436 (4 February 1999).

Mindell, A. (2000) *Quantum Mind: The Edge Between Physics and Psychology.* Portland, OR: Lao Tse Press.

Minkel, JR, (Feb 2009) "Strange but True: Superfluid Helium Can Climb Walls," *Scientific American*, 10.

MindMaster™: http://www.mindmaster.tv/sp/mindmaster.htm?hop=infobookde.

Mind of A Winner™: http://www.mindofwinner.com/subliminal-messages/.

Mishra, Satyendra Prasad, (1989) *Yoga and Ayurveda: Their alliedness and scope as positive health sciences.* India, Varanasi: Chaukhambha Sanskrit Sansthan.

Molina, R. A., *et al*, (2010) "Perspectives on $1/f$ noise in quantum chaos," *Journal of Physics: Conference Series 239 (2010) 012001.*

Molina-Terriza, Gabriel, *et al*, (December 2002) "Optical Vortex Streets," *Optics & Photonics News*, 56.

Monroe, Robert, (1971) *Journeys Out of the Body.* New York, NY: Doubleday.

Monroe, Robert, (1985) *Far Journeys.* New York, NY: Doubleday.

Montagnier, Luc, *et al*, (Dec. 23, 2010) "DNA waves in water," *Journal of Physics*: Conference Series, Vol 306, Issue 1.

Moore, James, (2004) "The Enneagram: A Developmental Study." Source: http://www.gurdjieff-bibliography.com/

Moore, Judith; Lamb, Barbara, (2001) *Crop Circles Revealed: Language of the Light Symbols*. Flagstaff, AZ: Light Technology Publishing.

Morgan, D. (2010 A, date when downloaded from the internet) *A Theoretical Framework for Hypnosis*.

Morgan, D. (2010, date when downloaded from the internet) *Principles of Hypnotherapy*.

Morgan, Marlo, (1991) *Mutant Message Downunder*. Summit, MO: MM Co.

Müller, Hartmut, (April 2001) "Telecommunications Free from Electric Smog!" Germany: *Raum&Zeit, (Special 1: Global Scaling)*, Vol 114, 119-126.

Murdock, D. M. (a.k.a. Acharya S): http://truthbeknown.com/

Murphy, Joseph, (1963 pdf) *The Power of Your Subconscious Mind*. Mansfield Centre, CT: Martino Publishing (2011 reprint).

Murphy, Michael, (1993) *The Future Of The Body: Exploration into the Further Evolution of Human Nature*. New York, NY: The Putman Publishing Group.

Muzzio, Isabel A., *et al*, (June 2009) "Attention Enhances the Retrieval and Stability of Visuospatial and Olfactory Representations in the Dorsal Hippocampus," *PLoS Biol 7(6)*: e1000140. doi:10.1371/journal.pbio.1000140.

Myss, C. (1996) *Anatomy of the Spirit: The Seven Stages of Power and Healing*. New York, NY: Three Rivers Press (Random House).

Nachalov, Yu V.; Sokolov, A. N., (c. 1991) "Experimental investigations of new long range actions," (paper on the internet).

Nakada, Tsutomu, (2000) Vortex model of the brain: The missing link in brain science? In: Nakada, T (ed.) Integrated Human Brain Science. Amsterdam: *Elsevier*, 3–22.

Nalbone, Karen, (1991) "Glossary of Objective Language of the Fourth Way," Lawrenceville, NJ: Private paper.

National Aeronautics and Space Administration (NASA), (April 7, 2015) "The Solar System and Beyond is Awash in Water,": https://www.nasa.gov/jpl/the-solar-system-and-beyond-is-awash...

Neimann, Gilbert, (Ed.), (1960) Between Worlds: An International Magazine of Creativity. Denver, CO: (Published for the Inter American University by Alan Swallow; now out of print.)

Newberg, Andrew; D'Aquili, Eugene, (2001) *Why God Won't Go Away: Brain Science and the Biology of Belief*. New York, NY: The Ballantine Publishing Group.

Newberg, Andrew B.; Newberg, Stephanie K., (2005) "The Neuropsychology of Religious and Spiritual Experience," (Chapter 11 in Paloutzian).

Newton, Michael, (2000) *Destiny of Souls: New Case Studies of Life Between Lives*. St. Paul, MN: Llewellyn Publications.

Newton, Michael, (1994/2003) *Journey of Souls: Case Studies of Life Between Lives*. St. Paul, MN: Llewellyn Publications.

Nicolescu, Basarab, (1991) *Science, Meaning, and Evolution: The Cosmology of Jacob Boehme*. New York, NY: Parabola Books.

Nicoll, Maurice, (1917) *Dream Psychology*. London, England: Oxford University Press.

Nicoll, Maurice, (2011 edition) *The New Man: An Interpretation of Some Parables and Miracles of Christ*. Guildford, England: White Crow Books.

Nicoll, Maurice, (1952) *Psychological Commentaries On the Teachings of G. I. Gurdjieff and P. D. Ouspensky* (Vols. 1 - 5) London, England: Vincent Stuart.

Niebur, Ernst, (2002) "Synchrony: a neuronal mechanism for attentional selection?" *Current Opinion in Neurobiology*, 12:190–194.

Nieper, Hans, (1983) *Dr. Nieper's Revolution in Technology, Medicine and Society*. Oldenberg, Germany: Druckhaus Neue Stalling.

NOAM: New Oxford American Dictionary.

Nobili, Renato. "Schrödinger wave holography in brain cortex." *Physical Review A* Vol. 32, No. 6, p.3618-26. Dec. 1985.

Noble, Andrew E., *et al*, (Apr. 8, 2015) "Emergent long-range synchronization of oscillating ecological populations without external forcing described by Ising universality," *Nature Communications*, DOI: 10.1038/ncomms7664.

Nugent, Fereshteh S., *et al*, (2008) "High-Frequency Afferent Stimulation Induces Long-Term Potentiation of Field Potentials in the Ventral Tegmental Area." *Neuropsychopharmacology*, 33, 1704–1712.

Norbu, Namkhai, (1992) *Dream Yoga And The Practice Of Natural Light*, Ithaca, New York: Snow Lion Publications. (The ebook edition, edited by Michael Katz, and scanned, proofed, and hyperlinked by Purusa, 2002.)

Novak, Peter, (1997) *The Division of Consciousness: The Secret Afterlife of the Human Psyche*. Charlottesville, VA: Hampton Roads Publishing Company, Inc.

Novella, Steven, (May 8, 2007) "Is the Brain Analog or Digital?" *Neurological Blog*: http://theness.com/neurologicablog/index.php/is-the-brain-analog-or-digital/

O'Brien, Justin, (1989) *Pioneer of Inner Space: Swedenborg.* Fryeburg, ME: J. Appleseed & Co.

O'Connor, Elizabeth, (1971) *Our Many Selves: A Handbook of Discovery.* New York, NY: Harper & Row, Publishers.

Ornstein, Robert, (1977) *The Psychology of Consciousness.* New York, NY: Harcourt Brace Jovanovich, Inc.

Oster, Gerald, (Oct. 1973) "Auditory Beats in the Brain," *Scientific American.*

Ostrander, Sheila; Schroeder, Lynn, (1970) *Psychic Discoveries Behind the Iron Curtain.* Englewood Cliffs, NJ: Prentice-Hall, Inc.

Ostrander, Sheila; Schroeder, Lynn, (1979) *Super Learning.* New York, NY: A Laurel Book.

Ouspensky, Peter D., (1934; 1997) *A New Model of the Universe: Principles of the Psychological Method in its Application to Problems of Science, Religion, and Art.* Mineola, NY: Dover Publications, Inc.

Ouspensky, Peter D. (1957) *The Fourth Way.* New York, NY: Alfred A Knopf.

Ouspensky, Peter D. (1949; 2001) *In Search of the Miraculous: Fragments of an Unknown Teaching.* New York, NY: Harcourt, Inc.

Ouspensky, Peter D. (2008) *Conscience: The Search for Truth.* Sandpoint, ID: Morning Light Press.

Ouspensky, Peter D. (1974—seventh printing) *The Psychology of Man's Possible Evolution.* New York, NY: Alfred A. Knopf.

Pagels, Elaine, (Jan. 26-30, 2004) "The Text of the Gospel of Thomas," Printed with permission, (Harry Camp Memorial Lecturer, Stanford Humanities Center: http://shc.stanford.edu). - Salem, OR: Polebridge Press. Pennington, Judith: https://www.institutefortheawakenedmind.com/

Pal, Madhabendra Nath, (1982) *Ayurveda: A View Through Modern Science.* Calcutta, India: Sree Saraswaty Press Limited.

Palhano-Fontes, Fernanda, *et al,* (Feb. 18, 2015) "The Psychedelic State Induced by Ayahuasca Modulates the Activity and Connectivity of the Default Mode Network," *PLOS* (pdf).

Paloutzian, Raymond F.; Park, Crystal L., (Eds.), (2005) *Handbook of the Psychology of Religion and Spirituality.* New York, NY: The Guildford Press.

Pappas, Catherine, (Mar. 5, 2012) "Viewpoint: New Twist in Chiral Magnets," *Physics* 5, 28.

Pascual-Leone A., *et al*, (2011). "Characterizing brain cortical plasticity and network dynamics across the age-span in health and disease with TMS-EEG and TMS-fMRI" *Brain Topography* **24**: 302–315.

Pavel, S. (1979). "Pineal vasotocin and sleep: involvement of serotonin containing neurons," *Brain research bulletin, 4*(6), *731-734.*

Pennisi, Elizabeth, (Sept. 13, 2018) "Plants communicate distress using their own kind of nervous system," *Science Magazine: http://www.sciencemag.org/news/2018/09/plants-communicate-*

Petoukhov, S. V., (1999) *Biosolitons - One Secret of Living Matter: The Bases of Solitonic Biology* (in Russian). Russia, Moscow: Department of Biomechanics, Mechanical Engineering Research Institute, *Russian Academy of Sciences.*

Piccoli T, *et al*, (2015) "The Default Mode Network and the Working Memory Network Are Not Anti-Correlated during All Phases of a Working Memory Task," *PLoS ONE,* 10(4).

Pletzer, Belinda., *et al*, (Jun 4, 2010) "When frequencies never synchronize: the golden mean and the resting EEG," *Brain Research*, 1335: 91-102.

Poil, Simon-Shlomo, *et al*, (July 18, 2912) "Critical-State Dynamics of Avalanches and Oscillations Jointly Emerge from Balanced Excitation/Inhibition in Neuronal Networks," *The Journal of Neuroscience*, 32(29): 9817–9823.

Pollack, Gerald H., (2001) *Cells, Gels and the Engines of life: A New Unifying Approach to Cell Function.* Seattle, WA: Ebner & Sons Publishers.

Pollack, Gerald H., (2013) *The Fourth Phase of Water: Beyond Solid, Liquid, and Vapor.* Seattle, WA: Ebner & Sons Publishers.

Popp, Fritz-Albert, (Fall 1998) "Biophotons and Their Regulatory Role in Cells," *Frontier Perspectives*, Vol. 7, No. 2, 13-22.

Popoff, Irmis, (1978) *The Enneagram of The Man of Unity.* New York, NY: Samuel Weiser, Inc.

Poponin, Vladimir, (Mar. 19, 1992) "The DNA Phantom Effect: Direct Measurement of A New Field in the Vacuum Substructure," *Tweston.net.*

Porges, Stephen W., (1995) "Orienting in a defensive world: Mammalian modifications of our evolutionary heritage. A Polyvagal Theory," *Psychophysiology, 32 (1995), 301-318. Cambridge University Press. Printed in the USA.*

Prigogine, Ilya, (1984) *Order Out of Chaos: Man's New Dialogue With Nature.* New York, NY: Bantam Books.

Psi Encyclopedia: https://psi-encyclopedia.spr.ac.uk/Quijano, Eduardo, (Nov. 16, 2007) "Head on Collision of Vortex Rings at Re-1071," *You Tube: "Vortex Ring Collision," https://www.youtube.com/watch?v=XJk8ijAUCil.*

Raghuraj, Puthige, *et al*, (2004) "Right uninostril yoga breathing influences ipsilateral components of middle latency auditory evoked potentials," *Neurological Science*, 25: 274-280.

Rahnev, Dobromir, (2013) "Entrainment of Neural Activity Using Transcranial Magnetic Stimulation," *The Journal of Neuroscience, 10 July 2013, 33(28): 11325-11326.*

Raia-Green, Courtenay, (2009) "Science, Magic, and Religion," (A UCLA Introductory History Course of 19 Lectures, available on You Tube).

Rajshree, Vajrapani Patil, (Apr. 27, 2010) "Medical Experts to Study How Yogi Survives Without Food and Water," *Medindia*.

Rama, Swami, (approx. 1990-1996) *Control Over Mind and Its Modifications: Sankhya and the Yoga Sutras*. Publication on the internet: http://www.swamij.com/yoga-sutras-10104.htm.

Rankin, Catharine H., *et al*, (Sept., 2009) "Habituation revisited: An updated and revised description of the behavioral characteristics of habituation," *Neurobiology of Learning and Memory*, Volume 92, Issue 2, Pages 135-138.

Rebbi, Claudio (Feb. 1, 1979) "Solitons," *Scientific American, Inc.*

Reddy, Ananda (2002) "Essentials Of Transformative Psychology," (A proposal to formulate the discipline of Indian Psychology, presented at a Conference in Pondicherry, September 30th - October 3rd, 2002. This event was sponsored by The Infinity Foundation: http://infinityfoundation.com/index.shtml).

Reichenbach, Baron Charles von, (1850) *Physico-Physiological Researches on the Dynamics of Magnetism, Electricity, Heat, Light, Crystallization, and Chemism in Their Relations to Vital Force*. New York, NY: J. S. Redfield, Clinton-Hall.

Reid, John Stuart, (Feb. 22, 2017) "Secrets of Cymatics," *You Tube* (https://www.youtube.com/watch?v=uMK3OVBjx2Q).

Rein, Glen (1989) "Effects of Non-Hertzian Scalar Waves on the Immune System," *Psychotronics Association Journal* (pdf).

Rein, Glen, (Fall-Winter 1998) "Biological Effects of Quantum Fields and their Role in the Natural Healing Process," *Frontier Perspectives*, Vol. 7, No. 1, 16-23.

Rochat, Philippe, (2003) "Five levels of self-awareness as they unfold early in life," *Consciousness and Cognition* 12, 717–731. (©2003 Elsevier Inc.)

Rohé, Fred (1975) *The Zen of Running*. Middletown, CA: Organic Marketing.

Roopun, Anita K., *et al*, (Dec. 2008) "Temporal Interactions between Cortical Rhythms," *Frontiers In Cellular Neuroscience*, 2(2): 145–154.

Roopun, Anita K., *et al*, (Apr. 2008) "Period concatenation underlies interactions between gamma and beta rhythms in neocortex," *Frontiers In Cellular Neuroscience*, Vol 2, 1-8.

Rosenfield, Israel, (1998) *The Invention of Memory: A New View of the Brain*. New York, NY: Basic Books.

Rossello, Josep L., *et al*, (Feb 2012) "Neural Information Processing: between synchrony and chaos." *Nature Precedings* : doi:10.1038/npre.2012.6935.1.

Rossi, E. (1986) *The Psychobiology of Mind-Body Healing: New Concepts of Therapeutic Hypnosis*. New York, NY: W. W. Norton & Company, Inc.

Rucker, Rudolf V., (ed), (1980) *Speculations on the Fourth Dimension: Selected Writings of Charles Hinton*. New York, NY: Dover Publications, Inc.

Russell, Walter B., (1926; 1974 edition) *"The Universal One: An exact science of the One visible and invisible universe of Mind and the registration of all ideas of thinking Mind in Light, which is matter and also energy."* Swannanoa, Waynesboro, VA: University of Science and Philosophy.

Russell, Walter B., (1947; 1994 edition) *"The Secret of Light."* Covington, GA: Newtown County Library.

Rutherford, W. (1984) *Pythagoras: Lover of Wisdom*. Wellingborough, Northamptonshire, England: The Aquarian Press.

Sanford, John A., (1987) *The Kingdom Within: The Inner Meaning Of Jesus' Sayings*.

Sannella, Lee, (1976) *The Kundalini Experience: Psychosis or Transcendence?* Lower Lake, CA: Integral Pub.

Saroka, Kevin, *et al*, (Dec, 2010) "Experimental Elicitation of an Out of Body Experience and Concomitant Cross-Hemispheric Electroencephalographic Coherence," *NeuroQuantology* Vol 8, Issue 4, 466-477.

Sato, Rebecca, (Dec. 12, 2007) "Human Cells Found to have Electric Fields as Powerful as Lighting Bolts," *The Daily Galaxy—Great Discoveries Channel*: http://www.dailygalaxy.com/my_weblog/2007/12/its-electrifyin.html.

Schartner, Michael M., *et al*, (April 19, 2017) "Increased spontaneous MEG signal diversity for psychoactive doses of ketamine, LSD and psilocybin," *Scientific Reports*, 1-12.

Schleip, Robert, "Fascia as an organ of communication," In: Schleip R, *et al*, (2012) "Fascia: the tensional network of the human body," Elsevier Ltd, Edinburgh, 77-112.

Schmid, Gary Bruno, (2016, downloaded from internet) "Conscious vs. Unconscious Information Processing in the Mind-Brain," *Zürich, Switzerland*: www.mind-body.info.

Schrödinger, Erwin, (1944) *What is Life? The Physical Aspect of the Living Cell*. (Based on lectures delivered under the auspices of the Dublin Institute for Advanced Studies at Trinity College, Dublin, in February 1943. Pdf version.)

Schroeder, Charles E., *et al*, (2009) "Aligning the Brain in a Rhythmic World," *ResearchGate*, 22-28.

Schueler, Gerald J. & Betty J., (2001) *The Chaos of Jung's Psyche*. An ebook on the web: http://www.schuelers.com/ChaosPsyche/table_of_contents.htm

Schutter, Dennis, J. L. G., *et al*, (Mar. 2012) "Cross-frequency coupling of brain oscillations in studying motivation and emotion," *Motivation and Emotion*, 36(1): 46–54.

Schwartz, Gary E.R.; Russek, Linda G.S., (Fall 1998) "The Origin of Holism and Memory in Nature: The Systemic Memory Hypothesis," *Frontier Perspectives*, Vol. 7. No. 2, 23-30.

Schweickert, Richard; Boruff, Brian, (1986). "Short-term memory capacity: Magic number or magic spell?" *Journal of Experimental Psychology: Learning, Memory, and Cognition* 12 (3).

Science Daily, (2009) "Long Distance Brain Wave Focus Attention," (Based on Gegorious 2009).

Sechrist, Alice Spiers, (1981) *A Dictionary of Bible Imagery (Compiled from the works of Emanuel Swedenborg)*. New York, NY: Swedenborg Foundation, Inc.

Seife, C. (2000) *Zero: The Biography of a Dangerous Idea*. London, England: Penguin Books.

Seymour, Percy, (1992) *The Scientific Basis of Astrology: Tuning to the Music of the Planets*. New York, NY: St. Martin's Press.

Sharamon, Shalila; Baginski, Bodo J., (1989) *Cosmo-Biological Birth Control*. Wilmot, WI: Lotus Light Publications.

Sharma, Arvind, (2004) *Sleep as a State of Consciousness in Advaita Vedanta*. Albany, NY: State University of New York Press.

Sheldrake, Rupert, (1989) *The Presence of the Past: Morphic Resonance and the Habits of Nature*. New York, NY: Vintage Books.

Shikhirin, Valeriy (Sept. 2005) "Development Prospects of Tore Technologies, Elastic Mechanics and 'Wonders' Worked By Them In Nature," (A paper (revised and updated) included into the Proceedings of the 2nd International Research and Application Conference "Tore Technologies" held September 21-25, 2005 at the Irkutsk State Technical University, Plenary Report, and pp. 3 – 41).

Shimamura, Arthur P., (Apr. 5, 2014) "Surrealism, Creativity, and the Prefrontal Cortex," *Psycholology Today*.

Shors, Tracey J., *et al*, (1997) "Long-term potentiation: What's learning got to do with it?" Behavioral And Brain Sciences (1997) 20, 597–655

Shrimali, Narayan Dutt, (1985) *Practical Hypnotism*. New Delhi, India: Pustak Mahal.

Sidorov, Lian, *et al*, (May 2012) "Biophysical Mechanisms of Genetic Regulation: Is There a Link to Mind-Body Healing?" *DNA Decipher Journal*, Vol. 2, Issue 2, pp. 177-205.

Singer, Wolf, (1999). "Neuronal synchrony: A versatile code for the definition of relations?" *Neuron*, 24, 49-65.

Singer, Wolf, (2013) "The Neuronal Correlate of Consciousness: Unity in Time Rather than Space?" *Scripta Varia* 121, Vatican City, Rome.

Singh, Jaideva, (1992) *The Yoga of Vibration and Divine Pulsation*. Albany, NY: State University of New York Press.

Sivananda, Sri Swami, (1990) *Mind: Its Mysteries and Control*. Himalayas, India: Yoga-Vedanta Forest Academy Press.

Skarda, Christine A. and Freeman, Walter J. (1990) "Chaos and the New Science of the Brain." *Concepts in Neuroscience* vol. 1, no. 2, p. 275. March 1990.

Skarka, V, *et al*, (July 18, 2010) "The variety of stable vortical solitons in Ginzburg-Landau media with radially inhomogeneous losses," *physics.optics*, arXiv:1007.3030v1.

Smith, Cyril W., (1998) "Is a Living System a Macroscopic Quantum System?" *Frontier Perspectives*, Fall/Winter, Vol. 7, No. 1, 9-15.

Soljacic, Marin, *et al*, (1999) "Self-Similarity and Fractals Driven by Soliton Dynamics," Invited Paper, Special Issue on Solitons, *Photonics Science News* 5(1), 3-12.

Spiegel, David, (ed), (1994) *Dissociation: Culture, Mind, and Body*. Washington, DC: American Psychiatric Press, Inc.

Sternheimer, Joel, (2002) "Method for the Regulation of Protein Biosynthesis," US Patent: US2002177186, 2002-11-28.

Stewart, Thomas Milton, (1978) *The Symbolism of the Gods of the Egyptians and the Light They Throw on Freemasonary*. London, England: A Lewis (Masonic Publishers) LTD.

Stigsby, Bent, *et al*, (1981) "Electroencephalographic findings during mantra meditation (Transcendental Meditation). A controlled, quantitative study of experienced meditators," *Electroencephalography and Clinical Neurophysiology*, 51, 434—442 (© Elsevier/North-Holland Scientific Publishers, Ltd).

Strassman, Rick J., (2001). DMT: *The Spirit Molecule. A Doctor's Revolutionary Research into the Biology of Near-Death and Mystical Experiences*. Rochester, VT: Park Street.

Strogatz, Steven H., (2003) *Sync: How Order Emerges from Chaos in the Universe, Nature, and Daily Life.* New York, NY: Hyperion.

Strogatz, Steven H., (1994) *Nonlinear Dynamics and Chaos.* Cambridge, MA: Perseus Books Publishing, LLC.

Strohl, James, (1998) "Transpersonalism: Ego Meets Soul," *Journal of Counseling and Development,* Fall, Vol. 76, 397-403.

Sumedho, Ajahn, (2014) *The Sound of Silence: The Anthology, Vol. 4* Hertfordshire, UK: Amaravati Buddhist Monastery.

Svoboda, Robert, (1984) *Lessons and Lectures on Ayurveda.* Albuquerque, NM: The Ayurvedic Institute.

Swedenborg, Emanuel. (1988) *Dictionary of Correspondences—Representatives and Significatives.* New York, NY: Swedenborg Foundation, Inc.

Swedenborg, Emanuel. (2000) *Heaven and Hell.* Translated by George F. Dole West Chester, PA: Swedenborg Foundation, Inc.

Swedenborg, Emanuel, (2009a) *Divine Love and Wisdom.* Translated from the Original Latin by John C. Ager. West Chester, PA: Swedenborg Foundation.

Swedenborg, Emanuel. (1758-2009b) *Arcana Coelestia: The heavenly arcana contained in the Holy Scripture or Word of the Lord unfolded, beginning with the book of Genesis.* Translated from the Original Latin by John Clowes Revised and Edited by John Faulkner Potts, Standard Edition, Vols. 1-12. West Chester, PA: Swedenborg Foundation.

Swedenborg Foundation: http://www.swedenborg.com/emanuel-swedenborg/explore/spiritual-world/

Szabo, Sandor, *et al*, (July 7, 2012) "The legacy of Hans Selye and the origins of stress research: A retrospective 75 years after his landmark brief, "Letter' to the Editor of Nature," *Informa Healthcare USA, Inc.*

Szegedy-Maszak, Marianne (Feb. 28, 2005) "Mysteries of the Mind: Your unconscious mind is making your everyday decisions," *U. S. News & World Report: Health & Medicine, 53-61.*

Talukder, Gargi, (Mar. 20, 2013) "Decision-Making is Still a Work in Progress for Teenagers," *Brain Connection: http://brainconnection.brainhq.com/2013/03/20/decision-making-is-still-a-work-in-progress-for-teenagers/.*

Tamburini, Fabrizio, *et al*, (Mar. 1, 2012) "Encoding many channels on the same frequency through radio vorticity: first experimental test," *New Journal of Physics* 14, 033001, 1367-2630.

Tamdgidi, Mohammad, (2009) *Gurdjieff and Hypnosis: A Hermeneutic Study.* New York, NY: Palgrave Macmillan.

Targ, Russell, (2004) *Limitless Mind: A Guide to Remote Viewing and Transformation of Consciousness*. Novato, CA: New World Library.

Tart, Charles T., (Mar. 30, 1997) "Six Studies of Out-of-the-Body Experiences," *Journal of Near Death Studies*.

Taylor, Eugene, (1999) *Shadow Culture: Psychology and Spirituality in America*. Washington, DC: Counterpoint.

Terhune, Devin Blair, *et al*, (2011) "Differential frontal-parietal phase synchrony during hypnosis as a function of hypnotic suggestibility," *Psychophysiology*, 48, 1444–1447. (Wiley Periodicals, Inc. Printed in the USA. Copyright © 2011 Society for Psychophysiological Research.)

Tesla, Nikola, (June, 1900) "The Problem of Increasing Human Energy with Special References to the Harnessing of the Sun's Energy," *Century Magazine*.

Tewari, Paramahamsa, (2007) *Discovering Universal Reality*. Editions India.

Thelen, Esther, (2001) "The dynamics of embodiment: A field theory of infant perseverative reaching," *Behavioral and Brain Science*, 24, 1-86.

Thompson, Jeffrey, (2009) "Epsilon, Gamma, HyperGamma and Lambda Brainwave Activity and Ecstatic States of Consciousness," Web article published by *the Center for Neuroacoustic Research*. Encinitas, CA.

Thomson, William, Sir (Lord Kelvin), (1869) "On Vortex Atoms," *Transactions of the Royal Society of Edinburgh* 15, 217-260.

Thornhill, Wallace; Talbot, David, (2007) *The Electric Universe*. Portland, OR: Mikamar Publishing.

Three Initiates, (1912, 1940) *The Kybalion: A Study of The Hermetic Philosophy of Ancient Egypt and Greece*. Chicago, Ill: The Yoga Publication Society, Masonic Temple.

Tiller, William, "Subtle Energies in Energy Medicine," *Frontier Perspectives* 4, no. 2 (Spring 1995): 17-21.

Tomes, Ray, (2012) "The Wave Structure of Matter," (text of a talk given to Friends of the Scientific and Medical Network: http://homepages.ihug. co.nz/~ray.tomes/.

Trevelyan, George, (1985) *Magic Casements: The Use of Poetry in the Expanding of Consciousness*. London, England: Coventure, L.T.D.

Trowbridge, Patricia. (1997) *Breaking the Memory Barrier: An Answer To Learning Problems*. Key West, Fl: Eaton Street Press.

Tsuda, Ichiro, (2001) "Toward an interpretation of dynamic neural activity in terms of chaotic dynamical systems," Behavioral and Brain Sciences, (2001) 24, 793–847.

Turow, Gabe, (2005) *Auditory Driving as a Ritual Technology: A Review and Analysis.* Religious Studies Honors Thesis, Stanford University.

Uhlhaas, Peter J., *et al*, (May 2011) "The Development of Neural Synchrony and Large-Scale Cortical Networks During Adolescence: Relevance for the Pathophysiology of Schizophrenia and Neurodevelopmental Hypothesis," *Schizophrenia Bulletin*, 37(3): 514–523.

Van den Bovenkamp, Frank. (March 2007) "Why Phi: a derivation of the Golden Mean ratio based on heterodyne phase conjugation," Source - http://science.trigunamedia.com/whyphi2007/Why%20Phi%20-%20a%20 derivation%20of%20the%20Golden%20Mean%20ratio%20based%20on%20 heterodyne%20phase%20conjugation.pdf.

Velmans, Max (ed), (1998) *The Science of Consciousness: Psychological, Neurological and Clinical Reviews.* London; New York; Canada: Routledge.

Viehweger, Rainer, (2012) *Understanding the Universe through Global Scaling®: looking at the world with fresh eyes.* Poole, UK: Quantum Health Ltd.

Vitiello, Guiseppi, (2009) "Coherent States, Fractals, and Brain Waves," *New Mathematics and Natural Computation*, 05, 245-264.

Wang, Jian, *et al*, (Jan 26, 2012) "Terabit free-space data transmission employing orbital angular momentum multiplexing," *Nature Photonics* 6, 488–496.

Waser, Andre, (2004) *The Global Scaling ® Theory: A Short Summary.* Private web publication.

Weiss, Harald; Volkmar Weiss, (2003) "The golden mean as clock cycle of brain waves" *Chaos, Solitons and Fractals* 18, No. 4, 643-652 - Elsevier Author Gateway (online version).

Welch, John R., (July–December 1990) "Llull, Leibniz, and the Logic of Discovery," *Catalan Review* IV, Nos. 1–2, 75–83.

Wen, Patricia, (July 5, 2001) "'Brain Fingerprints' May Offer Better Way to Detect Lying." *The Boston Globe.*

Werbos, Paul J., (Nov. 1999) "Can Soliton Attractors Exist in Realistic 3+1-D Conservative Systems?" *Chaos, Solitons & Fractals, Vol 10, Issue 11, Pages 1917–1946.*

Wernery, Jannis, (2013) "Bistable Perception of the Necker Cube in the Context of Cognition & Personality," A dissertation for the degree of Doctor of Sciences submitted to ETH ZURICH.

Werntz, D. A., *et al*, (1983) "Alternating cerebral hemispheric activity and lateralization of autonomic nervous function." *Human Neurobiology*, 1983, 2:39-43.

Werntz, D. A., *et al*, (1987) "Selective hemispheric stimulation by uninostril forced nostril breathing." *Human Neurobiology*, 6:165-171.

Westerland, M. (2007) "Healing with advanced hypnotherapy: A science of spirituality." www.rcpsych.ac.uk/college/specialinterestgroups/spirituality/publications.aspx.

Wheeler, Kenneth L., (2014) *Uncovering the Missing Secrets of Magnetism: Exploring the nature of Magnetism with regards to the true model of atomic geometry and field mechanics of rational physics & logic*. Richmond, MI: Dark Star Publications.

Wiener, Norbert, (2013) *Cybernetics: or, control and communication in the animal and the machine*. Cambridge, MA: MIT Press.

Wiener, Norbert, (1954) *The Human Use of Human Beings*. Cambridge, MA: MIT Press.

Wiener, Norbert, (1989) *The Human Use of Human Beings: Cybernetics and Society*. London, England: Free Association Books.

Wijk, B. C. M, *et al*, (2015) "Parametric estimation of cross-frequency coupling," *Journal of Neuroscience Methods* 243, 94–102.

Wilbur, Ken, (Nov. 20, 2006) "Ken Wilbur Stops His Brain Waves," You Tube video: https://www.youtube.com/watch?v=LFFMtq5g8N4.

Williams, R., (1979) *Biochemical Individuality: The Basis for the Genetotropic Concept*. Austin, TX: University of Texas Press.

Wilson, Colin, (1973) *The Occult*. New York, NY: Random House, Inc.

Windrider, Kiara; Sears, Grace, (2006) *Deeksha: The Fire From Heaven*. Makawao, Maui, HI: Inner Ocean Publishing, Inc.

Wise, Anna, (2004) *The High Performance Mind: Mastering Brainwaves For Insight, Healing, And Creativity*. New York, NY: Penguin Group Inc. (USA).

Wolf, Fred A., (1986) *The Body Quantum: The New Physics of Body, Mind, and Health*. New York, NY: Macmillan Publishing Company.

Wolf, Fred A., (1996) "On the Quantum Mechanics of Dreams and the Emergence of Self-Awareness," (In Stuart R. Hameroff, Alfred W. Kaszniak & A. C. Scott (eds.), 1996, *Toward a Science of Consciousness*. Cambridge, Massachusetts: MIT Press).

Wolff, Milo, (1990) *Exploring the Physics of the Unknown Universe: An Adventurer's Guide*. Manhattan Beach, CA: Technotran Press.

Wolff, Milo, (2008) *Schrödinger's Universe: Einstein, Waves and the Origin of the Natural Laws*. Parker, Co: Outskirts Press.

Wolfram, Stephen, (May, 14, 2002) *A New Kind of Science*. Wolfram Media, Inc.

Wolinsky, Stephen, Collaboration with Ryan, Margaret O., (1991) *Trances People Live: Healing Approaches in Quantum Psychology*. Falls Village, CT: The Bramble Company.

Wolinsky, Stephen, (1994) *The Tao of Chaos: Essence and the Enneagram*. Falls Village, CT: The Bramble Company.

Yang, Chun, *et al*, (Jan. 21, 2015) "Fascia and Primo Vascular System," *Hindawi Publishing Corporation* (Evidence-Based Complementary and Alternative Medicine), Article ID 303769, 6 pp.

Yates, Frances A. (1966) *The Art of Memory*. Chicago, Ill: The University of Chicago Press.

Yildirim, Serap, *et al*, (Jan. 2010) "Nasal Cycle in Schizophrenia: Left Nostril Dominance may be Associated with Cerebral Lateralization Abnormality and Left Hemisphere Dysfunction," *ResearchGate*.

Yirka, Bob, (May 15, 2015) "Microsoft study claims human attention span now lags behind goldfish," *Medical Xpress*.

Yogananda, Paramahansa, (1946; 2007) *Autobiography of a Yogi*. Los Angeles, CA: Self-Realization Fellowship.

Young, Arthur M., (1976) *The Reflexive Universe: Evolution of Conciouness*. United States of America: Delacorte Press.

Yüksel, Ramazan, *et al*, (Jan. 2014) "The Effects of Hypnosis on Heart Rate Variability," *International Journal of Clinical and Experimental Hypnosis*, 162-171.

Yukteswar, Swami Sri, (1901/2013) *The Holy Science*. Los Angeles, CA: Self-Realization Fellowship.

Yurth, D. G., (Dec. 5, 2000) "Torsion Field Mechanics: Verification of Non-local Field Effects in Human Biology," (Paper on internet).

Zeiders, Charles L. (2004) *The Clinical Christ: Scientific and Spiritual Reflections on the Transformative Psychology Called Christian Holism*. Birdsboro, PA: Julian's House.

Zephir (Blog): https://www.blogger.com/profile/06010623752049244967

Zukav, Gary, (1979) *The Dancing Wu Li Masters: An Overview of the New Physics*. New York, NY: William Morrow and Company.

Zvetkov, V. D., (1997) *Heart, Golden Ratio and Symmetry*, Puschino, Russian Academy of Sciences.

INDEX OF KEY TERMS AND AUTHORS

delta, 61, 62, 266
deoxyribonucleic acid, 32, 374
deselection, 87, 108, 119
destructive interference, 116, 121, 137, 309
Diamond, John, 176
digitized, 105
disempowerment, 293, 297
dissociation, 148-150, 270, 292
dissonance, 20, 138
dissonantly separate, 176
distant healing, 248
dividing attention, 36-37, 45, 262
Divine Mandate, 175
DLE. *See* death-like event
DMN. *See* default mode network
DNA. *See* deoxyribonucleic acid
DNA phantom effect, 220-222
doctrinal texts, 353
Doczi, György, 183, 328
dopamine, 76, 137, 155-156, 208
dreamed movement, 224
duality, 158, 238, 322,
dwell time, 322
dynamic equilibrium, 183, 319
ecstasy, 275, 279
ectoplasm, 253
EEG spectrum, 62, 115, 192, 199, 258-260, 266
Egne, Hakan, 55, 127
Einstein, Albert, 16, 100, 127
electro-dynamic fields, 223
electro-magnetic, 110-111, 209
electro-magnetic signature, 111, 209
electromagnetic signal, 210
electronic body, 39

Ellis, Amy Jo, 353
email transfer of DNA, 214
emergence, 131, 245, 340-341
emotional modalities, 350
EMS. *See* electromagnetic signal
encephalograph, 61
energy and information, 102-103
energy configured, 36
energy exchange, 161
energy expenditure, 135
energy management, 288
engrams, 60
enneagram, 34
entheogens, 346
entrainment, 133, 135, 148, 156, 202, 212-213, 259, 274, 298, 351, 352
ependyma, 69
epiphenomena, 232, 324
epsilon, 61, 62, 94
equanimity, 225
equilibrium, 181, 182, 227
Erickson, 97, 226, 267, 310, 311
Erickson, Milton. *See* Erickson
eschatology, 121
esoteric psychology, 235
esoteric symbols, 33, 34, 233, 307, 308, 337, 361
esotericism, 232, 235, 239
Essence, 63, 132, 134, 135, 139, 172-174, 286, 291, 301
Essence's signature, 330
eternal, 121
euphoria, 156
eustress, 143, 208
exceed speed of, 157
explosive, 81, 109, 162

love, v, 1-2, 5, 18, 30, 55-56, 59, 79, 125, 173, 275-276, 286, 289, 309-312, 315, 317, 333, 335, 348

love and compassion, 79, 276, 348, 352

love-and- understanding, 317

low pressure zones, 85, 105

LPCS. *See* laser photon correlation spectrometer

lucid dreaming, 24, 64, 246-247

Lull, Raymond, 325

lying, 176, 297

MacLean, Paul, 114

magnet, 118-119

magnetic resonance imaging, 70

magneto-electric, 110-111

magnitude of assembly, 327

Maharishi Mahesh Yogi, 22

mansions, 273, 314

mass action, 80, 355

master memory construct. *See* MMC

materiality, 101, 294, 316, 368-369

Maxwell, James Clerk, 251

MC, 4, 78, 83-84, 86-89, 88-89, 91-92, 108-109, 111, 119, 124, 133, 138, 142, 144, 148-149, 197, 201, 204, 208, 220, 224-227, 249, 285-286, 291-292, 297, 304, 305-306, 309, 311, 313, 318, 324-325, 329-332, 343, 347, 350, 356, 363, 370

meaning, 80, 144, 246, 283, 338

meditation, 261-264, 321

medium, 276-283

Melloni, Lucia, 92

memory, 209, 214, 216-217, 224, 265, 280, 313, 324, 330-336

memory construct. *See* MC

meridian system, 68, 105, 192, 346

metabolic rate, 16, 48, 98, 349

metabolism, 243

methodologies, 299, 338, 347

Meyl, Konstantin, 89, 102

Milewsky, John V., 111

mind pixel, 127-128

Mirabelli, Carlos, 281

mirror neurons, 298

MMC, 33-35, 180, 234, 290, 302, 304, 306, 309, 312, 327, 320, 330-332, 334-339, 342, 345, 349, 353, 358, 363-364

mnemonic, 35, 306

mnemonics, 171, 325-327

modalities to elicit WR, 348

modes of being, 150, 164, 330

molecular body, 27

molecular print, 86, 177, 308

molecular substrate, 139, 189, 220

monotheism, 358

Montagnier, Luc, 125, 209

moral principle, 316

motionless, 100, 275

motivation, 36, 87, 301

movement modalities, 348

movements, 349

movie-like, 95

MRI. *See* magnetic resonance imaging

mudras, 337, 348

Muller-Ortega, Paul E., 238

multiple personalities, 149, 291

Murphy, Joseph, 304

music, 351

mytho-symbolic, 268

natural resonant frequency, 89, 132, 177

NDE. *See* near-death experience

near-death experience, 313

Necker Cube, 321

negative and positive feedback, 164-170

negative emotions, 32, 36, 138, 309, 331

negative feedback, 149, 162, 177

negative space-time, 83, 96, 106, 110-111, 119, 123, 128, 102, 200, 207, 222-223, 227-228, 251-253, 227, 280, 285, 307, 312, 323-324

neocortex, 44, 111

neural regulation, 70

neuroglia. *See* glia

neuroglial system, 72

neuronal synchrony, 238

New Age, 18, 22

Newton, Michael, 269

Newtonian, 237

node, 101, 116

nodes, 85, 277, 327

noise, 195

non-conscious, 4, 23, 25-26, 36, 41, 45-46, 49, 58, 63, 87, 90, 92, 13-132, 134, 138, 141-42, 145, 149-150, 158-160, 167, 175, 178, 188, 190-193, 226, 246, 259-260, 267, 271, 295, 301, 304, 306, 320, 331-332, 349, 356

non-conscious, speed of, 190

nonlinear, 108, 185

nonlocal, 92

now, 110, 162

nucleus accumbens, 137, 155

OBE, ix, 12-13, 28, 37, 208, 220, 223, 225, 243, 294, 300, 313, 314, 347

objective compulsion, 232

objective compulsive disorder, 264

objective potentials, 120, 140, 315, 359

objective/subjective potentials, 55, 197

objective-subjective potentials, 359

obstacles, 285, 290

occipital lobes, 77

OCD. *See* objective compulsive disorder

oligodendrocytes, 67-68

OOPE, 20, 29, 32, 63, 87, 96, 115-116, 129, 133, 152-153, 158-159, 182, 170, 178,-179, 187, 189, 197, 208, 266-267, 290, 306, 318, 327, 335, 354, 357

orgone box, 256-258

Ostrander, Sheila, 28

other-directed agendas, 299

Ouspensky, 31, 144, 176, 233-237, 243, 293, 333

Ouspensky, Peter I... *See*

Printed in the United States
By Bookmasters